DOWNSIZING THE STATE

Dag MacLeod

DOWNSIZING THE STATE

Privatization and the Limits
of Neoliberal Reform
in Mexico

THE PENNSYLVANIA STATE UNIVERSITY PRESS
UNIVERSITY PARK, PENNSYLVANIA

Library of Congress Cataloging-in-Publication Data

MacLeod, Dag.
   Downsizing the state : privatization and the limits of
neoliberal reform in Mexico / Dag MacLeod.
      p.   cm.
Includes bibliographical references and index.
ISBN 0-271-02698-7 (alk. paper)
1. Privatization—Mexico. 2. Railroads—Mexico.
3. Airlines—Mexico. 4. Telecommunication—Mexico.
I. Title.

HD4013 .M32 2004
338.972'05—dc21   2003011651

Copyright © 2004 The Pennsylvania State University
All rights reserved
Printed in the United States of America
Published by The Pennsylvania State University Press,
University Park, PA 16802-1003

First paperback printing, 2005

The Pennsylvania State University Press is a member of
the Association of American University Presses.

It is the policy of The Pennsylvania State University Press
to use acid-free paper. Publications on uncoated stock
satisfy the minimum requirements of American National
Standard for Information Sciences—Permanence of Paper
for Printed Library Material, ANSI Z39.48–1992.

CONTENTS

List of Tables and Figures VI

Acknowledgments VIII

Abbreviations XI

1 Privatization and Competing Perspectives on Economic Organization 1

2 Public Ownership and the Rise of State-Led Development 33

3 Privatization and the Demise of State-Led Development 69

4 The Turbulent Privatizations of Aeroméxico and Mexicana de Aviación 109

5 Positive-Sum Games and the Sale of Telmex 149

6 The Transformation and Sale of the Mexican National Railroad 189

7 Economic Transformation and the Limits of Neoliberal Reform 231

Appendix 257

References 265

Index 291

LIST OF TABLES AND FIGURES

TABLES

1. Ideal Types of Exchange in the Economy   17
2. World Bank Loans to Mexico by Economic Sector, 1949–1980   57
3. Public-Sector Employment, 1980–2000   77
4. Revenue from Privatizations by Ministry, 1995–2000   88
5. Changing Ownership of Banks, 1990–2000   104
6. Distribution of Federal Investment in Transportation, 1952–1983   120
7. Principal Characteristics of Mexico's Two Largest Air Carriers, 1981–2000   124–125
8. Passenger Transportation in Domestic Scheduled Service, 1975–2000   128–129
9. Scheduled Commercial Airlines Operating in Mexico, 1980–2000   130
10. Passenger Transportation in Scheduled International Service, 1976–2000   131
11. Telephones in Service, 1880–2000   155
12. Sources of Telmex Revenue, 1976–2000   161
13. Telmex Long-Distance Rates, 1991–2000   183

14. Income and Transfers at FNM, 1965–1993   199
15. Employment at FNM, 1910–1996   202
16. Principal Characteristics of Newly Created Railroads   224
17. Interview Summary   259
18. List of Interviews   260–264

FIGURES

1. Number of Parastate Firms, 1930–1980   39
2. Balance of Parastate Sector as a Percentage of GDP, 1965–2000   44
3. Number of Parastate Firms, 1980–2000   72
4. Revenue from Privatization, 1984–2000   73
5. Sources of Revenue from Privatization, 1988–1993   83
6. Uses of Revenue from Privatization, 1988–1993   84
7. Parastate Firms' Share of Revenue Among Mexico's Fifty Largest Companies, 1984–1999   98
8. Private Mexican Firms' Share of Revenue Among Mexico's Fifty Largest Companies, 1984–1999   98
9. Foreign Firms' Share of Revenue Among Mexico's Fifty Largest Companies, 1984–1999   99
10. International Settlement Rate, 1993–2000   176
11. Fiber-Optic Network of Telmex's Principal Competitors   185
12. Rail Transportation, 1923–1999   198
13. Growth of Rail and Highway Infrastructure, 1952–2000   210
14. Foreign and Domestic Cargo Carried by Rail, 1990–2000   215
15. Restructured Rail System   222

ACKNOWLEDGMENTS

This book would not have been possible without my family, friends, and colleagues and the institutions that supported me throughout the process of research and writing. When I reflect on all the advice, companionship, encouragement, hot meals, photocopies and books, and financial and logistical support that I received, I have only two regrets: that I did not keep closer track of the many people who assisted me and that I was unable to incorporate every good idea into the final manuscript.

I need to begin by thanking two men who were, in many respects, my first teachers. Thomas M. Davies Jr. and Brian Loveman introduced me to the study of national development and encouraged my interest in Latin America during my undergraduate years at San Diego State University. By example, they taught me to appreciate good scholarship. Although it has now been more than ten years since I left San Diego State, I hope they will recognize their influence upon my intellectual development in this work.

Logistical support was vital during 1997 and 1998 as I shuttled between Mexico City, Baltimore, St. Louis, and Oakland. Cory Heyman, Kelly Schrum, and Mahua Sarkar all hosted me during this period. Melanie Arthur gave me the furniture necessary to write the first draft of Chapters 2 and 3 during my final semester at Johns Hopkins. Beth Simon, Tom Urbanski, Scout, and Makita welcomed me into their home and helped keep me sane during a productive and difficult semester in exile.

The fieldwork that constitutes the empirical basis for this book would not have been possible without the generous financial support of a Fulbright-García Robles scholarship, the facilities of the Departamento de Ciencias Sociales y Políticas at the Universidad Iberoamericana in Mexico City, and a travel grant from the Department of Sociology at Johns Hopkins University. James Goodyear of the Fulbright program at Johns Hopkins was instrumental in helping with the formulation of my original plan for the book. Margaret Keck and Franklin Knight at Hopkins reviewed the work in its early stages, and Patricia Fernández-Kelly provided helpful comments.

In Mexico City, Mari Hernández and Eduardo Andere assisted with the navigation of paperwork and other hurdles necessary to foreign residence. At the Universidad Iberoamericana, Esperanza Wilson of the Departamento de Ciencias Sociales y Políticas hosted me as a scholar in residence. A special debt is also owed to my dear friend Paul Franco not only for the logistical and institutional support that he provided at the Universidad Iberoamericana but also for his enduring friendship.

Although I promised confidentiality to many of the people I interviewed, others kindly agreed to permit me to use their names. In addition to those who are not named, I benefited immensely from conversations with Marco Antonio Leyva Piña, Andrés Caso Lombardo, Javiér Jiménez Espriú, David Gómez Díaz, Jaime Hernández, Carlos Hirsch, Rosana Ingle, Rafael Marino, Jorge Machado Dodera, Roberto Martínez, Concepción Rivera Romero, Jacques Rogozinski, Rafael García Rosas, Jorge Sandoval, and Jorge Silberstein. This work could not have been written without their help.

Fernando Cordero, Delphine Kachadourian, Pablo, and Adrian all enriched my stay in Mexico City. Richard Schauffler has provided inspiration and support, forgiving my distraction during the later phases of the writing process. Probably without knowing it, John French gave me renewed energy and helped me persevere during a difficult juncture in the fieldwork. My sister, Heather, helped to break through my isolation while I toiled in anonymity in St. Louis, and my brother, Max, has provided ceaseless support, becoming my number one cheerleader during the final push to finish the manuscript.

This book has also benefited from friends who have indulged me by reading and commenting on portions of it in various guises. I am especially grateful to Jennifer Bair, Paul Berks, Iona Mara-Drita, Kun-Chin Lin, Bruce Podobnik, and Sara Schatz for their comments on the work. Draft chapters dressed up as conference papers also received thoughtful comments from Giovanni Arrighi, Neil Brenner, Ben Brewer, Benedicte Bull, Ben Cashdan, Chris Chase-Dunn, Claudia Scholz, and Judith Teichman.

Beverly Silver and Alejandro Portes both offered thorough and detailed criticism of early drafts and challenged me to improve the manuscript and not shy away from difficult questions. Miguel Ángel Centeno and Nora Hamilton reviewed the manuscript for Penn State Press, pointing out additional areas for improvement. Sandy Thatcher has been thoughtful and gracious in leading me through the process of publishing my first book.

I am especially indebted to José Antonio Amozurrutia for the enthusiasm that he has consistently shown for the project. Toño helped me understand some of the most basic details of the privatization process and answered some of my most ignorant questions. He pointed me toward essential literature, shared his insights, challenged me to consider alternatives, prodded me to read more, and has read various drafts of the manuscript. The only thing that I value more than his thoughtful consideration of the intricacies of Mexican politics is his friendship.

And finally, I must thank the person without whom none of this would seem to matter much. Despite my extended absences and mental distraction, Terri continues to support and encourage me in every way possible. Without question, the only thing more challenging and more rewarding than completing this book has been her love.

ABBREVIATIONS

Afore	Administrador de Fondos de Retiro
	Retirement Fund Administrator
API	Administrador Portuario Integral
	Complete Port Administration
ASA	Aeropuertos y Servicios Auxiliares
	Airports and Auxiliary Services
ASPA	Asociación Sindical de Pilotos Aviadores
	Union Association of Pilot Aviators
ASSA	Asociación Sindical de Sobrecargos de Aviación
	Union Association of Flight Attendants
Banobras	Banco Nacional de Obras y Servicios Públicos
	National Bank of Public Works and Services
Canacintra	Cámara Nacional de la Industria de la Transformación
	National Chamber of Manufacturers
CCE	Consejo Coordinador Empresarial
	Business Coordinating Council
CDB	Comité de Desincorporación Bancaria
	Bank Divestment Committee
CFC	Comisión Federal de Competencia
	Federal Competition Commission
CFE	Comisión Federal de Electricidad
	Federal Electric Commission
CID	Comité Intersecretarial de Desincorporación
	Interministerial Committee for Divestment
CIGF	Comité Intersecretarial de Gasto y Financiamiento
	Interministerial Committee on Spending and Finance

Concamin	Confederación Nacional de Cámaras Industriales
	National Confederation of Chambers of Industry
Concanaco	Confederación de Cámaras Nacionales de Comercio
	National Confederation of Chambers of Commerce
Coparmex	Confederación Patronal de la República Mexicana
	Employer's Confederation of the Mexican Republic
CT	Congreso de Trabajo
	Labor Congress
CTM	Confederación de Trabajadores de México
	Mexican Confederation of Labor
CTTM	Compañía Telefónica y Telegráfica Mexicana
	Mexican Telephone and Telegraph Company
CWA	Communications Workers of America
EZLN	Ejército Zapatista de Liberación Nacional
	Zapatista Army of National Liberation
FNM	Ferrocarriles Nacionales de México
	Mexican National Railroad
IMF	International Monetary Fund
ITU	International Telecommunication Union
KCS	Kansas City Southern
LFEP	Ley Federal de Entidades Paraestatales
	Federal Law of Parastate Entities
MFD	Movimiento Ferrocarrilero Democrático
	Democratic Railroad Worker Movement
PAN	Partido de Acción Nacional
	National Action Party
Pemex	Petróleos Mexicanos
	Mexican Petroleum
PRD	Partido Revolucionario Democrático
	Democratic Revolutionary Party
PRI	Partido Revolucionario Institucional
	Revolutionary Institutional Party
SCT	Secretaría de Comunicaciones y Transportes
	Ministry of Communications and Transportation
SEMIP	Secretaría de Energía, Minas e Industrias Paraestatales
	Ministry of Energy, Mines, and Parastate Industry
SHCP	Secretaría de Hacienda y Crédito Público
	Ministry of the Treasury and Public Credit
SNTTAM	Sindicato Nacional de Técnicos y Trabajadores de Aeroméxico
	National Union of Aeroméxico Technicians and Workers

SPP	Secretaría de Programación y Presupuesto
	Ministry of Planning and Budget
STFRM	Sindicato de Ferrocarrileros de la República Mexicana
	Mexican Railroad Workers Union
STRM	Sindicato de Telefonistas de la República Mexicana
	Mexican Telephone Workers Union
Telmex	Teléfonos de México
	Mexican Telephone
TFM	Transportación Ferroviaria Mexicana
	Mexican Rail Transportation
TMM	Transportación Marítima Mexicana
	Mexican Maritime Transport
UACE	Unidad de Apoyo al Cambio Estructural
	Unit for the Support of Structural Change
UDEP	Unidad de Desincorporación de Entidades Paraestatales
	Unit for the Divestiture of Parastate Entities
UNT	Unión Nacional de Trabajadores
	National Workers Union

ONE

Privatization and Competing Perspectives on Economic Organization

Determining where and in the performance of which social functions to rely on automatic, spontaneous incentive and where on artificial incentive is one of the fundamental problems facing economic systems.

—*János Kornai,* The Socialist System

[T]he disappearance of power when one moves from the practical world of economics to its theoretical models is based on an easily exposed liberal sleight of hand, for it is the private-public partition that makes power invisible in the economy, and this partition ... is wholly arbitrary.

—*Samuel Bowles and Herbert Gintis,* Democracy and Capitalism

Privatization and the Roles of Markets and Governance in the Economy

Worldwide, more than a trillion dollars of revenue has been generated from the sales of public firms in the past two decades. The vast majority of the revenue from privatizations has come from advanced industrialized countries, predominantly Western European nations and Australia. Yet the impact of privatization may be most significant in the developing world, where public firms once accounted for a much larger share of domestic economies. In many developing countries, public firms dominated strategic sectors of national industry and played an essential role in promoting economic development, maintaining political stability and defending national sovereignty (Megginson and Netter 2001; *Privatisation International* 1998, 22, 24; Yergin and Stanislaw 1998).

Indeed, the importance of public firms in the developing world went far beyond their economic role. Beginning with the world depression of the 1930s and continuing through the post–World War II period, many developing countries built vast empires of state-owned firms, creating and expropriating companies in mining, oil, steel, electricity, rail, telecommunications, civil aviation, and almost every other sector of their economies.

Public firms were an integral part of the import-substituting industrialization model that was adopted to varying degrees throughout the developing world. As such, these firms supported specific political coalitions, providing high-wage jobs to workers, low-cost inputs to capitalists, political patronage to state bureaucrats, and subsidized goods and services to citizens. Public firms also defended many developing countries from economic domination by the multinational firms of industrialized countries and became potent symbols of national pride and developmental aspirations, occupying the commanding heights of both the economy and the polity (see Carrillo Castro and García Ramírez 1983, 148; Ramamurti 1996; Teichman 1995; Waterbury 1993; World Bank 1995a).

In this book, I examine one of the more remarkable cases of privatization in the developing world: Mexico's privatization program between 1983 and 2000. After more than fifty years of growing state intervention in the economy, Mexico abruptly abandoned public ownership in 1983 and reduced the number of state-owned firms from more than eleven hundred to barely two hundred by the year 2000. Mexico began privatizations earlier than most other developing countries and sustained the process over a longer period of time. Out of approximately $96 billion raised from privatization in the entire developing world between 1988 and 1993, more than one-quarter, $24.3 billion, came from Mexico. Between 1994 and 2000, Mexico continued to lead the developing world in privatization, opening basic infrastructure such as natural gas, the generation of electricity, satellite communications, ports, airports, and railroads to private investment (Rogozinski 1997, appendix 1; Spiller and Sales 1999; World Bank 1995a, 27).

The sweeping reforms introduced in Mexico since 1983 have been attributed to the "relative autonomy and insulation of the state vis-à-vis political pressures," "insulated change teams," and an "autonomous but embedded bureaucracy" (Petrazzini 1995, 6; Waterbury 1993, 3; Centeno 1994, 47). I argue that the designation of the Mexican state as "autonomous" distorts and oversimplifies the actual role of the state in carrying out privatizations. The application of the label *autonomous* to the Mexican state stems, in part, from a conceptual shift that has extended the meaning of *autonomy* from implying a concern with the independence of the state from elites to indicating state independence from *any* social pressure. Thus, the state's ability to overcome the interests of workers, peasants, and other popular-sector groups is sometimes said to constitute a form of autonomy.

Yet even where the pressure of elites upon the state is acknowledged, research on Mexico has still concluded that the Mexican state was autonomous by virtue of its coherent, effective bureaucracy. Here, too, I will present evidence that shows a more complex relationship between public and private actors. While the centralization of power in the Ministry of the Treasury was crucial to Mexico's privatization program, state actors also relied heavily upon the expertise and guidance of private consulting firms, international agencies and "agent banks." Rather than strengthening the bureaucratic capacity of the state, state managers actually subcontracted essential tasks to these other institutions, in effect privatizing part of the privatization process itself.

The question of state autonomy, however, is only one aspect of a larger question regarding the interrelationship between markets and governance in capitalist economies. In recent years, contributors to a growing body of literature in economics, political science, and sociology have emphasized that successful markets depend on social and political institutions that operate on principles quite different from those of the market.[1] Sociologists, in particular, have documented the myriad ways in which markets are *embedded* in social relations and nonmarket institutions that contradict the fundamental premises of neoclassical economics. These intellectual developments, however, seem to fly in the face of economic reforms around the world that are designed to unleash market forces: not only the privatization of public firms, but movements toward economic integration and deregulation that attempt to limit the extent to which nonmarket institutions determine economic allocations.

In this book I seek to bridge the gap between research in economic sociology that has demonstrated the poverty of neoclassical economic theory and the wave of economic restructuring that appears to conform to many of the principles underlying neoclassical economics. I do so first by building a theoretical model out of parts taken from economic sociology, classical political economy, and institutional economics. Rather than assume that nonmarket institutions invariably restrict the ability of actors to engage in market exchange, the model allows for an examination of the ways in which nonmarket institutions both limit *and facilitate* the operation of markets along the lines posited in neoclassical economics.

The argument that I make can be summarized as follows. The neoclassical conception of markets as atomized and asocial highlights an important

---

1. For a sample of the research by economists on the institutional underpinnings of markets, see Amsden 1989; North 1990; Wade 1990; and Williamson 1985. Literature on this topic in political science includes Goldthorpe 1984; Streek and Schmitter 1991. Sociologists who work in this area include Block (1990, 1994), Evans (1995), Fligstein (2001), Gereffi (1994), Granovetter (1985), Portes (1995a), and Zukin and DiMaggio (1990).

feature of social organization that economic sociology needs to acknowledge and incorporate into its research agenda. Capitalist markets are defined in part by a functional separation of actors from one another that permits them to operate in a manner that resembles neoclassical models. I borrow from institutional economic theory to show how legal and political structures—in particular *property rights* and *hard budget constraints*—create barriers to cooperation and collaboration among actors. These barriers enable and encourage exchange based on competition and utilitarian calculations of price and profitability not unlike that which neoclassical theory posits as the natural state of affairs. At the same time, I argue that there are limits to which an economy can function upon these principles, as illustrated by the concepts of *market failure* and *transaction costs*. I argue that there is an inherent tension between economic organization based on market exchange and the institutions that stabilize market exchange by correcting market failure or overcoming transaction costs. At certain historical junctures these tensions can no longer be reconciled, and new regulatory institutions and new markets are created.

Arguments regarding the structure of the economy beg the question of who the agents are that build markets and lead back to the issue of state autonomy. Here I argue that economic reforms in Mexico were the outcome of social struggles for and against the commodification of specific areas of the economy. Actors who stood to benefit from the extension of market principles to the production of certain goods and services pushed the state to separate, or "disembed," these areas of economic activity from the public sphere and the social obligations attached to nonmarket institutions. Actors whose interests were threatened by the creation of markets worked against this process (Polanyi 1957, 132). I argue that the "autonomy" of the state, understood as its capacity to implement policies, cannot be disentangled from the content of these policies and the coalitions and alliances that support and oppose them.

These theoretical formulations, however, are useful only insofar as they serve to guide empirical research. Therefore, I use the theory developed in this introductory chapter as a framework of analysis for examining the nuts and bolts of how markets were created in Mexico's privatization program. In Chapters 2 through 6, I draw on seventy-eight key-informant interviews conducted with fifty-eight unique sources to provide a detailed picture of the privatization process in Mexico between 1982 and 2000. These interviews are supplemented with data from government archives, legislation, press reports, corporate documents, trade publications, and secondary sources to foster an understanding of the reach and limits of Mexico's economic reform program (see the Appendix for details on the methodology).

In the remainder of this chapter, I outline key concepts that will be used for studying the relationship between markets and governance in the economy.[2] I show that the governance functions of the state actually belong to a broader category of nonmarket exchanges that are performed not only by the state but also by other social institutions within the economy, the most important of these being private firms. I argue that all market-based economic activity depends in some measure on nonmarket exchanges but that markets in a capitalist economy function in a manner that makes exchange *appear* independent of the social institutions upon which it depends. This model, then, requires that we examine how the chain of market and nonmarket exchanges is cobbled together so that private actors can appropriate resources from social institutions while, simultaneously, shielding these resources from public oversight or confiscation.

Chapter 2 provides a broad overview of the rise of state ownership in Mexico between 1917 and 1982. In this chapter I look at the legal and political foundations of public ownership, the number and types of firms incorporated into the state sector, and the role that public firms played in managing the state's relationship to labor, domestic capitalists, and the international economy. In Chapter 3, I examine the demise of the state-led development model from 1982 through 2000, looking at where and how the state privatized firms and introduced market reforms during the final three presidencies of the Revolutionary Institutional Party (PRI).

Chapters 4, 5, and 6 sharpen the focus of the study; here I examine the growth and decline of state intervention in specific industries within the transportation and communications sectors. These sectors are especially interesting for both theoretical and practical reasons. Theoretically, they are areas of the economy in which there is broad agreement about the legitimacy

---

2. Williamson (1975, 1985) refers to "markets and hierarchies" to contrast these two ideal types of economic action. I use the words *governance* or *coordination* in place of *hierarchy* because the word *hierarchy* suggests that governed economic activity is necessarily more coercive than exchange that occurs on the market. The extent to which one form of exchange is more or less coercive than another, however, is a question of empirical inquiry. For example, a democratically organized cooperative would be a less hierarchical form of economic organization than a market in which there is a high degree of resource inequality, even though the cooperative would be considered a "hierarchy" in Williamson's framework. Thus, the words *governance* or *coordination* are simply more accurate as a description of what firms, states, and other nonmarket institutions actually do in the economy and leave open the question of *how* these institutions govern economic exchange, whether hierarchically or not.

of state intervention. These sectors tend to produce positive externalities, with clear implications for national development, making the introduction of market reforms in transportation and communications more difficult than in other sectors of the economy. The practical justification for focusing on transportation and communications is equally compelling. Privatization within these sectors marked a major turning point in the process of selling state firms in 1988, included the single largest privatization in Mexico in 1990, and became the focus of intensified reform efforts between 1995 and 2000.

In Chapter 4, I examine the privatization of the Mexican airlines Aeroméxico and Mexicana de Aviación. Initiated at the end of the de la Madrid presidency, the sale of Aeroméxico, following its bankruptcy, was one of the first sales of a large firm that involved two of the three elements that would become key components of later privatizations: organizational restructuring of the firm and confrontation with labor unions. The subsequent privatization of Mexicana de Aviación added the third element that characterized this later phase of privatization: investment by foreign firms. By the early 1990s, however, intensified competition undermined profitability, and the two carriers attempted to blunt the effects of debilitating price wars by increasing the coordination of their activities. By 1995, both firms had gone bankrupt and were rescued by a government bailout.

In Chapter 5, I discuss the single largest privatization completed in Mexico: that of the telecommunications monopoly Telmex. The sale of Telmex was highly symbolic in that it allowed the Salinas administration to give economic restructuring a human face. The privatization of Telmex required even greater foreign investment than had Mexicana de Aviación and demanded a more complex restructuring of the firm. The privatization of Telmex also produced a more ambiguous outcome for the workforce, securing important guarantees for workers while also tying their interests more closely to that of the firm. While Telmex has remained profitable, its success appears to be the result of numerous restrictions on the influence of competition and price in the market.

The privatization of the Mexican National Railroad (FNM) is the subject of Chapter 6. The sale of FNM involved a complete transformation of the Mexican rail system. Throughout the twentieth century, the Mexican state consolidated the vast network of private rail companies into the operating structure of FNM. Following the debt crisis of the 1980s, this process was accelerated and the last remnants of private ownership were eliminated. Once consolidated, these assets could be evaluated and separated into individual, marketable rail companies. While the sale of Telmex provided trade-offs to workers in exchange for restructuring, the privatization of FNM depended on the iron grip of the official railroad labor union and the political power of the PRI to reward the leadership of the union.

Chapter 7 revisits the question of how states and markets may be separated from each other even while they remain interdependent. Drawing on examples provided in the case studies, in the concluding chapter I look first at the structural reorganization of economy and society in Mexico, then review the agents who supported and opposed the economic transformation and, finally, touch briefly on the political consequences of economic change. While the election of Vicente Fox Quesada in 2000 has reinforced the notion that economic and political liberalization go hand in hand, I argue that the election of Fox can be attributed more to the failure of economic reforms than to their success. While market reforms steadily eroded the living standards of large numbers of Mexicans, political reform became a key inducement used by the ruling party to reinforce its legitimacy. Although these reforms ultimately proved costly to the PRI, political reforms were implemented only after the scale and scope of state intervention in the economy had been significantly reduced, limiting the reach of a democratically elected government.

I encourage readers who are more interested in the empirical findings than in the theoretical implications to skip directly to Chapter 2. Although the organization of Chapters 2 through 6 is guided by the theoretical framework developed in the following pages, each of the empirical chapters should stand on its own.

Embeddedness, Atomization, and the Uncertain Boundaries Dividing Markets and Governance

ECONOMIC SOCIOLOGY AND NEOCLASSICAL ECONOMICS

Research in economic sociology has developed primarily as a response to neoclassical economics.[3] Neoclassical economic theory rests on the assumption of atomization in the marketplace. Even before economic actors can behave in a manner consistent with another foundational principle in neoclassical theory—utility maximization—they must exist as independent

---

3. The distinction is sometimes made between the new and the old economic sociology. See Granovetter 1990 for a discussion of the differences between the two. I use the term *economic sociology* to refer to recent research in this field and distinguish this from the old economic sociology only where there are important differences in interpretation or perspective.

agents. As the term suggests, atomization assumes an essentially asocial perspective on the economy. It implies that buyers and sellers are socially unrelated apart from the business they transact in the market and that markets operate without interference from the state or any other institution (Granovetter 1985, 484).

In the neoclassical perspective, the absence of social relations and nonmarket institutions allows actors to pursue their self-interest and, paradoxically, when individuals pursue their self-interest society benefits. Free exchange among individuals is said to naturally coordinate economic activity more efficiently than conscious efforts to manage the economy ever could. Drawing on the classical political economy of Adam Smith, neoclassical theory holds that markets tend toward equilibrium with price and competition harmonizing the countless transactions of unconnected individuals into the optimal production and allocation of the resources needed by society (Friedman 1982, 200; Hayek 1994, 55; Mercuro and Medema 1997, 13–15).

Economic sociology has challenged both the structural assumption of neoclassical economics regarding atomization in the marketplace and, to a lesser extent, the behavioral assumption regarding individual utility maximization. Empirical work in this field has shown that a wide range of institutions and social relations link actors in the marketplace and coordinate exchange, rendering the premise of atomization worthless. Similarly, this literature has uncovered numerous exceptions to the principle of utility maximization and shown these to be much more than random deviations from the neoclassical model (see Block 1990; Evans 1995; Fligstein 2001; Granovetter 1985, 1995b; Keister 2001; Portes and Sensenbrenner 1993; Uzzi 1996; White 1981; Zukin and DiMaggio 1990).[4]

Instead of atomization, research in economic sociology emphasizes the *embeddedness* of all economic action. Embeddedness refers to the web of social relations and nonmarket institutions that mediate exchange. These social relations and nonmarket institutions run the gamut of economic activity: from the psychological processes that prevent actors from calculating—let alone maximizing—their utility, to the politicolegal structures that define the boundaries of markets and states. In between, embeddedness may refer to patterns of interaction among the state and private actors, the regulatory functions provided by professional associations and civic

---

4. Research in economic sociology has also developed in response to efforts in institutional economics to account for social structures that deviate from the neoclassical premise of atomization in the marketplace. Institutional economists pointed out long ago that the most glaring omission of neoclassical theory is the business firm. See Coase 1937; Williamson 1985; and Granovetter's (1985) critique of Williamson.

organizations, shared conceptions of control that guide behavior in markets, kinship networks, or ethnic communities that restrain and facilitate economic exchange.

Although the dispute over the structure of markets—whether atomized or embedded—may appear obscure, the stakes in adopting one perspective over another could not be higher for economic policy. The premise of atomization in the marketplace serves as the foundation for arguments against state intervention in the economy. If the uncoordinated actions of individual buyers and sellers pursuing their self-interest is the key to the optimal allocation of resources, then state efforts to regulate, coordinate, direct, or otherwise interfere with the natural order of the market can only undermine economic efficiency. The most ardent proponents of the neoclassical perspective go further still, arguing that state intervention in the economy is roughly synonymous with coercion (see especially Friedman 1982; Hayek 1994).

Economic sociology, in contrast, has shown that innumerable social relations and political institutions already bring order to the market. Because economic action is already coordinated by nonmarket institutions, a wide range of choice is available to policy makers seeking to direct the economy toward specified goals. Turning neoclassical theory on its head, economic sociology argues that markets operate best when social relations and nonmarket institutions stabilize exchange, usually by *preventing* individuals from engaging in purely self-interested behavior and limiting the influence of price competition on allocations (Block 1990, 52, 67; Evans 1995, 26; Fligstein 2001, 23; Granovetter 1985, 1995b; Portes and Sensenbrenner 1993).

This theoretical orientation has led researchers studying the implementation of market-oriented reforms to emphasize the persistence of state intervention in the economy. Even in extreme cases of economic transformation such as those of Eastern Europe and China, where privatization and other neoliberal economic reforms would appear to signal a sharp break with historical patterns of state intervention in the economy, this literature claims that reform is characterized by "path dependence" and historical continuity.[5] As Andrew Walder (1994, 55) argues in an essay on economic reform in China, "[T]he distinctions commonly drawn between 'market versus plan' or 'public versus private ownership' are not always meaningful and may prove to be blunt tools to use in analyzing the organization of an economy" (see also Guthrie 1999, 22; Stark 1996).

---

5. The term *neoliberal* refers to a policy orientation that draws on neoclassical economics, usually formulated as minimal state intervention in the economy. See Balassa et al. 1986.

While this research provides a much needed corrective to the assumptions that underpin neoclassical conceptions of the economy, the formulation of the embeddedness perspective raises some serious problems of its own. To begin, the fact that nonmarket institutions shape economic exchange, however thoroughly documented, says nothing about *how* they shape exchange. Unlike the premise of atomization in neoclassical economics, there are no clear hypotheses that can be derived from the premise of embeddedness. By itself, this is not necessarily a failing of economic sociology. Indeed, research in this field has produced a rich body of empirical studies on economic organization precisely because it works inductively, examining the specific institutions that shape exchange in different contexts.

However, there is a general tendency in this literature to assume that social institutions invariably inhibit the influence of self-interest, competition, price, and profitability in determining economic allocations. Yet, given the emphasis in economic sociology on the malleability of economic organization, it is entirely possible that nonmarket institutions could construct exchange in a manner that *increases* the importance of market signals. Given the dominance of the neoclassical paradigm among policy makers, we need to take seriously the possibility that the "conception of control" in some markets may actually be the atomized, asocial market of neoclassical economics.[6]

A further concern has to do with the tendency in economic sociology to blur the boundaries between different forms of economic organization. This literature has challenged many of the distinctions that are commonly drawn between politics and markets, markets and hierarchies, capitalism and socialism, precapitalist and capitalist economies. Economic sociology tends to emphasize the common attributes of all forms of economic organization. But this perspective can obscure what is truly unique about capitalist economies. However blunt the distinction between private and public may be, private ownership of productive property is one of the defining features of capitalism. Private ownership of productive property is, moreover, the key policy goal of privatization. If the distinction between public and private is not meaningful, then it becomes difficult to understand what exactly is at stake in the process of privatization, let alone specify how markets differ from other forms of economic organization.

6. Even while arguing against neoclassical economics, Fligstein (1996b, 658) and Block (1994, 705) both suggest that actors could create markets with rules that resemble the neoclassical ideal. See also Evans 1995, 10; Guthrie 1999, 60; Swedberg, Himmelstrand, and Brulin 1990, 70; and especially Zukin and DiMaggio 1990, 14–23.

Finally, by emphasizing the stabilizing influence of institutions upon exchange, economic sociology produces a strangely static, at times functionalist, model of economic action. According to Harrison White (1981, 518), "Markets are self-reproducing social structures among specific cliques of firms and other actors who evolve roles from observations of each other's behavior." Following this line of reasoning, Neil Fligstein argues that "the purpose of action in a given market is to create and maintain stable worlds within and across firms": actors seek stability in the market because the problem of competition is endemic (2001, 35, 230; see also Fligstein 1996a, 663 n. 8). But where does competition come from in a world where actors are driven to create and maintain stable relations? In its zeal to show the failings of neoclassical economics, economic sociology has often removed some of the most important elements of economic action from its models—notably self-interest, price, profitability, and competition—leaving it with little to say about how markets may be destabilized, destroyed, or restructured. The sources of economic change and transformation then have to be reintroduced into these models as exogenous forces (Fligstein 2001, 84).

In the following pages, I develop a framework of analysis that draws on the strengths of economic sociology while attempting to address some of these weaknesses. While this framework maintains the focus of economic sociology on the social-structural foundations of markets, it leaves open the possibility that social institutions might actually construct exchange in a manner that appears atomized and asocial. I use the basic elements of the market found in neoclassical economics—self-interest, price, profitability, and competition—as indicators of ideal types not only to evaluate the ways that social relations and nonmarket institutions inhibit the operation of the market, but also to examine how market forces may, themselves, be socially constructed.

Notwithstanding the social relations inherent in all institutions, I argue that capitalist markets are a unique form of economic organization in that they require the separation of actors into functionally differentiated units capable of operating upon narrowly defined principles of exchange for the sake of gain. When markets function in this manner, allocations *appear* to operate independently of the political and social institutions upon which they depend. This does not imply that market allocations somehow magically become either apolitical or asocial. Indeed, if, as Lasswell (1950) claims, politics is about who gets what, when, and how, then the organization of markets in this manner is a supremely political process. Nor does it imply that the state has no role in the creation or maintenance of markets. It does mean, however, that the state delegates authority over essential aspects of social organization to private actors, fragmenting society into "political" and "economic" realms. Under these rules of exchange, in the normal course of events the state abstains from direct intervention in some of the most important aspects of social organization (Wood 1981).

Readers familiar with Marxist theory should recognize the Marxist influence on this formulation. Marx emphasized the inherent contradictions in capitalism between socially produced and privately appropriated value and referred to the appearance of an autonomously functioning market as "commodity fetishism." Despite the fact that the exchange of commodities "is a definite social relation between men . . . [it] assumes, in their eyes, the fantastic form of a relation between things" (Marx 1967a, 77). But while Marx sought to link commodity fetishism to the labor theory of value, I am more interested in exploring the institutional underpinnings that make the market appear autonomous: the specific organizational mechanisms through which markets mask the social character of exchange and the limits to this type of organization.

ECONOMIC SOCIOLOGY, MARXIST THEORY, AND KARL POLANYI

In his exegesis of Marxist theory, *The Limits to Capital,* David Harvey (1982, 141) asserts, "Capitalism cannot do without market coordinations and still remain capitalism." Research in economic sociology, in contrast, has demonstrated that markets operate very differently depending upon the institutions within which they are embedded. In this section, I attempt to identify what is unique about the "market coordination" that Harvey claims is indispensable to capitalism without abandoning the insight of economic sociology that markets can only be understood within their social context.

To begin, it is necessary to distinguish between capitalism and the market. Although the two terms are frequently used interchangeably, the market is only one aspect of a larger process that defines capitalism. Generally, capitalism is defined as a social system in which three conditions prevail: productive property is privately owned; wage labor is the primary means of marshaling labor power; and exchange relations are widely commodified, with money serving as a universal equivalent for the measurement and storage of value (Harvey 1982; Heilbroner 1985; Marx 1967a; Przeworski 1990).

In Marxist theory, a host of implications follows from identifying these as the central pillars of capitalism, the most important being that the production of *use value* is subordinated to the production of *exchange value.* That is, goods are not produced or distributed as needed by society but, instead, where individuals can profit. This concept is captured in Marx's most parsimonious formulation of the dynamic of capitalism, MCM'—money (M) is used to produce commodities (C) not as an end in itself but instead as a means earning more money (M').

A market is a point of contact where buyers and sellers meet for the purpose of exchange. But while markets have existed throughout recorded history, Marx argued that the operation of the market in a capitalist economy is unique. He described market exchange with both sarcasm and admiration for the peculiarity of the market in a capitalist economy where

> buyer and seller of a commodity, say of labour-power, are constrained only by their own free will. ... each enters into the relation with the other, as a simple owner of commodities, and they exchange equivalent for equivalent ... each disposes only of what is his own ... each looks only to himself. The only force that brings them together and puts them in relation with each other, is the selfishness, the gain and the private interests of each. (Marx 1967a, 173)

These highly schematic descriptions of capitalism and the market can be challenged on a number of levels. Economic sociologists would be entirely correct in pointing out that none of the elements that defines capitalism is absolute, because the circuit of MCM' is always embedded within specific social institutions. The power of capitalists to dispose of their property is limited in law and practice. Labor relations are similarly regulated by law, collective bargaining agreements, and social struggle. And money, dependent as it is on central banks and public faith, is only partially commodified. Indeed, some of the same criticisms aimed at neoclassical economics regarding its asocial conception of the economy have also been directed at Marxist theory (Block 1994; Granovetter 1985).

But Marx was emphatic that economic relations are not *actually* disembedded from the society in which they exist. Instead, his contention was that in a capitalist economy, exchange relations on the market *appear* to be asocial. Indeed, Marx's reply to Fernando Galiani could just as easily be directed at recent work in economic sociology: "When, therefore, Galiani says: Value is a relation between persons ... he ought to have added: a relation between persons expressed as a relation between things" (1967a, 79 n. 1).

And although markets are necessary to the operation of capitalism, they represent only one phase through which goods pass in the process of expanding value. Goods also pass through nonmarket institutions in which the social relations among actors are perfectly clear. One of these institutions—as essential to a capitalist economy as the market—is the firm. The importance of the firm to Marx is that it does not actually function according to principles of market exchange even though it is organized for the express purpose of capturing profits on the market. Indeed, firms facilitate the generation of profits precisely by suspending the rules of market exchange: by coordinating the activities of actors, inhibiting (or, at the very least, channeling) their self-interest and drawing on gains from the detailed division of labor.

The insight that Marx provides, then, is that capitalism is defined by the *combination* of both market and nonmarket exchanges. Moreover, while these two principles of social organization are interdependent, they are also deeply antagonistic. Although capitalism is predicated on private ownership, the growth and development of markets tends to make economic life more social than ever. Capitalist development extends the division of labor, draws increasing numbers of people into relations of interdependence, and requires the creation of institutions that can provide the authoritative legal and political framework on which markets depend. The growth of firms extends the reach of the planned economy, linking workers and capitalists in the collective endeavor of production.

In an economic system as intensely social as capitalism, the market is critical, then, because it facilitates the integration of actors under rules of exchange that deny the very social relations that make exchange possible. In a capitalist market, actors can bring their resources into relation with one another without sacrificing their control over those resources. Organized in this way, markets produce a form of impersonal resource allocation that constrains actors, whether collective or individual, to focus on narrowly defined gain and self-interest (Meyer, Boli, and Thomas 1987, 21).[7]

This formulation of the structure of the capitalist economy suggests that states and other nonmarket institutions must strike a delicate balance between these contradictory principles of social organization. Social institutions are necessary to capitalism in that they integrate actors, providing the social structure that economic sociology has demonstrated makes exchange possible. At the same time, social institutions must also fragment actors to a sufficient degree that their nonmarket connections to other actors do not impede the type of rational calculation that is the hallmark of capitalism (Arrighi 1994; Harvey 1982; Piore 1996).

---

7. In a similar vein, contemporary Marxists have argued that while individual states may exercise a degree of control over their national economies, it is their participation in the international economy—over which no single state has control—that sustains capitalism at a global level (see Arrighi and Silver 1999; Chase-Dunn 1989, 135; Wallerstein 1974, 15).

These distinctive types of exchange may be distinguished from one another using the concepts of exchange and use value. I employ these terms as ideal types to emphasize primarily the *purpose* or *motivation* behind production, but also to highlight the mechanisms that structure economic action. *Exchange value* refers to production and exchange that are motivated primarily by the immediate increase of money value, exchange for the sake of gain. *Use value* refers to production and exchange that are directed by something other than exchange for the sake of gain.

While this formulation borrows from Marx, it is hardly unique to the Marxist perspective. It is analogous to Max Weber's (1968, 85) distinction between formal and substantive rationality in economic action as well as Karl Polanyi's ideal types of exchange, though my definition of use value encompasses both the redistributive and the reciprocal exchanges that Polanyi uses to contrast market exchange.[8] It is in this distinction between different types of exchange that recent work in economic sociology departs most sharply from the "old" economic sociology of Karl Polanyi. In his chronicle of the emergence of capitalism in Europe, Polanyi argues that social institutions were used to create precisely the type of asocial, atomized exchange that neoclassical theory posits as the natural state of affairs and economic sociology claims cannot exist: "A self-regulating market demands nothing less than the institutional separation of society into an economic and political sphere" (1957, 71).

Polanyi's description of "fictitious commodities" is especially valuable for understanding how the concepts of use and exchange value supplement the model of the new economic sociology. Consistent with the new economic sociology, Polanyi emphasizes the embeddedness of the economy, in particular of labor, land, and money: "The commodity description of labor, land, and money is entirely fictitious." And yet Polanyi goes on to argue that

> it is with the help of this fiction that the actual markets for labor, land, and money are organized; they are being actually bought and sold on the market; their demand and supply are real magnitudes; and any measures or policies that would inhibit the formation of such markets would *ipso facto* endanger the self-regulation of the system. The commodity fiction, therefore, supplies a vital organizing principle in regard to the whole of society affecting almost all its institutions in the most varied way, namely, the principle according

---

8. I have no intention of entering into the labyrinth that Marx and his followers have constructed around the precise meanings of these terms. For a sample of the debates regarding use and exchange value, see Baudrillard 1981, 130–42; Lukács 1971; Žižek 1989.

to which no arrangement or behavior should be allowed to exist that might prevent the actual functioning of the market mechanism. (72, 73)

Emphasizing these distinctive characteristics of capitalist economies is not to argue that markets are disembedded in the sense that exchange could take place without social regulation or state intervention. To the contrary, Polanyi pointed to the growth in state intervention that was necessary to impose these conditions upon exchange relations (1957, 140). And he went on to argue that markets constructed in this manner were inherently unstable. The havoc wrought by enforcing the laissez-faire ideal provoked countermovements among various sectors of society seeking social protection from the ravages of the market.

Markets in a capitalist economy may be disembedded, then, to the degree that they channel social action into price-based competition. This appears to require breaking the social structure into units capable of acting as independent agents. As Harvey (1982, 18) argues, "Exchange of commodities presupposes the right of private proprietors to dispose freely of the products of their labour. . . . This means that 'juridical individuals' (persons, corporations, etc.) must be able to approach each other on an equal footing in exchange, as sole and exclusive owners of commodities with the freedom to buy from and sell to whomsoever they please."

INSTITUTIONAL ECONOMICS AND THE MECHANISMS THAT STRUCTURE EXCHANGE

The difficulty of employing this perspective is that it operates at such a high level of abstraction. In this section, I seek to ground these broad theoretical considerations in more manageable concepts, drawn from institutional economics. The concepts of *property rights* and *hard budget constraints* illustrate the importance of exchange value, or disembedded economic action in a capitalist economy. They point to the mechanisms that structure exchange relations so that price, profitability, and competition exert a dominant influence on the allocation of resources. In contrast, the concepts of *market failure* and *transaction costs* point to the impossibility of structuring an economy in which all resource allocation is guided by these principles. Market failure and transaction costs point to the necessity of resource allocation that is guided by considerations of social use value, guided by coordination and planning and is, thus, embedded in social relations that restrict the influence of the market. This analytical framework is presented schematically in Table 1.

TABLE 1 Ideal types of exchange in the economy

Principle Guiding Economic Action	Exchange Value (Disembedded)	Use Value (Embedded)
Mechanisms		
Macroeconomic	Hard Budget Constraints Actors are cut off from resources outside the market and adjust behavior to accommodate market allocation.	Market Failure States provide goods and services that are socially necessary but independently unprofitable.
Microeconomic	Property Rights Actors retain rights of exclusion and part with their property only when profitable.	Transaction Costs Boundaries of firms extend to overcome coordination problems.
Indicators	• Goods distributed on the basis of price; • Actors' survival dependent on profitability in the market; • Relationships among actors in the market guided by competition; • Actors are free to pursue individual gain.	• Goods distributed on basis other than price; • Actors sustained through nonmarket transfers unrelated to profitability; • Relationships among actors guided by cooperation and coordination; • Actors subordinate individual gain to goals of larger group.

Property Rights

According to North (1981, 21), "[T]he essence of property rights is the right to exclude." At first glance, this definition appears to suggest that *property rights* refers to the use toward which owners put their property. The importance of exclusion to property rights, however, does not derive from the use that owners have for their property, not directly. Rather, the importance of exclusion to property rights is that it allows owners to use their resources for the purpose of increasing exchange value in the market: to use property as capital. As Kornai (1992, 63) argues, the enforcement of property rights "excludes nonowners from making arbitrary use of property without the owner's consent."[9]

Note the very particular meaning of *arbitrary* here. In this formulation, the arbitrary use of property refers to actors using resources for some reason *other* than profit making. Defining arbitrary in this manner transforms instrumental values of exchange for the sake of gain into ultimate values toward which actors should strive in the market. Under this definition, if a well-fed owner of a bakery cannot exclude a penniless but hungry nonowner from the products of the bakery, then the property is at risk of being used in an arbitrary manner. Owners part with their property only when nonowners have something to offer in exchange, usually money.

In contrast, an essential aspect of public goods—goods that are not produced for their exchange value—is the fact that nonowners cannot be excluded from their use, the lighthouse being the classic example. The capacity to exclude others, however, should not be understood only in the narrow sense of those cases where exclusion is technically difficult. Kinship and other forms of group affiliation create obligations that may make the exclusion of nonowners difficult as well. Indeed, even though public ownership may predominate in sectors of the economy where it is technically problematic to exclude others, the simple fact of public ownership makes exclusion more difficult by subjecting production decisions and performance goals to a broader range of criteria and input than may exist for a private firm.

Both North and Kornai, then, point to the central paradox of property rights. "For one to be able to speak of property at all, there has to be a social mechanism that enforces the assertion of the property rights" (Kornai

---

9. Although economic sociologists employ the concept of property rights, consistent with the embeddedness perspective, they tend to minimize the importance of exclusion to property rights, emphasizing instead the *uses* to which property is put. According to Walder (1994, 55), "Ownership, and by extension contract, is the right to exclude others, *but this right is always restricted by regulation or law*" (emphasis added). See also Fligstein 1996b, 658.

1992, 62; North 1981, 21). This mechanism is, of course, the state. Yet while the rights of property rest on the principle of exclusion, the legitimacy of the state rests on the principle of inclusion. The state must claim to represent general interests, not particular ones. Consistent with neoclassical economic theory, the ideology that supports the role of the state in enforcing the exclusion of nonowners from using the property of others claims that the defense of private property ultimately improves the general welfare (see Meyer, Boli, and Thomas 1987; Przeworski 1980).

Another paradox of property rights should be highlighted here. Although property rights refer fundamentally to exclusivity, property within a capitalist economy loses its exchange value if it is withheld too long from society. Capitalists may pressure other actors in society by denying access to productive assets. However, for property to be productive, capitalists must put their property into relation with other actors and their property. Most important, they must bring their property into contact with workers who sell their rights over their own labor power and consumers who may have some discretion to choose one product over another (Marx 1967a, 151).

Hard Budget Constraints

A second concept from institutional economics that illustrates the necessity of generating exchange value in a capitalist economy is referred to as *hard budget constraints*. According to Kornai (1992, 143, 144) the budget constraint

> refers to a firm's external conditions: What will its environment tolerate? ... Does the firm operate in a social environment that recognizes it will fail and exit from the economy in the case of continuing losses and financial catastrophe? Or will the environment be unable to accept this and bail it out? One manifestation of the softness of the constraint is a manifestation of the [sic] fact that a firm's survival and expansion do not depend on the market; it is decided in the frames of bureaucratic coordination and financial bargaining with the authorities.

As with property rights, the principle of hard budget constraints establishes instrumental, profit-maximizing behaviors as the basis upon which exchanges must take place. In a capitalist market, not only allocations but survival itself must be determined by formally rational calculations of price and profitability. To the extent that actors deviate from these rules, to the extent that their activities fail to generate the profits necessary for their independent survival, they must be denied nonmarket assistance and forced to adjust their behavior to conform to market allocations.

Although Kornai applies the principle of hard budget constraints to firms, the concept may be extended to other actors in the market: individuals, households, and states all operate within varying degrees of hard or soft budget constraints. Furthermore, state subsidization is hardly the only means by which a soft budget constraint may be created. Any actor with resources in reserve can engage in some arbitrary activity and ignore losses that would undermine the less affluent. A firm may choose to subsidize loss-making activity with the revenue that it generates in profitable segments of the market, funneling money to a "loss leader" that may not yield profit over the short run, if ever. By so doing, a firm applies soft budget constraints to its own activities.

The precise level of profitability that a particular market may demand of actors, the time frame over which actors must show profitability, and even the range of activities to which hard budget constraints may apply are all highly variable. Market conditions can change. Loans may be restructured or contracts renegotiated without the financial bargaining with the authorities that Kornai claims are characteristic of soft constraints. Despite the variability of budget constraints over time and across markets, however, they represent the structural limits to actors engaging in nonmarket exchange. They are unique to capitalist markets not simply because firms or individuals are unable to enlist the state's support of some activity that is independently unprofitable. Rather, hard budget constraints apply when actors are cut off from nonmarket resources and forced to accept the allocations provided on the market.

Like the principle of exclusion in property rights, hard budget constraints are also paradoxical. They are a market principle that needs to be understood in relation to nonmarket allocations, in particular, to the ability to gain access to resources through nonmarket channels. And, although hard budget constraints imply individual failure within the market, the rationale that justifies the failure of some actors rests on the notion that imposing hard budget constraints ultimately promotes the general welfare. Not all actors or activities can survive in a properly functioning market. Instead, hard budget constraints are said to eliminate inefficient producers and, thus, improve the general welfare by rewarding efficient producers and raising the aggregate level of productivity.

Market Failure
The classic reference to state intervention in the economy is Adam Smith's widely cited passage from *An Inquiry into the Nature and Causes of the*

Wealth of Nations. In it, Smith asserts that the duties of the sovereign include "erecting and maintaining those public institutions and those public works, which, though they may be in the highest degree advantageous to a great society, are, however, of such a nature, that the profit could never repay the expense to any individual or small number of individuals and which it therefore cannot be expected that any individual or small number of individuals should erect or maintain" (1981, 723).

Although this passage is said to identify *market failure* as the justification for state intervention (Feigenbaum and Henig 1997, 339), it actually contains two analytically distinct elements. One element of Smith's argument refers to the technical question of whether a good or service can be produced for a profit—whether the benefit will outweigh the cost. The other element of the definition rests on how we define "advantageous to a great society." While the question of profitability refers to the exchange value of production, whether or not a good is "advantageous to a great society" is a determination of ultimate values or the use for which goods are produced.

Distinguishing between the two parts of Smith's argument is especially important in understanding the limitations of the utility of the concept of hard budget constraints. For markets to allocate resources on the basis of hard budget constraints, actors that are not operating profitably must adjust their behavior or fail (Wiles 1962, 20). Yet certain loss-makers continue to operate and do not fail. As Smith points out, some loss-makers are propped up because they perform functions within the economy upon which private exchange depends and, therefore, are taken up by the state.

Loss-making activity may, of course, be sustained for purely political reasons that are unrelated to maintaining productive activities. The state may subsidize capitalists in any number of ways, just as it may support workers through social welfare benefits. It is precisely this type of arbitrary, or politicized, activity that Kornai criticizes in the socialist model and that leads him to argue that the state must allow firms to fail in order to ensure economic efficiency. Yet he does not establish the limits to this process. Which firms cannot fail? What unprofitable activities must be undertaken by means other than for-profit institutions?

Any effort to answer this question merely highlights additional contradictions of the market mechanism. First, it points to the need for some institution to make political decisions regarding the redistribution of resources to provide collective goods. Second, it challenges the exclusivity of property rights because sustaining unprofitable activities implies the existence of some mechanism such as taxation that extracts resources from private owners and transfers them to other sectors of society.

Moreover, the question of necessary losses goes beyond the technical requirements of economic exchange—infrastructural projects, the provision of weights and measures, or the enforcement of contracts—to losses necessary for the legitimacy of the political system and the social relations that sustain market exchange (Bourdieu 1980, 123; North 1981, 44). The amount and type of failure that a society can tolerate, the safeguards or forms of social insurance that would undermine the market mechanism and the legitimacy of the political institutions that support it, are not at all clear. Nor is there any certain range of time or set of goods over which losses may be sustained for the purposes of greater long-term efficiency or social welfare. As Kornai acknowledges, "There is no universally valid dividing line between private and public goods. The dividing line varies through history, depending on the state of technology and the social conditions" (1992, 62 n. 1).

Transaction Costs
Another line of inquiry into why certain goods are provided through market exchange and others are allocated by alternative mechanisms focuses on *transaction costs*. Arrow (1969, 48) sees market failure as a subcategory of transaction costs that "in general impede and in particular cases block the formation of markets." North sees transaction costs as an essential element of microeconomic theory but also employs the concept to deal with macroeconomic issues and with more general questions of institutional development (1981, 5; 1990, 28). While there is no dominant school of thought on the nature of transaction costs, I focus here on transaction costs in the microeconomy and especially the "private ordering" examined by Williamson in order to distinguish this concept from that of market failure.

Williamson (1985, 19) argues that transaction costs are "the economic equivalent of friction in physical systems." This "friction" is defined, in large part, by uncertainty that alters the structure of incentives, making the arm's-length transactions of neoclassical economic theory problematic. Williamson goes on to specify three attributes of the contracting process that increase the level of uncertainty—bounded rationality, opportunism, and asset specificity—all of which create problems for economic exchange that is premised on the utility-maximizing action of atomized actors. This conceptualization is useful in that it draws our attention to the fact that different goods have different characteristics that can influence the way in which they are exchanged (Kollock 1994; Perrow 1990). Furthermore, by conceptualizing the firm as a governance structure, transaction cost analysis acknowledges the political nature of the economic power inherent in the organization of production.

At the same time, the internalization of exchanges into more comprehensive governance structures—whether private, as with firms, or public, as with states—creates social relations that inhibit the self-interest of individual actors and defy the exclusivity of property rights. The long-term relationships implied by the private ordering of economic activity may stabilize exchange and alleviate certain risks that are inherent in the market. But they can also grant associates, partners, subcontractors, or employees increased leverage over the organization of production and increase the importance of managing these relationships (Perrow 1990). Because nonmarket institutions are legitimated by nonmarket principles, these institutions may also become the locus of social struggle, with actors making claims on resources not in terms of market exchange but, rather, because of their status as members of a group, employees of a firm, or citizens of a state.

Moreover, in order to realize long-term gains, institutions generally spread costs and revenues over time and space, thus transforming the parameters of hard budget constraints. In the process of stabilizing exchanges, the very units to which different principles of exchange are applied change. Exchanges that in one part of institutional economic theory should fail as a consequence of hard budget constraints become part of a larger social unit that is sustained by nonmarket exchanges under transaction cost analysis.

The notion that nonmarket institutions are constructed to overcome uncertainty in the market raises the question, as with market failure, of the extent to which actors can be protected from the force of the market without impairing the mechanism of hard budget constraints. If hard budget constraints depend on the possibility of failure and exit from the market, transaction cost economics suggest that a wide range of economic activity may be shielded from these forces and internalized into the larger structure of a firm.

DEFINING THE CAPITALIST MARKET: SOCIAL CONNECTION
AND FRAGMENTATION

The purpose of highlighting the contradictions between different components of institutional economics is not to dismiss institutional theory. To the contrary, I believe that the contradictions between different aspects of this literature accurately reflect contradictions that are built into the economic structures and processes of capitalist economies. Moreover, these elements of institutional theory help to ground more abstract concepts of exchange and use value, showing the mechanisms by which social relations are structured within and outside markets. Placed in relation to one another in this way, they provide a useful analytical framework for examining how markets operate.

Studying privatization within this framework allows for an exploration of the interrelationship between the embedded exchanges on which economic sociology focuses and the arm's-length transactions of the market that neoclassical economics assumes. The coordinated production that takes place in the firm is best understood when we take into account the external environment of the firm, which may very well consist of competitive, arm's-length transactions among firms. Similarly, the redistributive, nonmarket allocations that states establish through the provision of public goods, entitlement programs, subsidies, and other forms of nonmarket exchange must be assessed in light of the larger, competitive environment of the international economy in which states operate (Ruggie 1983).

To be clear, this is not an argument about the relative efficiency of one form of economic organization over another. Rather, the argument is that economic exchange under capitalism depends on a delicate balance between two opposing principles: connecting actors through nonmarket institutions while simultaneously separating, or fragmenting, them into groups that are capable of exchanging on a market basis. Markets depend on nonmarket institutions that connect actors and restrict their ability to operate on principles of naked self-interest. At the same time, if the institutions that link actors for the purpose of stabilizing exchange become dominant, they can destroy the very foundation of market exchange.

The question that remains to be answered is the following: Why does the market become the governing institution for any allocations at all? If one of the salient characteristics of markets is the real risk of failure—as the discussion of hard budget constraints suggests—and actors seek to evade market allocations through the construction of nonmarket institutions, why are markets not overwhelmed entirely by other forms of economic organization? Who, after all, would seek the uncertainty of the market over the security and predictability of an embedded relationship?

In order to answer this question, I return to the question of state autonomy. I relate state autonomy to the concepts of hard budget constraints and property rights to show that actors seek to create markets so that they can exclude other actors from property. This does not imply that actors seek to create markets indiscriminately. To the contrary, we should expect to find actors seeking to impose market allocations on certain exchanges while simultaneously seeking to avoid market allocations in others. This is a key area of conflict in the process of privatization. It involves extracting firms from some of the political coalitions and social obligations within which they are embedded without exposing them to the dangers of unrestrained competition.

Taking the State Back Out? State Autonomy and Neoliberal Reform

Research on economic transformation in the developing world frequently employs the concept of state autonomy to explain the introduction of major economic reforms. In an overview of the politics of economic reform, Stephan Haggard and Robert Kaufman (1992, 7) point to the crucial role of political institutions in shaping adjustment decisions and argue that "the central theoretical debate in this area concerns the importance of state autonomy for the initiation and consolidation of fundamentally new policy directions." In his study of developmental states, Peter Evans (1995, 12) also argues that autonomy is indispensable for states seeking to promote economic transformation: "The internal organization of developmental states comes much closer to approximating a Weberian bureaucracy. Highly selective meritocratic recruitment and long-term career rewards create commitment and a sense of corporate coherence. Corporate coherence gives these apparatuses a certain kind of 'autonomy.'"

In the implementation of neoliberal economic reforms, state autonomy is said to matter because it enables the state to overcome what has been called the "orthodox paradox." According to Miles Kahler (1990, 55), the "[o]rthodoxy [of neoclassical economics] has not dealt successfully with the paradox of using the state—its only instrument—to change policy in a less statist direction." Citing the orthodox paradox, Haggard and Kaufman (1992, 25) argue that "for governments to reduce their role in the economy and expand the play of market forces, the state itself must be strengthened." Evans (1995, 27), too, cites the orthodox paradox, noting that "when liberalization, privatization, and other policies associated with neoliberalism were implemented, it was in fact state managers who formed the core of the 'change teams' that made change possible" (see also Grindle 1996; Waterbury 1993).

Much like research in economic sociology, the literature on state autonomy serves as an important corrective to conventional wisdom and neoclassical analyses of the economy. This research highlights the central role of the state in economic transformation and provides a useful conceptual framework for analyzing the state. It emphasizes the importance of understanding both the internal organization of the state and the external environment in which states operate. My critique of the state autonomy literature is divided into two parts, both of which are necessary for understanding the process of privatization in Mexico.

First, recent literature on state autonomy suggests two different meanings of the term that need to be clarified. One strand of this literature focuses on autonomy as the *capacity* of the state to implement reforms while another strand focuses on the *insulation* of the state from social pressure. Both of these interpretations, however, frequently treat the content of economic reforms as if it were unrelated to either the external forces acting upon the state or the ability of the state to implement these policies. In contrast, I will seek to relate state capacity to the content of reform measures and the varying abilities of nonstate actors to resist or support economic reforms. This involves looking at state autonomy relationally. With rare exceptions, the state's ability to act "autonomously" from one social group simply means that it is acting in the interest of some other social group.

The second problem with the state autonomy literature is related to the first. By failing to examine the specific interests that are served by economic reforms, this literature tends to suggest that autonomy enables the state to harmonize the collective interests of society as a whole. The problem of economic reform is reduced to one of collective action in which neoliberal reforms involve concentrated, short-term costs that make it difficult for policy makers to implement these policies but also diffuse, long-term benefits that make reform attractive (Haggard and Kaufman 1992, 18; Schneider 1999, 55). While the collective action problem provides an appealing theoretical framework for understanding the barriers to implementing reforms, the question of who wins and who loses in the process of economic reform needs to be examined empirically.

I will argue that in Mexico the consequences of neoliberal reforms were precisely the opposite of those assumed in the collective action problem. Privatization and neoliberal reform in Mexico generally produced concentrated benefits for a relatively narrow segment of society while imposing diffuse costs on large swaths of the Mexican population. The real challenge of implementing reform had to do with gaining control over the unwieldy apparatus of the state in order to transfer public assets to private actors.

Self-interest explains part of the motivation of actors who implemented reforms as well as those who benefited from them. Fifty years of growing state intervention in the economy created enormous opportunities for anyone who could take hold of specific pieces of the state and bring them under private control. Mexico's largest and most powerful capitalists were not shy about demanding just such a reform program. But self-interest in the Mexican context was also framed within the neoclassical paradigm by a group of powerful actors trained in neoclassical economics (Babb 2001). An essential element of these actors' conception of control was the ideal-typical market in which price, profitability, and competition exert a dominant influence upon allocations.

STATE AUTONOMY: CAPACITY AND INSULATION

The concept of state autonomy has undergone a subtle but significant shift in recent years. Early attempts to define and use this concept came from a Marxist tradition. Debates surrounding the autonomy, "relative autonomy," or "potential autonomy" of the state have been summarized elsewhere and it is not necessary to review them here except to note that this literature emphasized the function served by the state in reproducing capitalist relations of production.[10] Thus, in its earliest formulations, state autonomy was concerned almost exclusively with the class basis of the state in capitalist economies and referred to independence of the state—either structurally or instrumentally—*from capital*. Marxist writers generally concluded that the state could not act independently of dominant groups, although their reasoning differed.

In dialogue with the Marxist tradition, Theda Skocpol (1979, 27, 29) introduced a focus on the institutions of the state—its administrative, and coercive, organizations—in order to examine the potential autonomy of this apparatus from elite interests. Looking at the state as an "autonomous structure—a structure with a logic and interests of its own," Skocpol identified two dimensions along which to assess the potential autonomy of the state, one internal, the other external. Internally, the state "first and fundamentally extracts resources from society and deploys these to create and support coercive and administrative organizations ... these fundamental state organizations are at least potentially autonomous from direct dominant-class control." Externally, "a state's involvement in an international network of states is a basis for potential autonomy of action over and against groups and economic arrangements within its jurisdiction" (29, 31).

More recent formulations of the concept reveal two related meanings of autonomy with very different implications. Evans's (1995) research on developmental states adapts Skocpol's framework of analysis—as well as his own earlier work (see Evans, Rueschemeyer, and Skocpol 1985)—and refers to the external-internal dimensions of the state as embeddedness and

---

10. For useful overviews of these debates, see Domhoff 1967; Hamilton 1982; Miliband 1969; Poulantzas 1972, 1976; and Skocpol 1979.

autonomy, respectively. Unlike Skocpol's focus on the interstate system, for Evans the external dimension of state autonomy refers to relations between the state and private capital, primarily domestic capital. *Autonomy*, then, refers exclusively to the internal dimension of state organization, to a state's bureaucratic capacity, which allows it to operate on a logic distinct from that of private accumulation.

The other meaning of *autonomy* that can be found in recent literature in which the term is used retains the focus on state independence. However, contributors to this literature tend to identify state autonomy with the independence of the state from *any* social group, not only capital. Haggard and Kaufman (1992, 8), for example, highlight the importance of state autonomy to the initiation of policy reform, noting that reform is "more likely when technocratic decision makers enjoy some degree of independence from particularistic interests." The meaning of autonomy shifts drastically here from its roots when "particularistic interests" are defined to mean "the pull of distributive politics:" implicitly those of popular sector groups (see also Callaghy 1990, 263; Centeno 1994, 47).

This use of the concept is problematic for two reasons. First, a state may actually succeed in resisting the demands of specific groups most effectively when it penetrates these groups rather than insulating itself from them. In Mexico, for example, the capacity of the ruling party to implement policies detrimental to labor depended upon its ability to moderate the demands of labor through organizations that were tightly linked to the state. Not only did the state manage to avoid being captured by these groups, these groups were actually captured by the state. Conversely, the ability of elites to influence state policy has depended in some measure on their autonomy from the state, that is, on their independent control over resources that gives them the capacity to resist the state and pressure it from outside the state apparatus.

The second problem with this use of the concept of state autonomy is its tendency to treat as equivalent all social pressures acting upon the state. The ability of the state to ignore the interests of workers and peasants is treated as if this were the same as a state defying more powerful groups in society. Sometimes this emphasis is explained by the fact that popular sectors and their representatives form an integral part of a ruling coalition (Petrazzini 1996; Waterbury 1993). Yet when a state demonstrates its independence from some particularistic interest, even if it is acting in its "own" interests, its policies will invariably benefit some other group in society.

This suggests a need to look at state autonomy not in an absolute sense but, rather, relationally. Achieving independence from one group generally requires that the state anchor itself more firmly in the support of another social group. Thus, gaining independence from capitalists may force the state into an alliance with peasants, workers, the military, or international capital (see Hamilton 1982, 284–86; Skocpol 1979, 31). The specific correlation of forces and balance of independence from and dependence on different groups itself depends on the resources that these groups possess and how these resources are deployed to influence the state.

AUTONOMY AND SOCIAL STRUGGLE

State autonomy, understood as the capacity of the state to contradict the interests of specific groups in society, may be usefully compared to the concept of hard budget constraints. Recall that in order for markets to serve as the ultimate arbiter of survival, states must resist the temptation to intervene in the economy to prop up insolvent firms. In my elaboration on the concept of hard budget constraints above, I argue that nonstate actors can also impose hard budget constraints on other actors, including the state. By denying resources to other actors except on market terms, nonstate actors contribute to the construction of hard budget constraints. They demonstrate not only their autonomy from one another, but their very identity as discrete agents controlling independent resources.

Thus, while it is useful to examine whether the state can resist external pressures, it is equally important to investigate whether nonstate actors can resist the pressures of the state in its efforts to influence their actions. Indeed, this is the critical question for states that seek to implement polices that contradict the market. The discussion of neoclassical theory and institutional economics above provides some clues about where we should begin looking for indications of this type of autonomy.

Private property is the resource that gives capitalists independence from the state. In a properly functioning market, private actors withhold their resources except where they perceive that exchange is in their own self-interest. Similarly, states withhold their resources from private actors: they create hard budget constraints by refusing to prop up the unprofitable activities of any particular firm while at the same time subsidizing those institutions and public works that, though unprofitable, "may be in the highest degree advantageous to a great society" (Smith 1981, 723).

In Skocpol's work, the ability of states to operate independently of private actors revolves, in part, around their control of the administrative and coercive apparatus of the state. In theory, the resources necessary to operate the administrative and coercive apparatus of the state depend on the extraction of resources from the private economy. Because public goods are unprofitable, states have an interest in the smooth functioning of the private economy to maintain their own operations. In practice, however, public ownership is not always tied to unprofitable firms. Indeed, the focus of Hamilton's study of state autonomy in Mexico was the nationalization of the profitable petroleum industry. Not only did the appropriation of foreign oil holdings in Mexico contradict the interests of an important segment of capital, it simultaneously secured a base of resources for itself independent of domestic capitalists.

Similarly, while private property is said to depend on the legal framework provided by the state for its protection, private actors can skirt the authoritative framework provided by the state. Especially in developing countries, the ability of capital to exit the economy through capital flight increases the significance of its voice. Thus, the enforcement of property rights by *other* states can grant private actors the power that, in theory, depends on their own state.

Public ownership of firms, then, presents both dangers and opportunities. The nationalization of profitable firms may provide the state with resources independent of private capital. It may also alienate the private sector by depriving capitalists of opportunities for profit. At the same time, when market failure affects basic social and economic infrastructure such as finance, communications and transportation, energy, or the food supply, the problems of private actors quickly become the problems of states. Even when market failure affects nonessential sectors of the economy, social mobilization and protest can force states to alter their policies. If social upheaval makes it impossible for commerce to take place, then state spending to contain social unrest becomes as necessary to supporting market exchange as the provision of standardized weights and measures or the enforcement of contracts.

Social struggle and the crises that these provoke, then, need to be incorporated more directly into the analysis of the introduction of neoliberal reforms. Although crisis occupies the background of most studies of economic restructuring, it is generally presented as external to the reform process, something like a physical barrier beyond which actors cannot stray without endangering their existence. Economic crises, however, are not the social equivalent of plate tectonics. Instead, economic crises are caused by actors who create speculative bubbles, introduce new technologies, deny other actors resources, engage in mass protests, or take any other actions that undermine existing political and economic arrangements.

Once again, this perspective draws on the work of Karl Polanyi and his study of the rise—and subsequent demise—of the laissez-faire ideal.

According to Polanyi, the creation of the markets of the late nineteenth and early twentieth centuries consisted of a "double movement":

> [O]ne was the principle of economic liberalism, aiming at the establishment of a self-regulating market, relying on the support of the trading classes, and using largely *laissez-faire* and free trade as its methods; the other was the principle of social protection aiming at the conservation of man and nature as well as productive organization, relying on the varying support of those most immediately affected by the deleterious action of the market. (1957, 132)

Changing the rules of economic engagement, reorganizing the public sector, and transferring assets from the state into private hands present enormous opportunities for some actors while threatening the very survival of others. It is in the struggle between groups affected by these changes that we should seek to understand the process of economic transformation. It is to these struggles that we now turn.

TWO

Public Ownership and the Rise of State-Led Development

Extensive state ownership—of steel, textile, and sugar mills, of the railroads and telegraph system, of the petroleum and electric-power industries, of chemical plants, of banks and auxiliary financial institutions, and of dozens of additional economic activities—created a need and simultaneously an opportunity for competent young men. Twenty-five years of state management of the oil industry and the railroads, for example, has brought into existence a broadly competent clique of "oilmen" and "railroaders."
—*Frank Brandenburg,* The Making of Modern Mexico

[I]t was in the 1950–1960 interval that one detects a growing awareness among virtually all observers that Mexico had become one of the handful of so-called underdeveloped nations to effect the transition to sustained, more or less self-generating economic expansion.
—*William P. Glade Jr. "Revolution and Economic Development"*

In this chapter I examine the growth and development of Mexico's state-owned enterprise sector, also known as the parastate sector.[1] I begin with an overview of the legal foundations and political institutions on which public ownership depended, look at the different forms of state ownership, and examine the scale and scope of Mexico's parastate sector from the end of the Mexican Revolution until 1982. I then explore the contribution of different actors to the growth of the parastate sector, the role of parastate firms in managing the state's relationship to these actors, and the administrative structures that were developed to control the growing state sector.

By the early 1980s, the growth of the state's role in the economy appeared to have overwhelmed the market as the principal mechanism for allocating resources on a number of important measures. On the eve of the economic crisis of 1982, state spending was approaching one-half of Mexico's gross

---

1. There is no direct translation for the word *paraestatal*. The *Pequeño Larousse Ilustrado* (1988) defines *paraestatal* in the following way (my translation): "said of the entities that cooperate with the state without forming part of its administration." In the Mexican context it has a more specific meaning, which is elaborated on below.

domestic product (GDP). Parastate companies operated in almost every sector of the economy, and the five largest companies in the country were state-owned. Eight of the largest twenty-five firms in Mexico were state owned and captured 57 percent of the total sales revenue of the top twenty-five companies (Aspe 1993, 157; *Expansión* 2000).

And yet, despite the vast reach of the parastate sector and the extension of the state's role into almost every niche of economic activity, the public sector had not by any means eliminated the force of market allocations over the economy. International markets in particular continued to exert an important influence over the domestic economy. International markets provided an escape valve for Mexican capitalists, allowing them to move capital abroad when they felt threatened by the national development plans of the state. In addition, the Mexican state's ability to extend its ownership of and support for parastate firms became increasingly dependent on access to foreign capital and the price of petroleum on international markets. When oil prices fell and foreign lending dried up, Mexico's economy was thrown into deep crisis.

The Revolutionary Institutional Party (PRI) used parastate firms to help it deal with three main challenges in its efforts to promote economic development. First, it had to respond to the demands of popular-sector groups whose members periodically rose up in protest against the widening gap between the ideals of the revolution and the reality of their daily lives. Second, it sought to appease domestic capitalists who controlled economic resources that were vital to national development. And third, it needed to manage its relationship with the international economy, to cushion the impact of sometimes volatile external markets and prevent the domination of the national economy by foreign interests.

Until the late 1960s, parastate firms assisted the state party in balancing the competing interests of these three forces. The inability of any one of the factions within or outside the PRI to completely overcome the interests of the other forced a degree of collaboration and accommodation that allowed for high levels of sustained economic growth during the 1950s and 1960s, in what came to be known as the "Mexican miracle." But the state's very success in holding each of these forces in check would soon undermine the state-led development project.

In 1968, a student uprising challenged the government to address unmet social needs and was violently repressed. Business groups that had prospered behind state protection became more independent during the 1970s, took on more foreign debt, and exercised their options to send capital abroad. And volatility in international markets created conditions that allowed the Mexican state to defy domestic capitalists on whom it had previously depended for resources. Although 1982 marks the end of the project of state-led development, the foundation on which this project was built had already begun to crumble in the late 1960s.

An Overview of the Growth of Mexico's Parastate Sector

THE LEGAL AND POLITICAL FOUNDATIONS OF PUBLIC OWNERSHIP

The Mexican Revolution (1910–17) is alternately described as the first social revolution in world history and as a bourgeois, political revolution. It is characterized as a social revolution because "the expansion of political consciousness and the entry of previously excluded social forces into politics are central characteristics of all social revolutions" (Middlebrook 1995, 1). At the same time, it has been portrayed as essentially bourgeois because it merely reformed existing property relations rather than seeking to abolish private property (A. Córdova 1972).[2]

Understanding the precise nature of the Mexican Revolution is important in examining the growth of the parastate sector because the social forces unleashed by the revolution largely determined Mexico's subsequent development strategy. Disagreement over its meaning stems in part from the fact that the revolution was a hybrid. While it had some bourgeois origins, it also included social revolutionary elements, in the mobilization of peasants and organized labor. It was also anti-imperialist, though for different reasons among different sectors of society.

Despite the entry of previously excluded social forces into politics, all observers agree that the Mexican Revolution ended with a peculiar form of incorporation of popular sectors. According to Córdova,

> The middle-class supporters of the revolution invented populism, not so much in struggle against the oligarchic system (by 1914 that had already been annihilated as a political power) as, precisely, in struggle against the independent peasant movement that Villa and Zapata commanded. Mexican populism, therefore, had a counter-revolutionary core: it tried to avoid the transformation of the mass movement into a social revolution. (A. Córdova, 1972, 32)[3]

---

2. See also Aspe and Beristain 1984, 18; Centeno 1994, 6; and Hamilton 1982, 61–63.
3. All translations are mine unless otherwise indicated.

With the winning coalition representing radical agrarians, landholders, anarchosyndicalist labor organizations, and small and large capitalists, the Constitution of 1917 came to embody the contradictory goals of these competing factions. Drawing on the liberal Constitution of 1857, the Constitution of 1917 guarantees individual rights and provides for a separation of powers. At the same time, it also includes sweeping provisions for state intervention in the economy and ultimately constitutes the legal foundation on which Mexico's parastate sector was built (Ruiz Dueñas 1988, 116; Villarreal 1988).[4]

Articles 25, 26, 27, 28, and 90 deal most explicitly with the role of the state in the economy. They give the state responsibility for directing, coordinating, and orienting economic activity and national development (Articles 25 and 26); raise public interest above the rights of private property (Article 27); and prohibit monopolies except in specific "strategic" activities and entities that are reserved for the state and enumerated in Article 28, such as the coining of money, the postal service, and basic petrochemicals. Article 90 blurs the line between public and private sectors by establishing the basis for parastate firms in the operation of the economy and declaring that federal administration will be "centralized and parastate" (see Fernández Varela 1991, 31–35; Tamayo 1988a, 636).[5]

Despite the creation of this legal framework providing for a wide range of state intervention in the economy, the development of Mexico's public sector in the 1920s and early 1930s was unexceptional. As Mexico recovered from its devastating revolution, the growth of the state took place principally in the areas of finance and public works. In the 1920s the state sought to stabilize the banking system with the creation of the National Banking Commission and a central bank, the Bank of Mexico. During the same period, the National Roads Commission was created to integrate Mexican territory, and a National Irrigation Commission and National Bank for Agricultural Credit were created to promote agricultural development. Typical of the interventions of all states that have market economies,

---

4. Fernández Varela (1991, 31) asserts that the Constitution was fundamentally liberal, with the exception of Article 123, which granted a variety of social rights to workers: "The statist project appears through the modifications suffered over the course of seventy-five years along with the diverse regulatory laws of the principal articles." At the same time, Fernández Varela points to others who argue that "two different national projects can be derived from the Constitution: one liberal and the other statist" (see also Hamilton 1982, 61).

5. Articles 27, 28 and 123 deal with what Polanyi (1957, 72) calls "fictitious commodities": land, money, and labor, respectively. They assert the right of the state, rather than the market, to allocate these for public rather than private ends.

the Mexican state created regulatory institutions and developed a legal framework primarily to facilitate the smooth operation of the private economy (Ruiz Dueñas 1988, 116; Shelton 1964).

The realization of the radical potential of the Constitution and the construction of a large sector of state-owned firms would first require the consolidation of political power. The founding of a state party in 1929 was crucial to this process. The National Revolutionary Party—predecessor to the PRI—provided an umbrella organization for interest representation and began to lay the foundations of political stability that would allow for a more active state role in the economy.

The timing of the formation of the state party was especially significant because Mexico's capitalist class remained relatively weak and fragmented in the years following the revolution. Mexico was still a predominantly rural country; its territory was poorly integrated; political power was splintered among local and regional bosses; and the commanding heights of the economy were dominated by and directed toward foreign markets, mostly those of the United States. In 1930 slightly more than 70 percent of the economically active population was employed in the primary sector with only 15 percent involved in industrial activities (Petricioli 1988, 841). Regional strongmen were still able to circumvent the authority of the federal government and impose local taxes and tariffs. Railroads, petroleum, electricity, telephones, and mining were all controlled by foreign capital.

In the absence of any other organized interest in Mexican society capable of promoting national development, the state party became the focus of both economic and political organization. At the same time, the ruling party came to incorporate into its structure the same disparate elements that shaped the revolution and the Constitution. In a power struggle with former president, and *jefe máximo*, Plutarco Elías Calles, President Lázaro Cárdenas (1934–40) mobilized workers and peasants to overcome the more conservative elements of the revolutionary family. Cárdenas encouraged and facilitated the organization of workers, peasants, and middle-class groups into peak associations; each of these would represent its sector's interests within the state party. Cárdenas rallied all these forces around the project of national development and the liberation of the domestic economy from foreign domination (see A. H. Chávez 1979; A. Córdova 1974; Middlebrook 1995, 26).

The extent to which the PRI ultimately penetrated almost every organized interest in Mexico is captured by Pellicer de Brody and Reyna (1978, 31), who assert that by the 1950s, "[i]t would not be possible to study the action of popular organizations in Mexico ... without remembering the firm control that the state had achieved over them which had led them to become confused with the state apparatus." Bailey (1988, 89) argues, "The PRI lacks an identity separate from the government in terms of structure or purpose." When the state extended its presence in the economy, then, by definition, so too did the PRI.

But if the state was the party and the party was the people, there was one important exception. Business groups were never formally incorporated into the PRI. The exact relationship between the PRI and the private sector is subject to some debate (Centeno 1994; Hamilton 1982; P. H. Smith 1979). However, there is considerable evidence that the most powerful private-sector groups enjoyed close, informal access to the leadership of the state-party while simultaneously maintaining their independence (Maxfield 1990, 36–45; Tello 1984, 54; Torres 1984, 36). Working outside the PRI and controlling economic resources that were independent of the state would ultimately grant these groups greater leverage over the policies of the government than they could have exercised from within the party.

Although declarations by the newly formed state party on the role of the state in the economy were often vague, the Cárdenas administration pointed to both external and internal problems plaguing the economy. The former revolved around the domination of the Mexican economy by foreign interests, the latter around the incapacity or unwillingness of Mexican capitalists to meet the developmental needs of the country. These two themes were clearly sounded when President Cárdenas expropriated foreign petroleum holdings in support of Mexican sovereignty and petroleum workers who had been locked in conflict with the companies. Placing an exclamation point on the nationalization of the petroleum industry, the 1938 Declaration of Principles and Program of Action of the Party emphasized the need to "liberate the country from outside economic influence" and to "organize the economy of the country on the principle that production and distribution are oriented toward the true satisfaction of popular needs" (Tamayo 1988a, 640).

Yet apart from the petroleum nationalization—which to this day retains overwhelming public support, making it almost untouchable politically—the commitment to state ownership varied considerably from one administration to the next. Presidents Ávila Camacho (1940–46) and Alemán (1946–52) insisted that economic development was first and foremost the responsibility of the private sector. But their belief in the primacy of private enterprise did not prevent these presidents from making important investments in public firms. When markets failed to produce essential goods and

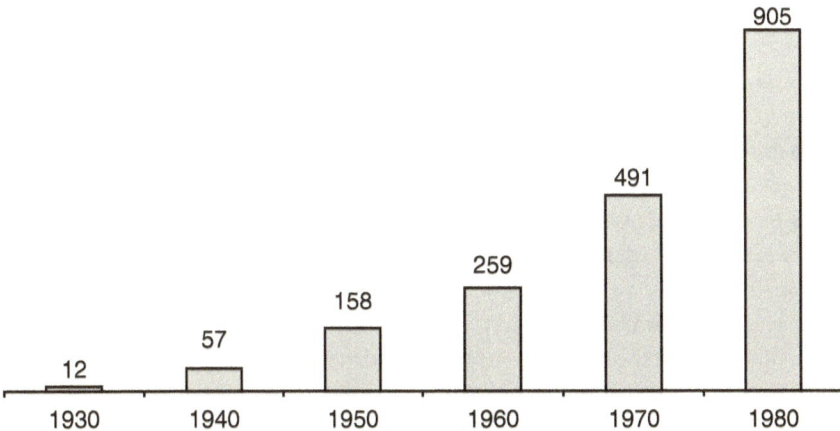

FIG. 1 Number of parastate firms, 1930–1980
Sources: Years 1930–70 from Banamex (1991); data for 1980 from Salinas (1994, 91).

services needed for development, both Ávila Camacho and Alemán made public investments in basic infrastructure and even extended the state's role in the economy by entering into the production of industrial and agricultural inputs to facilitate development (Torres 1984, 28, 103).

Except for populist outbursts by Presidents López Mateos (1958–64) and Echeverría (1970–76) and the desperate bank nationalization of López Portillo (1976–82), direct confrontation with national capital was rare. Instead, the state appeared most willing to intervene proactively in those sectors where Mexican capital was not established—in particular in capital-intensive industries such as steel, automobiles, and shipbuilding; in resource extraction, especially mining and petroleum; and in basic infrastructure dominated by foreign companies: electricity and later telecommunications. The state also acquired a range of firms through defensive measures, less focused on economic development than on maintaining political stability.

Thus, while the political orientation of a particular administration may have mattered in terms of the speed with which the state-owned enterprise sector grew, as Figure 1 shows, the absolute size of the parastate sector increased throughout the entire period from 1930 to 1980. As I discuss below, the parastate sector grew steadily until the 1970s and then expanded more rapidly in response to popular unrest and, later, as a result of resources provided by the petroleum boom. During this entire period of expansion, however, every administration struggled to control and direct the activities of the growing number of parastate firms.

THE SCALE AND SCOPE OF PUBLIC OWNERSHIP

Different studies citing aggregate data on the growth of the public sector often contradict one another. The very definition of what to include in the parastate sector has changed over time and we should read the numbers provided by official sources with great caution. Rather than a precise quantification of state ownership, the figures presented here are intended to convey a sense of the general direction in which the parastate sector developed between the early 1930s and 1982. To begin, parastate firms are defined by four principal forms of property in which the state participates: (1) ownership of stock by the state as either a majority or minority shareholder, (2) decentralized organisms, (3) "deconcentrated" organisms, and (4) special funds known as *fideicomisos,* established for a specific purpose, ranging from the promotion of Mexican cinema to the construction of rural roads or urban housing.

Participation by the state as a majority or minority shareholder in a firm does not necessarily alter the firm's organizational structure. It does, however, create an obligation that the firm serve the public interest (Acle Tomasini and Vega Hutchison 1986, 74–78; Carrillo Castro and García Ramírez 1983, 16, 17). Decentralized organisms cannot be created without the approval of Congress but operate with a degree of autonomy. They are governed by a board of directors and sometimes generate their own resources. Although they are formed by the executive and the legislature, they operate outside the hierarchy of the central administration of the state (16). Deconcentrated organisms may be established directly by presidential decree. Because they do not have either a board of directors or their own resources, they are more directly controlled by line ministries (Bailey 1988, 83; Interviews 50, 72).[6] *Fideicomisos* may be created at the complete discretion of the president.

One important area of economic activity operated by the state but excluded entirely from the parastate sector is education. Mexico's federal government provides primary, secondary, and postsecondary education. Combined, the educational institutions include the largest segment of organized workers in the entire state sector and the largest union in all of Latin America. However, because education is controlled directly by the

---

6. I use the term *line ministry* to refer to those individual ministries that oversee a particular functional area of the economy—e.g., transportation and communications, agriculture and hydraulic resources, energy—in contrast to the Ministries of the Presidency, Treasury, and Planning and Budget, that deal with the overall resources and organization of the state.

All citations of interviews refer to the numbers in the interview list in Table 18.

central administration of the federal government through the Ministry of Education, schools fall outside of the parastate sector (Cook 1997).

Although data on the exact date of acquisition or creation of specific parastate firms are sketchy, most studies agree on two points. First, there was an increase in the rate of growth of the parastate sector beginning in 1970 with the Echeverría administration. Much of this growth appears to be related to the creation of a large number of *fideicomisos*. Second, despite some contradictory data, most authors settle on 1982 as the zenith of the parastate sector; in this year 1,155 firms operated with state participation (Aspe 1993, 156; Bailey 1988, 67; Rogozinski 1997, 111; Teichman 1995, 29). According to one set of official figures, in 1982 the Mexican state was a majority shareholder in 744 firms, operated 231 *fideicomisos* and 102 decentralized organisms, and was a minority shareholder in 78 firms (SHCP 1994b, 9).[7]

Apart from the increase in the rate of expansion of the public sector in the 1970s, however, clear patterns in the growth of the parastate sector are hard to find. The lead official for selling parastate firms under the Salinas administration, Jacques Rogozinski, estimates that half of Mexico's state-owned firms were failing at the time that they were incorporated into the public sector. President Salinas's minister of the Treasury, Pedro Aspe, also points to the incorporation of bankrupt firms into the parastate sector in the second half of the 1970s and argues that the growth in the number of state-owned firms was not the result of an industrial strategy but, rather, was "a massive rescue operation designed to protect employment" (1993, 158; Boeker 1993, 5).

From the 1930s through the early 1950s, the state appears to have focused most of its attention on providing basic infrastructure for transportation as well as inputs for agriculture and heavy industry. Without question, the expropriation of all foreign petroleum holdings in 1938 and their consolidation into the state-owned company Pemex was the single most dramatic expansion of the parastate sector. Pemex is one of the largest companies in the world, and its sales place it at the top of the list of Mexico's largest firms every year, with three to four times the revenues of the second largest company (*Expansión* 2000).

More often, the state entered into the direct production of goods and services through the gradual acquisition of stock and the "Mexicanization"

---

7. Casar and Peres (1988, appendix 1) provide dates for the creation of a subset of parastate firms. Unfortunately, this data does not indicate whether the firm was created as a private enterprise and incorporated into the public sector at a later date or if it was created as a public firm. For other limitations of this data, see Casar and Peres (1988, 32).

of an industry. The Mexican National Railroad (FNM), originally created in 1908 to coordinate and administer the holdings of failing railroads, was brought under operational control of the government in June of 1937 by the Cárdenas administration. During the following fifty years, FNM slowly acquired all of the country's rail lines (see Chapter 6). The Federal Electric Commission (CFE) was also created during the Cárdenas administration. Although it produced only a small percentage of Mexico's total electricity needs in 1937, over time CFE consolidated under its control all foreign holdings of electricity-generating capacity (Ruiz Dueñas 1988, 160; Alcudia 1988, 786).

In the 1940s, the state created a national iron and steel company, Altos Hornos de México, and founded the fertilizer company Guanos y Fertilizantes de México. The state also began entering into joint projects with the private sector during this period. In the 1950s the state invested in a company to build rail cars in support of the national railroad and took control of the automobile and truck company Disel Nacional, a joint venture between private Mexican and Italian capital that was nearly bankrupt (Bennett and Sharpe 1985, 141). In addition, the state entered into the production of newspaper and textile machinery in the 1950s and acquired the airline Aeroméxico in an effort to pacify a labor dispute at the company (see Chapter 4).

In 1960 the state completed the project it had begun with the creation of CFE by acquiring the last remaining shares of foreign-owned power and light companies (Bennett and Sharpe 1985, 35) and created regional light and power companies for Pachuca, Toluca, and Mexico City in 1963. It also expanded its efforts to use public firms for the promotion of transportation and communications services during the 1960s. Throughout the 1940s and 1950s, public support for transportation was focused primarily on ground transportation through FNM and large-scale highway construction projects. In the 1960s, the state began organizing air and water transportation by creating Airports and Auxiliary Services (ASA), shipbuilding and repair companies for the Gulf and Pacific Coasts, and cable-laying companies to support the growth of telecommunications infrastructure.

Also during the 1960s, the state extended its presence in the provision of basic consumer goods, especially foodstuffs. The National Export and Import Company (Ceimsa), created in the late 1930s to oversee the import and export of food, entered into the retail distribution of food in the 1940s (Pellicer de Brody and Reyna 1978, 26). In 1961, Ceimsa was liquidated and replaced by the National Company of Staple Goods (Conasupo). The state began incorporating sugar mills into the parastate sector during the

1960s as price caps made sugar production unprofitable (Tamayo 1988a, 699). In the late 1960s and 1970s, the state broadened the scope of parastate operation still further by entering into the processing of secondary petrochemicals and completing the acquisition of the majority of the stock of the telecommunications monopoly, Mexican Telephone (Telmex).

In 1965, twenty of the largest parastate firms were incorporated into the federal budget in an attempt to gain control over their spending and income. These firms were primarily in energy, transportation, and social welfare and included Pemex; CFE; seven local electric companies; FNM; three regional state-owned railroads; the flagship airline, Aeroméxico; the National Lottery; the two social security administrations; the national institute for housing, Conasupo; and the agency that administers federal toll roads and bridges (Carrillo Castro and García Ramírez 1983, 64; Ruiz Dueñas 1988, 119).

The addition of these twenty firms to the balance sheet of the federal government in 1965 shows that the economic weight of these parastate firms was roughly the same as the remainder of the public sector. Public-sector spending in 1964, without any parastate firms on the books, was 12.8 percent of GDP. The following year, public-sector spending appears to double, hitting 26.3 percent of GDP, because of the inclusion of these twenty firms. Although public-sector spending and parastate sector spending rose in tandem in the early 1970s, parastate sector spending began to fall as a percentage of total state spending in 1976. Public spending climbed to a peak of 47 percent of GDP in 1982. At that time, the parastate sector represented less than 30 percent of the total, or 14 percent of GDP (Aspe 1993, 114, 157).

Tamayo argues that parastate firms were financially troubled because they were slow to raise prices to account for inflation. For example, the price per kilowatt hour of electricity declined in real terms at a rate of 3.5 percent annually between 1962 and 1984; in 1982, regular and diesel gasolines were sold on the domestic market by Pemex at prices that were, respectively, 40 and 72 percent lower than international prices; the prices charged for hauling cargo by rail did not change between 1962 and 1975; and telephone rates were not raised between 1955 and 1975 (Tamayo 1988a, 731–38). Yet, as Figure 2 shows, income from the twenty parastate firms incorporated into the federal budget—including profitable firms such as Pemex as well as the chronically unprofitable FNM—always matched spending increases. Between 1965 and 1982, the budget deficit of the parastate sector exceeded one percent of GDP in only one year, and in eleven of the eighteen years the parastate sector showed a slight surplus. During the

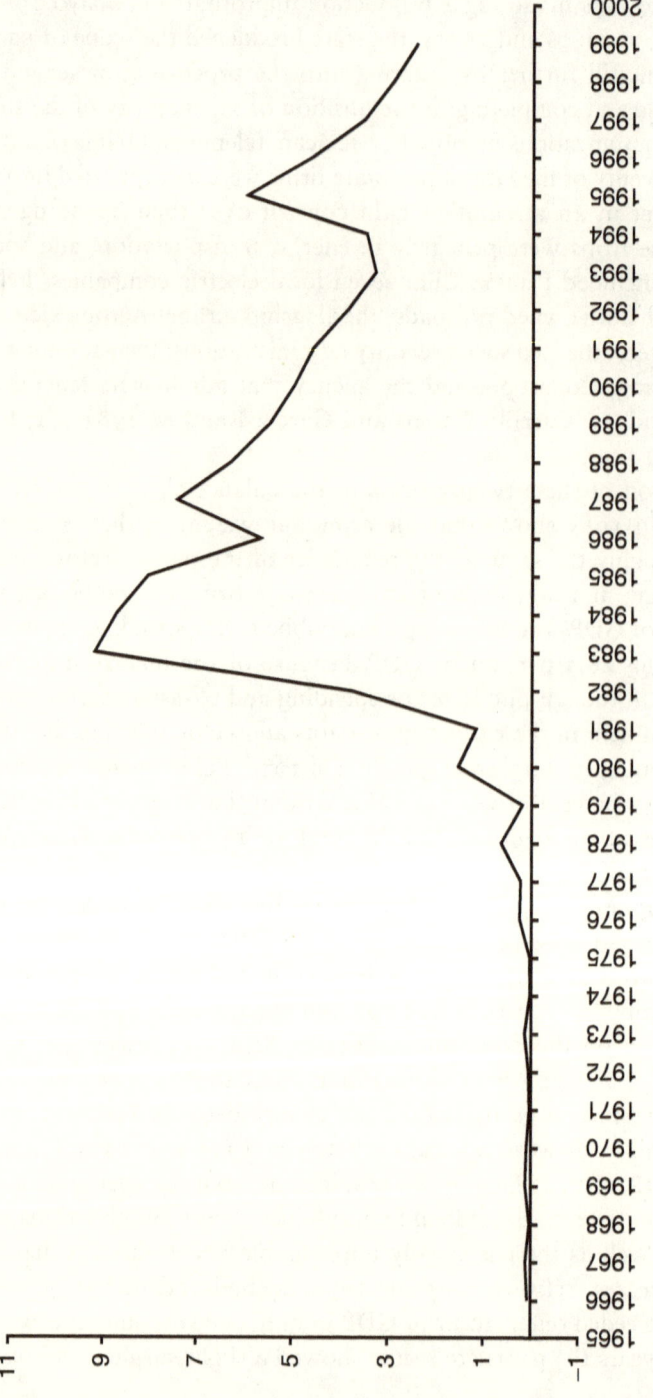

Fig. 2 Balance of parastate sector as a percentage of GDP, 1965–2000
Sources: Years 1965–92 from Aspe (1993, 157); years 1993–2000 calculated from INEGI (2002, table 1).

crisis of the 1980s, the state began running large surpluses to help pay down the debt (see Chapter 3).

Furthermore, parastate firms were an important source of tax income and foreign exchange for the federal government. Perhaps nowhere is the metaphor of the state as a "black box" more appropriate than in the balancing of the income and spending of parastate firms. Profitable public firms paid large sums of money into the general treasury that were then transferred through cross-subsidies to unprofitable firms or to general revenues. Public firms were actually milked to make up for the short-fall of tax revenues from the private sector. Between 1970 and 1980 the contribution of parastate firms to the federal treasury almost quadrupled, growing from 8 percent of total tax revenue to 29 percent (Carrillo Castro and García Ramírez 1983, 152).

The transfer of resources from the parastate sector forced even profitable firms to resort to external financing during the 1970s. Pemex, in particular, took on massive foreign debt while subsidizing other parts of the federal government and other parastate firms through the taxes it paid. One study estimates that Pemex increased its net foreign borrowing by $2.7 billion in 1980, the same year that it paid taxes of more than $7 billion to the federal government (Philip 1984, 45 n. 1). By the late 1970s, the parastate sector had surpassed the private sector as the principal source of foreign exchange in Mexico, with more than half of all export earning coming from public firms and Pemex alone accounting for 46 percent of foreign revenue (Carrillo Castro and García Ramírez 1983, 153).

As the number of public firms grew, so did the number of workers employed in the parastate sector. In 1970, public firms employed 411,000 people, or 3 percent of the economically active population. By 1980, public firms had more than doubled their absolute number of employees to 879,000 while the percentage of the economically active population employed by parastate firms increased to 4.7 (Carrillo Castro and García Ramírez 1983, 157).

The final lurch in the growth of the parastate sector came in August 1982, when outgoing president José López Portillo nationalized the country's banks. During the 1970s, economic policy oscillated between expansionary efforts to accelerate growth and meet popular demands for public services, and austerity imposed by the International Monetary Fund (IMF). By the late 1970s, however, the second petroleum shock left international financial markets awash in liquidity while discoveries of untapped oil reserves made Mexico an exceptionally attractive candidate for credit.

But by 1982, conditions had changed drastically. Declining petroleum prices and a steep hike in interest rates engineered by the U.S. Federal

Reserve led to widespread speculation against the peso. Faced with diminishing foreign reserves, a suspension of foreign lending, and reports of private bankers transferring hundreds of millions of dollars out of the country in a single day, President López Portillo nationalized the country's banking system. Along with the fifty-six private credit institutions that the expropriation added to the parastate sector, the bank nationalization also made the state a shareholder in all the firms in which the banks held stock (Aspe 1993, 115; Maxfield 1992, 82; Tello 1984, 47).[8]

Public Firms and State Formation

Understanding why Mexico acquired specific firms and how they operated requires an examination of the different social forces with which the state had to contend as well as the organization of the state bureaucracy. Between 1930 and 1982, the Mexican state used public firms to balance the competing pressures of three forces: popular sectors, especially organized labor in strategic sectors of the economy, domestic capital, and the international economy all exerted an important influence on the growth and development of Mexico's parastate sector.

CONTROLLING POPULAR SECTORS

Although the Mexican political structure under the PRI was often characterized as "corporatist," state-labor relations in Mexico failed the test of corporatism in at least one important area. Corporatism involves the organization of interest representation into peak associations that monopolize negotiations between the state and a sector of society. In contrast to this ideal type, Mexican labor unions were divided into a number of different groups each with a distinct relationship to the state party. The central pillar of organized labor within the PRI is the Mexican Confederation of Labor (CTM), composed primarily of small, generally weak unions. Larger, stronger, national-industry unions operating in strategic sectors of the economy are usually independent of the CTM (see Dubb 1998; Middlebrook 1995, 341; Schmitter 1974; Zapata 1989).

The PRI attempted to control organized labor through at least two distinct mechanisms. While the CTM organized the disorganized into a base of

---

8. For a range of estimates on the number of firms that the state acquired through the bank nationalization, see Carrillo Castro and García Ramírez 1983, 109; Centeno 1994, 193 n. 58; Ortiz Martínez 1994, 24; Tello 1984, 107, 108.

political support that could be bought at a relatively low price, state intervention and the imposition of pro-government leaders, known as *charros*, disorganized the organized. *Charros* attempted to manipulate workers in the stronger unions through a combination of incentives and constraints. Pro-government leaders sought to limit union demands in important sectors of the economy but also offered improved benefits to workers in exchange for loyalty to the ruling party. When all else failed, *charros* resorted to election fraud, intimidation, and violence against dissident union members to maintain control over unions (see La Botz 1992; Middlebrook 1995).

As the state progressively increased its ownership in strategic sectors of the economy, some of the most powerful unions came to represent workers in parastate firms. National industrial unions are found in mining, railroads, petroleum, telecommunications, electricity, and the airlines. Despite the heterogeneity of the parastate sector as a whole, the industrial organization of parastate firms in strategic sectors of the economy shared a number of characteristics: "In these sectors high rates of unionization prevailed, well above the national average, organized around the most important national industrial unions" (Bensusán 1990, 13). In addition, labor unions in these firms exercised a monopoly on the interest representation of workers and, in some cases, after a brief experience of democracy, ended up with state-imposed leaders in charge (see also Dubb 1998, 1; Hamilton 1982, 112).

These similar patterns of development in unions representing workers in state-owned sectors of the economy merit further consideration. Most of these unions were strong and many were organized at the national level *before* the firms became part of the public sector. In 1930 approximately 130,000 workers were organized in the railroads, mining, petroleum, and electric-power industries alone. Their unions were often the least malleable and most independent among Mexico's working class. The historical control of strategic sectors by foreign capital also contributed to the militance of these workers (Hamilton 1982, 111; Middlebrook 1995, 108).

Thus, the growth of public ownership in these sectors of the economy appears to be related both to the domination of the sectors by foreign interests and to the potential power of unions to make demands on the political system that were incompatible with the PRI's development strategy. Conflict between labor and capital in strategic sectors of the economy could wreak havoc on the economy and created clear incentives for the Mexican state to intervene and even expropriate foreign holdings. A struggle between electrical workers and the British-owned Mexican Light and Power Company shut down almost the entire industry of Mexico City for ten days in 1936 and no doubt contributed to the decision to create the Federal Electric Commission in 1937 (Hamilton 1982, 149). Financial problems appear to have been more important than labor conflict in determining the 1937 nationalization of the Mexican National Railroad (see Chapter 6). However, labor conflict was at the heart of the dispute between the Mexican state and foreign petroleum companies in 1938 and it was during a strike by the pilots' union in 1959 that President López Mateos nationalized Aeroméxico (see Chapter 4).

State ownership of the commanding heights of the economy, then, was important not only for allowing the state to direct resources to developmental priorities; it also allowed the state to retain a tight grip on workers, whose independent base of power could make them powerful political opponents. This mechanism of control within the public sector was far from perfect. Unlike private owners, who could discipline workers by disinvestment or, as foreign oil companies had, by shifting production to another country, the development plans of the state depended on low-cost production in strategic sectors of the economy. Attempts by democratic factions within parastate-sector unions to break the grip of the official union leadership were crushed, often violently, by the state. At the same time, the formal structures within which official unions operated came to incorporate a wide range of benefits and rights to participate in the organization of work and even in the management of state firms (Teichman 1995, 67).

In order for public firms to lead economic development, the state needed to limit and redirect the demands of workers. For example, union leaders at FNM repeatedly argued that the railroad needed to raise its rates to enable it to invest in new equipment and prevent continuing losses. Obviously, raising rates would also make it easier to pass on pay raises to union railroad workers. But the state management of FNM refused to raise rates, arguing that price increases would hurt small mining companies that depended on the subsidized rates. Mining was an important source of foreign exchange for Mexico (see Chapter 6).

It is not clear whether the price that the state ultimately paid to buy off workers—in terms of low motivation, high nonwage benefits, inflexible work rules, and repeated strikes and slowdowns—was less than it would have paid had it allowed public firms to raise their rates to meet workers' demands. But the parastate sector permitted the state to make concessions to workers that did not have an immediate monetary impact. Subsidized food and health benefits, for example, could be provided by other parastate firms. Between 1950 and 1960, workers in petroleum, electricity, and the railroads received a growing percentage of their compensation in the form of nonwage benefits (Pellicer de Brody and Reyna 1978, 169).

Labor unrest was never entirely subdued in the industries that were taken over by the state. Shortly after the petroleum expropriation, the Cárdenas administration found itself in the uncomfortable position of having to refuse many of the same union demands that it had supported when workers were locked in struggle with international petroleum companies. During the 1940s and 1950s, the state sent troops to occupy the oil fields and take over the railroads in response to labor unrest. In the late 1950s, the state weathered a series of strikes in education and a railroad strike with serious political overtones that threatened to spread to other sectors of the economy (Hamilton 1982, 248; Torres 1984, 32).

Even with their expansion into retail distribution of food, it appears that public firms served primarily to benefit private firms. Parastate firms provided cheap communications and transportation infrastructure, and subsidized electricity and petroleum, most of which went to promote private firms in light manufacturing, retail trade, and some heavy industry. They also provided a valuable revenue base that allowed Mexico to maintain some of the lowest tax rates in the world (Tello 1979, 45). But by the late 1960s, it was clear that parastate firms had not addressed a wide range of social needs.

A protest movement led by students in 1968 brought the limitations of Mexico's development model into sharp relief. Despite the acclaim that Mexico had received for its success in promoting sustained growth with low inflation, by the end of the 1960s, 35 percent of the population older than six years old lacked any formal education, while only 22 percent had completed primary school; only 59 percent of the population between six and fourteen years old attended primary school. Almost 60 percent of the housing in Mexico lacked indoor plumbing and only 24 percent of the population belonged to one of the social security institutions. In rural Mexico, still lower percentages of Mexicans had access to basic services, while higher percentages lived in poverty (Tello 1979, 16, 17).

The student movement of 1968 also managed to attract support from important segments of organized labor, including railroad, electric, and petroleum workers. Although the protest movement was violently repressed, it made a lasting impression on the state party. President Echeverría sought to address some of the demands of the movement by increasing the use of parastate firms outside strategic and priority sectors. President Echeverría created a large number of *fideicomisos* that were devoted to the promotion of social welfare, such as funds for the promotion of rural community development and affordable housing (Brachet-Marquez 1996, 167; Tello 1979, 43).

President Echeverría also sought to increase the growth rate by expanding investments in basic industry, including steel, energy, and petrochemicals. In addition, he sought, and achieved, increased prices and rates for goods and services produced by the parastate sector, though a parallel effort to implement tax reform failed, forcing the government to rely still more heavily on the parastate sector for general revenues (Tello 1979, 76). As noted above, the number of workers employed by parastate firms jumped during the 1970s. But it is not clear that popular sectors regained their trust in the government following the repression of 1968. Moreover, inflation was eroding gains in wages and benefits. As the economy entered into recession in 1976, Echeverría devalued the peso by almost 60 percent—the first devaluation in more than two decades—and Mexico entered into an austerity agreement with the IMF to resolve a growing balance-of-payments deficit (176).

Following negotiations with the IMF, state spending declined in 1977 and the López Portillo administration actually reduced the number of public firms. But by 1978, the spike in oil prices, combined with increased investment in petroleum, were beginning to pay dividends, allowing the López Portillo administration to prepay its obligations to the IMF and move back toward expansionary economic policies (Carrillo Castro and García Ramírez 1983, 108, 109).

APPEASING CAPITAL

Like organized labor, Mexican capitalists can be divided into groups that depended on the state and those with an independent power base. Small and medium-sized businesses dependent on the state's protection entered into a cross-class alliance in support of the project of state-led development. This group, sometimes referred to as the Cárdenas Coalition, included peasant and worker organizations mobilized by President Cárdenas, middle-class workers, small and medium-sized businesses represented by the National Chamber of Manufacturers (Canacintra), and the ministries of state that dealt with these groups, in particular, the Ministries of Agrarian Reform, Agriculture, Hydraulic Resources, and Transportation and Communication (Maxfield 1990, 35; Hamilton 1982).

Arrayed against the Cárdenas Coalition was an alliance of private bankers, the head of Mexico's central bank, the Bank of Mexico, and after

1946, the most important ministry of state in regard to financial matters: the Ministry of the Treasury and Public Credit (SHCP) (Maxfield 1990, 34).[9] Unlike the national industrial labor unions, the Bankers Alliance, as this group has been called, retained a power base independent of the state through its control of scarce capital. Moreover, allies within the government provided this group with the organizational links to the state that were needed for them to influence policy more directly. Where consultation through the SHCP and the Bank of Mexico failed to produce the desired results, capital flight was always an effective means of persuading the state to change its policies.

Even as the state acquired a growing number of companies, it still lacked the resources necessary to industrialize the economy. Although the state operated a number of investment banks and channeled lending from international agencies through a state development bank, Nacional Financiera (Nafinsa), it never came close to monopolizing control over money.[10] Milking public firms for tax revenues did not help the situation. Somehow, the state needed to marshal the resources of private capitalists and channel their investments into economic sectors that were deemed priority or strategic. Establishing a clear division of labor between public and private sectors and striking the appropriate balance between inducements and constraints, however, was never easy.

One of the principal benefits of public firms for the private sector was the subsidization of basic inputs, including gasoline, electricity, steel, fertilizers, and basic foodstuffs for workers. Although it is difficult to calculate the exact amount that businesses would have paid for essential inputs in the absence of parastate firms, Tamayo (1988a, 738) estimates that the subsidies from parastate firms to the private sector amounted to more than $6 billion in 1982 alone. But while public investment and the growth of the parastate sector created opportunities for some, it invariably posed a threat to others. Members of the Bankers Alliance, in particular, viewed the expansion of the state's role in the economy with considerable suspicion.

Thus, while a number of commentators point to the excellent relationship between the PRI and Mexico's major business groups until the early 1970s, state efforts to direct economic development repeatedly put members of the Cárdenas Coalition at odds with the Bankers Alliance. For

---

9. Hamilton (1982) makes a similar distinction between what she calls the progressive and the conservative alliances.

10. Nafinsa is a complicated institution that performed a wide range of functions not normally associated with any single institution. For a useful treatment of the structure and activities of Nafinsa, see Blair 1964.

example, efforts to ensure sufficient credit for agriculture and basic infrastructure under the Cárdenas administration established legal distinctions between "productive" and "unproductive" investments. Under these regulations, private banks were required to provide a minimum amount of loans for these sectors. State investment in priority sectors and requirements that private credit be made available to these sectors obviously created opportunities for private industries that were supplying inputs to agriculture and infrastructure. The response of those whose money would be channeled into these sectors, however, was a round of speculation against the peso and capital flight (Maxfield 1990, 62; Medina Peña 1994, 117; see also Anderson 1963; Shelton 1964).

The real question was never whether the state should intervene in the economy. Rather, the concern of business groups had to do with where, how, and in whose interest the state intervened. Thus, while business groups attacked the government for creating a program to distribute subsidized food in the early 1950s, they saw no contradiction in turning around to complain that state austerity was choking off economic growth. Similarly, while business groups supported trade barriers against foreign competitors, they simultaneously sought exemptions from tariffs on inputs to their products (Pellicer de Brody and Reyna 1978, 26; Torres 1984, 95).

Even during the years of greatest accommodation between the Cárdenas Coalition and the Bankers Alliance, from 1940 to 1970, repeated efforts by the state to direct private investment into productive activities met with mixed success. Private bankers complained that the state was unfairly competing against them when it sought to raise money in 1949 by issuing ten-year savings bonds (Torres 1984, 129). Early in the administration of President Ruiz Cortines, (1952–58), capital flight convinced the president to abandon plans to cut state spending and return to a policy of subsidies and fiscal incentives for private business. President López Mateos made a similar policy correction after declaring that his government was of the "extreme left within the Constitution" and openly supporting the Cuban Revolution. Accusations that he was a socialist were followed by an acceleration of capital flight. The response of the López Mateos administration was "an extraordinary effort on the part of the Mexican government to demonstrate its continued esteem for, and appreciation of, the critical role of business in Mexico's economic life" (Vernon 1963, 122; Brachet-Márquez 1996, 148).

By the early 1960s, however, the structure of the private sector and the state's dependence upon it for capital were both beginning to change. The years between 1954 and 1970 are sometimes referred to as the period of "stabilizing development." It was during this time that Mexico achieved high levels of sustained growth with low inflation—an average annual growth rate of 7.03 percent with average annual inflation of only 3.83 percent—leading many analysts to conclude that Mexico was on the verge of joining the nations of the advanced industrialized world. Among the keys to stabilizing development was the use of parastate firms to provide subsidies to domestic production, and tax exemptions to private industry even while a tight lid was kept on public spending (Hernández Rodríguez 1989, 47).

In order to capture the resources needed for investment in basic infrastructure and to stimulate private investment, then, the state maintained high interest rates, to attract foreign capital. The protected internal market brought considerable foreign investment to Mexico in the form of finance capital as well as in manufacturing. But the model also created a captive market in which the private sector had little incentive to export its goods or to improve their quality. It also led to a growing concentration of resources.

The years of uninterrupted growth had concentrated a large fraction of total financial resources in the hands of an ever smaller number of financial institutions. In 1950, 75 percent of all financial resources in Mexico were held by forty-two banks. By 1970, only eighteen banks controlled 75 percent of these resources, with the largest two banks controlling more than 40 percent (Tello 1984, 29). Simultaneously, ownership and control of industry were becoming more concentrated. In 1955, more than 14,000 firms produced 82 percent of national output. By 1970, barely 5,000 firms accounted for more than 90 percent of industrial production, with the largest 2,050 firms alone generating more than 80 percent of total production (Hernández Rodríguez 1989, 56).

In the 1970s, these tendencies toward concentration of capital, especially of financial capital, sharpened still further. In an effort to regulate the growth of holding companies, known as *grupos,* the state passed legislation in 1970 that formalized this form of business organization. Finance and mortgage banks strengthened their position in the economy by coordinating their activities with industrial firms. Legislation passed in 1974 encouraged private banks to tap into international capital markets, to make up for the dearth of financing available in pesos. And in 1976, the legislature passed a banking reform law that encouraged still greater coordination among banks and private firms that needed financing by permitting banks to engage in all forms of financial activities (Tello 1984; Maxfield 1992, 79).

The 1976 legislation permitting universal banking was intended to help smaller savings banks join forces with larger finance banks to promote "greater dispersion of resources in the national system" (Tello 1984, 31). In fact, it facilitated greater concentration in Mexico's financial sector. By 1980, the ten largest banks in Mexico controlled more than 85 percent of all resources in the country's financial market. The top two banks, Bancomer and Banamex, held about 49 percent of the resources of the entire banking system. Although the state owned two of the ten largest banks, the combined resources of these two institutions barely surpassed 10 percent of Mexico's total financial resources. Public financial institutions were important for providing credit and passing on loans from international agencies, but they captured very few of the resources available on the national market. Even loans from the public banking system to public firms often found their way into private banks as parastate firms deposited money lent from public banks into accounts held with private banks (Tello 1984, 32, 33).

These developments in the structure of ownership in Mexico were soon matched by a reorganization of major business organizations in opposition to the policies of President Echeverría. As the Echeverría administration sought to respond to the unrest of the 1968 student movement, a guerrilla movement threatened political stability with a series of kidnappings and assassinations of business leaders. Business leaders began to lose faith in the state's capacity and willingness to protect even their most basic interests (Hernández Rodríguez 1988, 60).

In response, the private sector began to challenge the government more openly. A new generation of more confrontational leaders was emerging among Mexico's major business organizations, culminating in the creation of the Business Coordinating Council (CCE) in 1975. The CCE was an umbrella organization for every major business group in the country except those connected with agriculture and provided a link among specialized business groups as well as multisector business groups. Although much of the private sector had once depended on the protection of the state, many in the business community now felt that the government no longer served their interests, and they felt no compulsion to continue working with the state (Ledesma 1992, 46, 47; Rubio 1989).

DEFENDING NATIONAL SOVEREIGNTY

Mexico's relationship to the international economy is dominated by its interaction with the United States. After feckless attempts to shape the outcome of the Revolution, the United States sought to limit the scope of the change that took place in its wake. The U.S. government supported American companies seeking reparations from the Mexican government and payment on debt held by the Mexican state (FNM 1987, 45, 46). In 1923 the

U.S. government withheld its recognition of the administration of Álvaro Obregón (1920–24) until it received assurances that Mexico would not apply Article 27 of the Constitution retroactively to the holdings of U.S. nationals. Article 27 declared that natural resources were the inalienable property of the nation. In addition, it granted the state the right to expropriate private property in the public interest and prohibited foreign capitalists who operated under concessions in Mexico from calling on their home government for protection in disputes with the Mexican state (Fernández Varela 1991; Hamilton 1982, 61).

In effect, the United States made recognition of Mexican sovereignty conditional upon Mexico's recognition of the rights of foreign capital. Almost immediately after reaching an agreement with the United States, however, the Mexican government began creating more restrictions on the operation of foreign capital in the country. In 1925, the administration of Plutarco Elías Calles (1924–28) passed the Alien Land Law and the Petroleum Law. The Alien Land Law prohibited foreign ownership of any land within fifty kilometers of a Mexican border and prohibited majority ownership in land-development companies, while the Petroleum Law was an attempt to make foreigners who had owned their property prior to the Revolution request "confirmatory concessions" (Hamilton 1982, 71; Meyer 1977).

Conflicts between Mexico and the United States over economic policy continued to erupt from time to time. However, more than any specific dispute, it was the general instability of international markets that exerted the most important external influence on Mexico's state-led development model. Throughout Mexican history, foreign markets have been highly volatile and the postrevolutionary period was no exception. Commodity prices—and with them, export earnings—fell during the world depression, then rose during World War II, only to collapse again following the war and then rise again during the Korean War. Foreign investment in the domestic economy was no more reliable. Foreign owners often repatriated their profits instead of reinvesting in Mexico. Investment, especially in basic infrastructure, tended to be concentrated in urban centers or in enclaves that were more closely linked to the U.S. economy than to that of Mexico, neglecting outlying areas that most needed the investment (Newfarmer and Mueller 1975).

The substitution of imports with domestic production began in practice as a response to the world depression before economic theorists identified these policies as a model of industrial development. As specific policies for substituting imports with domestic manufacturing evolved, however, it became clear that the state needed to do more than simply exclude foreign capital from the domestic economy. Mexico also needed to attract foreign investment into priority sectors. The challenge for the Mexican state, then, was to exclude foreign capital from sectors of the economy where it had failed to meet Mexico's developmental needs while attracting capital into areas where it could be beneficial (Medina Peña 1994, 116).[11]

The conflict between President Cárdenas and foreign petroleum companies illustrates the difficulty of striking a balance between establishing national control over the economy while, simultaneously, attracting the foreign investment needed for development. Control over petroleum was not merely of symbolic importance, although there was widespread support for national control of the industry. Petroleum was also an important source of foreign exchange and would be needed by Mexico literally to fuel its industrial development. At the same time, the Mexican state could hardly afford the "prompt, effective, and adequate indemnization," that the United States demanded that Mexico provide to expropriated foreign oil companies (Meyer 1977, 188).

The development of the petroleum sector under state ownership would require foreign technology, and the expertise of foreign engineers and managers, in addition to foreign capital. For years after the petroleum nationalization, the U.S. State Department prohibited U.S. government financing for development in that sector unless Mexico agreed to readmit U.S. companies (Torres 1984, 177). But with war looming in Europe, the Roosevelt administration was not in a position to push Mexico too hard on behalf of U.S. petroleum companies. The need to secure the southern border as well as the need for raw materials from Mexico would force the United States to adopt a more conciliatory approach toward the Cárdenas administration.

Although the United States and multilateral lending agencies still sought to promote private ownership in the 1960s and 1970s, World Bank lending policies became more supportive of public firms (Torres 1984, 182, 183). Lending from the bank could not match the capital available from private lenders, but the bank did provide about $4.2 billion in loans between 1949 and 1980. Table 2 shows World Bank lending to Mexico during this period. All but about 13 percent of the loans were channeled through Nafinsa. Until 1960, all bank lending went to support the development of electricity, railroads, and roads. In the 1960s, lending expanded into agriculture and in

---

11. Brandenburg (1964, 215) asserts that Mexico largely succeeded in expelling foreign capital from strategic sectors of the economy while harnessing it to the goals of the state during the postrevolutionary period.

TABLE 2 World Bank loans to Mexico by economic sector, 1949–1980 (millions of nominal dollars)

	1949–60	1961–70	1971–80
Energy	124.8	220.0	125.0
Transportation			
Railroads	61.0	0.0	325.0
Roads and Highways	25.0	130.0	210.0
Ports	0.0	0.0	20.0
Aviation	0.0	0.0	25.0
Agriculture	0.0	117.5	1,755.0
Tourism	0.0	0.0	114.0
Unspecified "Industry Development"[a]	0.0	0.0	247.0
Steel	0.0	0.0	165.0
Fertilizer	0.0	0.0	130.0
Development Finance[b]	0.0	0.0	151.5
Water Supply and Sewage[c]	0.0	0.0	255.0
Mining	0.0	0.0	40.0
TOTAL	210.8	467.5	3,562.5

Sources: World Bank annual reports, various years. Missing data for 1965, 1967, and 1970.
[a]Also includes unspecified "Industry, Small and Medium Scale."
[b]Development finance includes generic "Development Finance" as well as projects labeled "Urban Development" and "Rural Development" except where "Agriculture" is specified as recipient of loan. Does not include "Industry Development."
[c]Does not include "Irrigation" projects, which are included under "Agriculture."

the 1970s, it began supporting other sectors: tourism, steel, mining, and ports (see Babai 1988, 256; Blair 1964).[12]

It was more difficult to channel private trade and investment. Following the Cárdenas presidency, Mexico's policy toward foreign capital often wavered between one displaying an ideological preference for laissez-faire and another stressing the practical necessity of protection. Efforts to open the economy to foreign trade repeatedly collapsed in the face of mounting trade deficits as Mexico invariably imported more than it could sustain with its exports. A trade agreement negotiated between the United States and Mexico in 1942, for example, soon produced trade imbalances that eroded Mexico's foreign reserves. The reduction in demand for raw materials at the end of World War II and a sharp rise in imports created a balance-of-trade crisis in 1947 that forced the Alemán administration to withdraw

12. In the first years of its operation, the World Bank was known as the International Bank for Reconstruction and Development.

from the treaty, increase tariff barriers, and create a system of licensing for imports (Torres 1984, 40, 226–29).

As the state increased its control over basic infrastructure, heavy industry, and raw materials, foreign investment moved increasingly into manufacturing. In 1940, less than 7 percent of foreign investment was concentrated in manufacturing. By 1970, this figure was more than 70 percent (Newfarmer and Mueller 1975, 49). Many of Mexico's smaller firms, represented by Canacintra, criticized the entry of foreign capital into manufacturing. Even larger and more conservative business groups organized around the National Confederation of Chambers of Industry (Concamin) and the National Confederation of Chambers of Commerce (Concanaco) accepted the need for at least provisional protectionist measures (Torres 1984).

But as the state struggled to determine the appropriate role of foreign industrial capital in the country, more important changes were taking place in Mexico's relationship to international financial capital. During the 1960s, the Mexican state increasingly came to rely on foreign borrowing to offset a growing balance of payments deficit. The stabilizing-development model actually depended on a current-accounts deficit, and between 1960 and 1970 the deficit in Mexico's current-account balance grew from $340 million to slightly more than $1 billion. During the same period, Mexico's external debt grew from $864 million to $3.2 billion (Bennett and Sharpe 1985, 34; Reynolds 1978; Villarreal 1977, 84).

With high levels of growth, social stability, and free convertibility of the peso from the mid-1950s through the late 1960s, the Mexican state had no trouble attracting the foreign investment it needed to offset the current-accounts deficit. But as social unrest began to threaten stability in the late 1960s and early 1970s, Mexico's increasing dependence on inflows of foreign capital became riskier, even after oil deposits continued to attract foreign investors.

In the 1970s, foreign lending to Mexico began to "dollarize" the economy. While foreign currency in the Mexican banking system amounted to only 5.7 percent of total financial resources in 1970, by 1980 foreign currency in Mexican banks, both private and public, had grown to 15.4 percent of total financial resources. As the petroleum boom of the late 1970s gained strength and Mexico sought to expand its production capacity, domestic banks were unable to meet the financing needs of either the private or the public sector. High levels of liquidity in international financial markets also made foreign debt cheaper than peso-denominated debt (Tello 1984, 47, 48). Foreign public debt, which was $6.8 billion in 1972, or less than one-half of one percent of GDP, more than tripled, to $20.8 billion in 1976, or 5.6 percent of GDP. By the end of the López Portillo presidency, in 1982, public foreign debt would more than double again, growing to $58 billion as government spending grew from 38 percent of GDP in 1976 to 47 percent of GDP in 1982 (Aspe 1993, 114, 115).

The state was not alone in its apparent inability to live within its means or restrain its foreign borrowing. The private sector also went on a borrowing binge during the late 1970s. Although starting at a lower absolute level, private-sector foreign debt actually grew more rapidly than that of the public sector during the López Portillo administration, more than tripling from $5 billion in 1976 to $18 billion in 1982. But much of the money that entered the country in foreign loans took a round trip, returning to the United States and other foreign destinations in the form of capital flight. Estimates from the Bank of Mexico show capital flight growing from less than $1 billion in 1978 to $10.9 billion by 1981 (Aspe 1993, 115; Tello 1984, 50).

The bank nationalization was not the logical conclusion of Mexico's state-led development policies. Instead, it represented a failure of the country's effort to harness private interest to national developmental goals. In his national address announcing the expropriation of the banking system on September 1, 1982, President López Portillo pointed to the divergence between the pursuit of profits and meeting Mexico's basic needs: "Speculation and coupon clipping translate into a multiplication of wealth for a few without producing anything and of necessity come from the simple plunder of those who produce. In the long run, they inevitably lead to ruin" (cited in Tello 1984, 129).

PRESIDENTIALISM AND BUREAUCRATIC STRUCTURE
IN THE PARASTATE SECTOR

One of the most enduring propositions regarding Mexico under the PRI is that the president enjoyed almost absolute powers during his *sexenio* (six-year term in office). The president dominated the military and the public bureaucracy. He commanded state and local organizations that were connected to the PRI and thus controlled government relations with the media, business leaders, and foreign powers. This assessment of presidential power is generally based on the lines of authority; the powers of appointment (especially of a successor); an absence of real checks and balances from the legislature or the courts; and the simple fact that, historically, presidents have wielded exceptional power during their terms in office (Bailey 1988,

36; Camp 1996, 156; Cornelius, Gentleman, and Smith 1989, 8; Garrido 1989, 421–23).

The growth of the parastate sector in Mexico has generally been seen as enhancing the power of the executive. As the number of public firms increased, so did the patronage available for the president to dispense in the form of contracts, jobs, and public works (Centeno 1994, 82). The evolution of the bureaucratic structures governing the parastate sector, however, reveals that the increased amount of patronage at the president's disposal did not translate automatically into increased capacity to wield that power. Instead, as the number of parastate firms grew, the structure for governing them became increasingly cumbersome, requiring an almost constant development of new strategies for bringing these firms under the control of the executive and for harnessing their activities to the goals of the federal administration.

Three distinct phases of administrative and institutional control over the parastate sector have been identified (Carrillo Castro and García Ramírez 1983, 37–39). During the first phase—1925–46—the state acquired and created firms primarily to meet the new requirements of the 1917 Constitution, stabilize the market, secure greater economic independence, and meet basic demands for social justice. During the second phase—1947–58—the state attempted to gain control over the growing number of parastate firms. Beginning with the passage of the Law for Federal Government Control of Decentralized Organisms and Firms with State Participation, in 1947, the executive branch attempted to use the Ministry of the Treasury to exercise greater oversight and control of public firms (*Diario Oficial* 1947).

The third phase—1959 to 1982—was characterized by efforts to implement systems of planning and control and can be further divided into two subperiods. During the first—1959 to 1976—the state's efforts to control the activities of parastate firms involved centralization and control over the income stream of parastate firms. Between 1977 and 1982, planning and control were achieved through decentralization of authority to line ministries within the federal government and by establishing goals and criteria of evaluation based on the global national development strategy (Carrillo Castro and García Ramírez 1983).

Throughout all these periods, the struggle between representatives of the Bankers Alliance and the Cárdenas Coalition within the federal bureaucracy continued. Under the reforms of 1947, the SHCP gained the power to review and veto the budgets of public firms, place restrictions on spending, promote organizational innovations, and designate a member of the executive body of parastate firms. The SHCP was also granted the right to propose the liquidation of public firms if they failed to "realize useful functions" or when they competed with private firms that "duly fulfill[ed] their obligations" (Carrillo Castro and García Ramírez 1983, 44). These powers were expanded still further with the creation of a National Commission of Investment under the direction of the SHCP in 1948 (*Diario Oficial* 1948). However, while the SHCP presided over the commission, other ministries, including the Ministry of Energy, Mines, and Parastate Industry (SEMIP) also had permanent representation on the commission.[13] This created serious problems of coordination, leading to the dissolution of the commission only a year after it was established (Carrillo Castro and García Ramírez 1983, 44, 45).

The system governing parastate firms was then fragmented. The SHCP gained control over parastate financial firms through the National Banking Commission and the National Insurance Commission. At the same time, the SEMIP was given control of the Mexican National Railroad and later the Federal Electric Commission (Carrillo Castro and García Ramírez 1983, 46; Ruiz Dueñas 1988, 118, 119). One of the biggest problems in coordinating the parastate sector was the considerable strength that these firms wielded through their control of economic resources. In some cases, directors of parastate firms were more powerful than the ministers of state who were supposed to monitor and control the activities of the parastate firm (Carrillo Castro and García Ramírez 1983, 47).

In 1954, control of public firms was placed in the hands of an Investment Commission, directly under the control of the president. Carrillo Castro and García Ramírez (1983, 48) claim that the Investment Commission was "the most successful attempt to date to achieve greater rationalization of the investments of the federal, public sector." Yet within four years still another scheme of control over parastate firms was devised. The 1958 Law of State Ministries and Departments created a new Ministry of the President to coordinate the activities of the parastate sector along with the SEMIP and the SHCP. These three ministries were charged with the planning, budgeting, control, and oversight of the parastate sector and were given the misnomer the "triangle of efficiency" (Ruiz Dueñas 1988, 119).

Part of the control problem was related to the structure of relations between parastate firms and the federal bureaucracy. In parastate firms, as in private ones, ownership and control are functionally separated from each

---

13. The Ministry of National Goods and Administrative Inspection was renamed the Ministry of National Patrimony by 1958 and in the 1970s was renamed the Ministry of Energy, Mines, and Parastate Industry (SEMIP). For the sake of simplicity, I use the acronym SEMIP to refer to this ministry even before it took its final name.

other (Acle Tomasini and Vega Hutchison 1986, 36). The principal-agent dilemma in the parastate sector, then, refers to the potential independence of managers within parastate firms to pursue goals distinct from those of the central administration. One of the most common goals of individual managers was quite simply individual enrichment. External supervision of the parastate sector by the SEMIP and the creation of various committees for oversight were attempts to identify and limit the use of public firms for personal profit. Throughout the 1960s, new commissions were created and laws passed to bring the parastate sector under the control of the executive, always with mixed success. Most of these attempts were designed to gain control over the spending of public firms and, in particular, to avoid the siphoning of public money into private hands (Acle Tomasini and Vega Hutchison 1986, 39).

Gaining fiscal control over the parastate sector, however, was especially problematic because the political and economic roles that the firms played were distinct. To maximize their potential as sources of political patronage, parastate firms needed to permit some opportunities for personal enrichment. Without this a job as a director or manager of a public firm would hardly constitute a proper reward for political loyalty. To maximize their contribution to national development, however, parastate firms were supposed to provide efficient production. High rates of turnover in the management—especially upper management—of parastate firms in conjunction with the six-year presidential cycle made the goal of efficient production especially difficult to meet (Acle Tomasini and Vega Hutchison 1986, 78, 79).

Revisions to the Law for Federal Government Control of Decentralized Organisms and Firms with State Participation were passed in 1965, and again in 1970, formalizing the structure of control by the triangle of efficiency. Yet this macro level of oversight still depended on the analysis and evaluation of budgets and programs by the governing bodies of these firms. In an attempt to bypass the management of public firms and get an independent source of information, the SEMIP implemented a system of external audits, though it is unclear to what extent these audits were capable of penetrating the bureaucracy of the firms (Carrillo Castro and García Ramírez 1983, 64–71). Thus, while Acle Tomasini and Vega Hutchison contrast the "actual" control over parastate firms exercised by the central government with the "formal" control exercised by the management of parastate firms, control over the information about the day-to-day activities of firms provided a very real source of power to line ministers and directors of parastate firms.

The contradictions between the interests of efficient production and personal enrichment were further aggravated by the structure of the triangle of efficiency. The SHCP's position within the Bankers Alliance made it generally hostile to the expansion of the parastate sector or increased investment by existing parastate firms. The SEMIP, in contrast, existed for the express purpose of administering and overseeing the parastate sector. Although specific firms such as Pemex were actually more powerful than the ministry itself, there were many firms over which the SEMIP could exercise control and extract benefits.

The role of the Ministry of the Presidency within the triangle of efficiency is less clear. It is likely that this ministry, as the direct representative of the president, took a position with regard to parastate firms that changed from one administration to the next. Positional interests aside, within the triangle of efficiency, "the Law of Ministries and Departments of State, the Law of Control and numerous presidential agreements granted concurrent, even overlapping and duplicate functions to each of them [the ministries].... These three dependencies established instruments and measurements—sometimes together, but most of the time each one by itself" (Carrillo Castro and García Ramírez 1983, 62).

Beginning in 1975, executive decrees granted parastate firms greater administrative autonomy through the creation of Committees for the Promotion of State Socioeconomic Development (Ruiz Dueñas 1988, 120). Parastate firms were also "sectorized" during this period: placed under the control of the line ministry in whose sector the firm operated. For example, state-owned hotels were placed under the control of the Ministry of Tourism, railroads under the Ministry of Communications and Transportation. At the same time that administrative control over parastate firms was decentralized, central control over the finances of parastate firms was tightened. A series of administrative reforms in 1976 began to bring the finances of individual ministries—and, by extension, parastate firms—under the control of the president by exercising greater control over budgets.

Under the new system, line ministries would develop their own budgets based on their evaluation of the needs of the sector. The budgets of all the line ministers would then be aggregated and passed to the SHCP, which would determine how much of the federal budget would be allocated to each ministry. Once these decisions had been made, the budget was disaggregated and passed back to the line ministries, which would allocate resources within their sector to match the requirements of the SHCP's budget. The Intersecretarial Committee on Spending and Finance served to monitor and coordinate the activities of each ministry, while the Organic Law of Public Administration and the Law of Budgeting, Accountability, and Public Expenditures further strengthened the hand of the executive over the finances of the bureaucracy (Centeno 1994, 86, 87; Interviews 55, 69).

Ironically, it was precisely during this period of heightened scrutiny of the finances of parastate firms that the federal government as a whole began taking on large amounts of external debt. For the first time, "the administrative reforms of 1976 and the accompanying plans allowed an elite to centralize control over policy making and the general agenda of the government while apparently attempting to decentralize the bureaucracy" (Centeno 1994, 87). As I will show in the following chapter, this process had not yet concluded. Individual ministries and executives of parastate firms retained a power base by virtue of their control over the basic information on the activities of parastate firms as well as their ability to execute the decisions required for the operation of parastate firms. While Presidents de la Madrid and Salinas would both argue for the need to decentralize the bureaucracy, they would actually centralize power still further in order to privatize parastate firms.

Discussion

The Mexican Revolution shattered the social, legal, and political foundations of the Mexican economy. In the process of rebuilding these foundations, new institutions were developed to link the rights of private property to broader goals of social welfare and economic development. The creation of these new institutions involved continual negotiation and renegotiation not only of the domestic social pact between labor and capital but also of Mexico's relationship with the United States and the international economy.

From the mid-1930s to 1982 the PRI used public firms to engineer a stalemate among the diverse and competing constituents within Mexico. Until the early 1970s, that stalemate was generally reinforced by Mexico's position in the global economy. Public firms filled a gap that national capitalists could not, and that the state preferred that international capitalists would not. They served to legitimate the revolution by imposing Mexican sovereignty on strategic sectors of the economy and promoting growth.

This development strategy was not always coherent. It consisted of tactical maneuvers, swings in support of whichever group could bring the most consistent pressure to bear on the state within the constraints of Mexico's class structure and position in the international economy. But decisions to create or appropriate a particular firm appear to have been based less on ideological predilection than on pragmatic calculations of the costs and benefits of public ownership. Indeed, the parastate sector continued to grow *despite* the ideological preferences of presidents such as Ávila Camacho and Alemán (see Jones and Mason 1982).

The structure of public ownership was characteristic of this pragmatic strategy of state-led development. Although the growth of state intervention in the economy throughout this period appears typical of a "command economy," the Mexican state was often incapable of commanding at all. Instead, the growth of the public sector in Mexico reflected the standoff among different sectors of society that had become institutionalized within the public bureaucracy. Capitalists got cheap inputs; workers won a wide range of formal controls over the process of production and economic benefits, even while they were subject to the political control of the PRI; public bureaucrats increased their access to sources of patronage; politicians cultivated constituencies; and consumers got inexpensive goods and services, however shoddy.

The repeated attempts to bring the parastate sector under control seem to confirm Evans's assertion that an important problem for economic development is not an excess of bureaucracy but, rather, a lack of it. Public firms were created and grew in an effort to direct economic investment toward sectors in which the profit motive was insufficient to attract private capital. Once these firms were in place, the state faced the challenge of reining in the utility-maximizing behaviors of workers and managers in public firms and bringing these institutions under the control of the executive.

Some see the inefficiency of the public sector as inherent in the structure of public ownership (Kornai 1992, 71–75).[14] In the absence of market signals, allocations are made arbitrarily, without regard for the actual supply of or demand for goods and services. In light of the historical experience of developing countries, however, the growth of state intervention in much of the developing world was itself a response to the vagaries of the international economy during the interwar period (Sanderson 1992). The Mexican state attempted to impose a different rationality on parts of the economy, and for almost forty years it did a reasonably good job of allocating resources toward growth, even if it failed abjectly to improve the level of welfare for the majority of its citizens.

Yet even as the Mexican state increased its presence in and regulation of the domestic economy, it never completely escaped the force of the market.

---

14. Jacques Rogozinski (1997, 27–29), general director of the UDEP under President Salinas, cites Kornai to explain the failures of the parastate sector in Mexico.

Mexico still required private investment to meet its developmental goals. International markets—commodity markets on which Mexico sold its products, markets for technology and capital equipment, and especially financial markets—placed important constraints on the Mexican state. Foreign markets also provided the option of capital flight to many domestic capitalists. Mexican capitalists could rely on the United States as a safe haven for investment when they feared that the Mexican state might not show sufficient respect for their property rights.

This is not to claim that the international market was the pristine realm of free exchange found in neoclassical models. To the contrary, an oil cartel and the United States Federal Reserve Bank may have exerted the strongest influence on the Mexican economy in the late 1970s and early 1980s. But what makes these international influences on Mexico unique is that they escape the control of the Mexican state. While Mexico was integrated into the international economy, foreign markets provided the opportunity for capitalists to detach their resources from the oversight and regulation of the state.

The social bargains within which the Mexican economy had grown for more than 30 years began to break down in the late 1960s and early 1970s. During this period, domestic capital became more concentrated, more closely linked to finance capital, and more independent of the state. The rift between large capitalists and the state widened as popular sectors made new demands on the state and international financial markets allowed the state to bypass domestic capitalists for its resource needs.

But none of the economic indicators that are cited as signs of imbalance in the economy led automatically to the crisis. Under different circumstances—a sustained rise in world oil prices, continued stagnation and low interest rates in the United States—Mexico might have managed to buy its way out of the crisis with growing petroleum exports. Private capitalists bet against this, contributing to the collapse of the model. When the U.S. Federal Reserve Bank allowed interest rates to spike and world oil prices began to fall, these bets paid handsomely.

Thus, Mexico's parastate sector did not collapse under its own weight. Rather, it crumbled as one part of a larger structure of social compromises and political bargains that was falling apart. In the early 1980s, the international environment within which domestic actors had operated changed dramatically and it became impossible to maintain the arrangements through which the Mexican state had controlled labor, appeased capital, and defended its economic sovereignty.

When speculation undermined the economy in the early 1980s, one of President López Portillo's close advisors, Carlos Tello, argued that the bank expropriation would help to resolve the dispute over economic policy in favor of the Cárdenas alliance. But the nationalization ultimately added the burden of massive foreign debt held by private firms to the already considerable debt of the state. The need to pay off that debt would later be used as an argument to justify dismantling Mexico's parastate sector (Maxfield 1990, 144).

THREE

Privatization and the Demise of State-Led Development

[P]ublic firms have permitted a progressive ability to manage the economy on behalf of the state, in accordance with the popular will . . . the ability to direct the economy is a basic element of the stability of social institutions.
—*Miguel de la Madrid*, La empresa pública en México factor de desarrollo económico y social del país

One of the best ways to prepare the private sector to operate efficiently as the new owner of public firms is to protect it as little as possible.
—*Jacques Rogozinski*, La privatización de empresas paraestatales

By incorporating the diverse elements of the revolutionary family into the ruling party, the Mexican state was able to balance competing, often contradictory interests of society. Parastate firms proliferated as a means of subsidizing domestic capitalists, blocking the entrance of foreign capitalists, and controlling the activities of labor in strategic sectors of the economy. By 1982, the structure of the parastate sector clearly reflected this labyrinth of compromises and concessions, industrial promotion and protection, labor suppression and subsidization. The growth of the parastate sector consistently outstripped the capacity of the central administration to actually control the activities of these firms and direct them toward developmental goals. An unwieldy amalgam of more than one thousand firms ranging from the petroleum behemoth Pemex to a bicycle factory and a soccer team all operated within a state that was suddenly deprived of the international credit on which it depended.

Mexico's parastate sector was not merely deeply embedded in the political and social institutions of Mexican society. In addition, the country's political, economic, and social structures reflected a specific accommodation with the international economy. Revolutionary nationalism provided the ideological justification for the nationalization of foreign holdings in mining, petroleum, electricity, railroads, and telecommunications. Control over these sectors provided the state with some of the resources it needed to direct economic development without having to directly challenge domestic capitalists.

Through the post–World War II period, foreign capital began returning to Mexico, primarily in two forms: in the manufacturing sector to take advantage of the protected domestic market and, increasingly during the early 1960s, as financial capital. Foreign debt allowed the state to postpone economic reforms while it sought to meet the demands of popular sectors whose members rebelled against their exclusion from the gains of stabilizing development. Tax reform was delayed and employment in public firms expanded during the petroleum bonanza of the 1970s.

This arrangement collapsed in August 1982, when representatives of the Mexican Ministry of the Treasury announced to their counterparts at the U.S. Treasury and Federal Reserve Board that the Mexican state was bankrupt. The nationalization of the banking system by outgoing President López Portillo shattered the confidence of Mexico's elite and led to a major rupture between the Mexican state and domestic capital. Forced to choose among the various factions that made up the revolutionary family, the de la Madrid administration wasted no time in establishing whose interests could be sacrificed. As the state implemented a wrenching austerity program to regain the favor of national and international capital, the traditional base of the PRI—workers, peasants and the middle class—lost a decade of development. Parastate firms, relics of Mexico's state-led development strategy, would play a critical role in the transformation of the Mexican economy.

Given the legal foundation on which the parastate sector rested, the historic difficulties of bringing parastate firms under control, and the wide range of constituencies built into public ownership, how did the Mexican state rid itself of this vast empire of public firms? In the following sections I answer this question, working my way from the inside of the parastate sector out. First, I provide an overview of privatizations in the three presidencies since 1982, paying special attention to the bureaucratic maneuvering that made such a radical transformation possible. I then examine the effects of privatization on the PRI's principal domestic constituencies: labor and capital. I conclude by looking at how privatization has changed Mexico's relationship to the world economy.

Selling the State

"One cannot privatize oneself," insisted a high-level official who had participated in the privatization program under President Salinas (1988–94) and continued to work on privatization within the Ministry of Communications and Transportation (SCT) under the Zedillo administration (1994–2000). The official emphasized the importance of being close to the action. However, he also pointed to the need to remain operationally separate from the entity that is being privatized (Interview 65).

In fact, the Mexican state did privatize itself rather successfully. Between 1982 and 2000, the number of parastate firms fell from 1,155 to 202. During this period, revenues from privatizations exceeded $30 billion (see Figs. 3 and 4). In order for the state to privatize itself, however, it first had to transform the public bureaucracy. As shown in the previous chapter, presidential power to dispense positions within the parastate sector was not the same as actually controlling the operation of public firms. Indeed, as a source of patronage, positions in public firms were valuable only as long as lower-level public officials within the bureaucracy enjoyed some discretion in managing the resources of the firms. When it became clear during the de la Madrid administration that the very source of political power and patronage—the parastate firm—might actually be taken away, officials within the bureaucracy quickly developed strategies for resisting privatization. Implementing neoliberal reforms would require the creation of new institutions to overcome this resistance.

The process of privatization occurred in three connected but analytically distinct phases, roughly corresponding to the de la Madrid, Salinas, and Zedillo *sexenios*.[1] These three phases can be distinguished from each other by (1) the state agencies that were used to privatize public firms and (2) the characteristics of the firms that were privatized. During both the de la Madrid and the Salinas administrations privatizations were accomplished through a centralization of control within the Treasury Ministry, which culminated in the creation of the Unit for the Divestiture of Parastate Entities (UDEP). This centralization of power enabled officials within the SHCP to isolate opponents of the privatization process and undercut their capacity to resist privatization by depriving them of the resources of the parastate firm. During the Zedillo administration, authority over privatizations was decentralized to special teams drawn from the SHCP and placed in strategic locations within specific ministries.

While the de la Madrid administration presided over the privatization of small firms in peripheral sectors of the economy, the Salinas administration launched a frontal assault on the heart of the parastate sector by selling large firms that operated in "priority" and "strategic" sectors of the economy.

---

1. A *sexenio* is the president's six-year term in office.

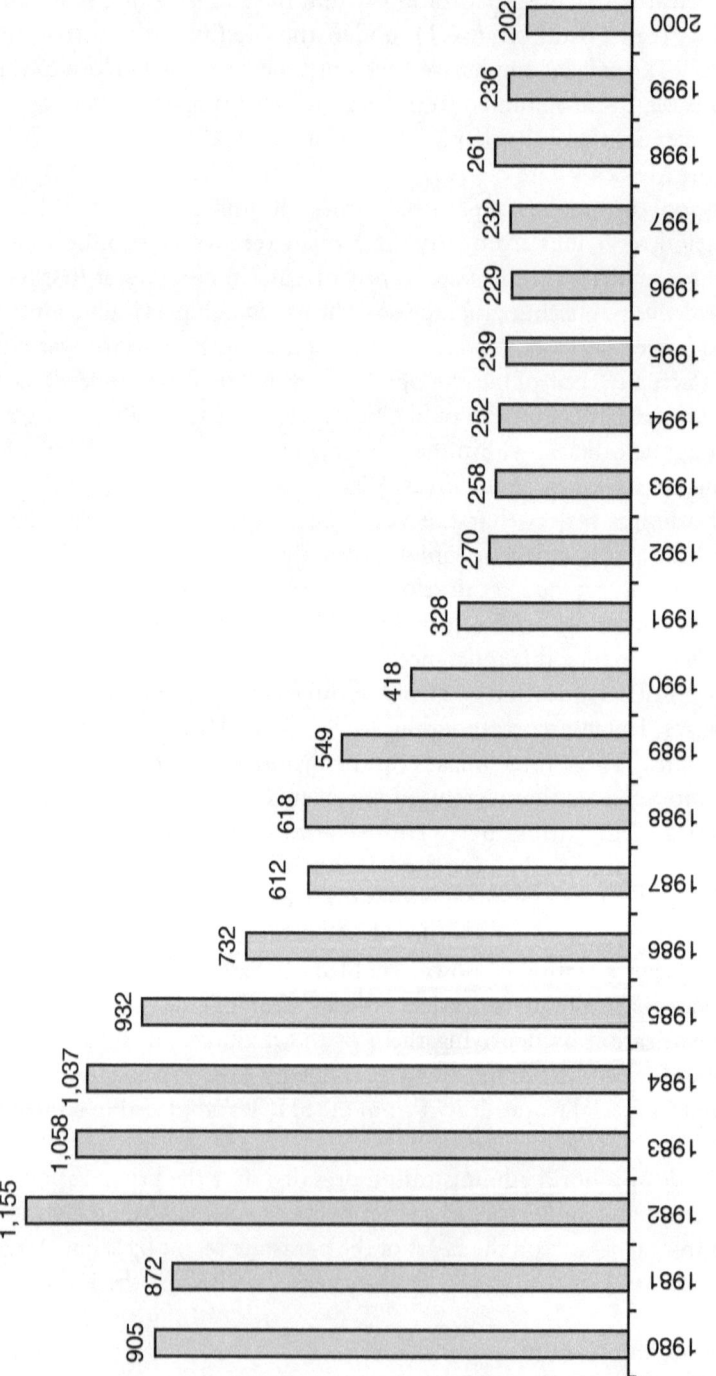

FIG. 3 Number of parastate firms, 1980–2000

Sources: Years 1980–87 from Salinas (1994, 91); years 1988–93 from Zedillo (1998, 81); years 1994–2000 from Fox (2002, 235).

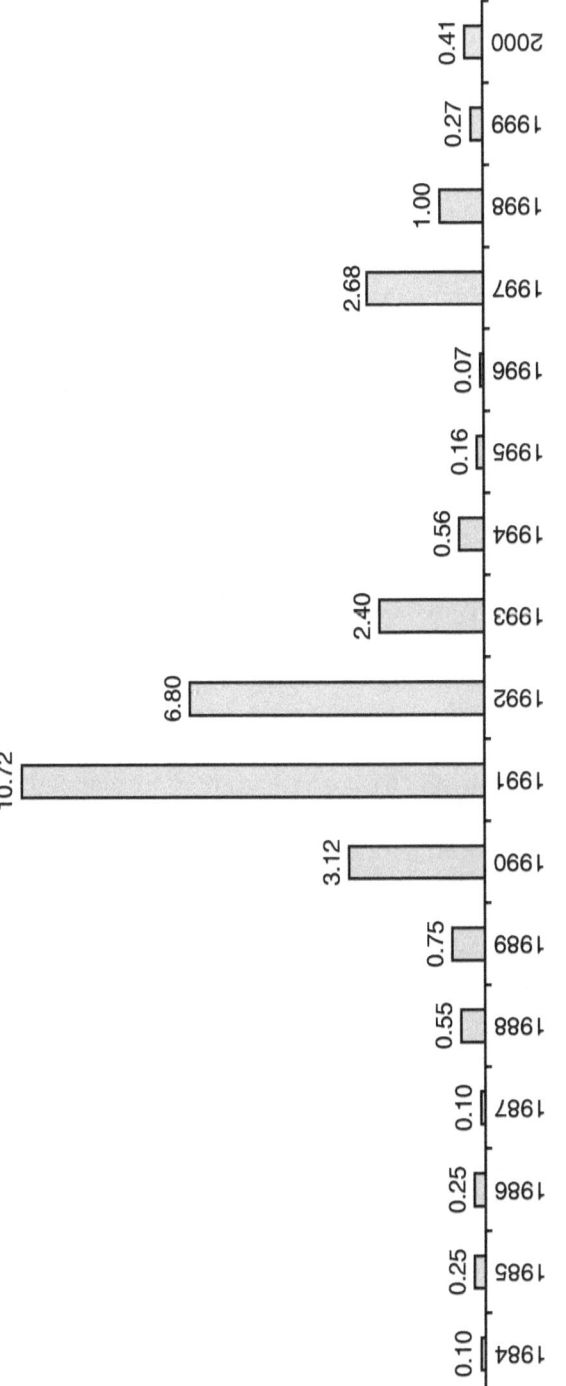

Fig. 4 Revenue from privatization, 1984–2000 (billions of nominal dollars)

Sources: Figures for 1982–88 estimates based on Tandon (1994, 60); 1989–94 from Rogozinski (1997, appendix 1). Figures for 1995–2000 from SHCP (n.d.).

The Zedillo administration maintained the focus upon privatization in core sectors of the economy; however, it generally accomplished this by selling concessions and introducing market mechanisms into sectors of the economy that remained "public."

THE EASY PHASE OF PRIVATIZATION: 1983–1988

President Miguel de la Madrid took office on December 1, 1982, three months to the day after outgoing President López Portillo announced his decision to nationalize the country's banking system. The nationalization of the banks swelled the size of the parastate sector, adding companies in which the banks held shares in addition to the banks themselves. It also infuriated important elements of the private sector, enlarging the rift between these groups and the state (Hernández Rodríguez 1986; Maxfield 1989). President de la Madrid had reaffirmed the central role that the state would play in economic development during the presidential campaign of 1982. He reiterated this position in the National Development Plan, which points to the role of the parastate sector in "strengthening the mixed character of the economy," "participating in the supply of socially necessary goods and services," "supporting the integration of the productive apparatus," and "regulating market activity in order to reduce the negative effects produced by income inequality" (SPP 1983, 6.3.3.1–6.3.3.5). The debt crisis and the need to reassure private investors, both foreign and domestic, would change all that. Privatization became a key component of the new strategy to pay back the debt, reform public finances, and regain the trust of the private sector.

Privatizations were always preceded by legal reforms. A flurry of legislation was passed in the first two months of the de la Madrid presidency, culminating with an amendment to the Constitution on February 3, 1983. During this period the Congress amended the Organic Law of Federal Public Administration; the Planning Law; and the Regulatory Law of Public Service, Banking, and Credit. Amendments were added to the Constitution, with state "rectorship" over national development added to Article 25 and placed next to the state's role of planning, directing, coordinating, and orienting national economic activity. Article 28 of the Constitution was amended to formalize the nationalization of the banks by making the provision of banking service the exclusive province of the state. In addition, satellite communications and railroads were added to the list of "strategic" areas over which the state exercised exclusive rights (see *Diario Oficial* 1983; Fernández Varela 1991, 32; Ruiz Dueñas 1988, 127).

A second set of legal reforms was adopted midway through the de la Madrid *sexenio* in an attempt to rationalize the administration of parastate firms. The most important measure adopted during this period was the 1986 Federal Law of Parastate Entities (LFEP), drafted by the powerful Ministry of Planning and Budget (SPP). The SPP had been at the center of efforts to rationalize public administration since its creation in 1976 and was a stronghold of reformers within the government. The LFEP attempted to provide for greater vigilance and control over the parastate sector. It mandated the presence of representatives from the Treasury Ministry and SPP on the governing bodies of parastate firms (Article 9), and it linked the continued existence of parastate firms to the achievement of the broader performance goals of the public sector.

The LFEP states that public participation in firms that are neither "strategic" nor "priority," as defined in Article 28 of the Mexican Constitution, will be subject to "alienation" (SPP 1986). The LFEP established early draft guidelines for privatization, creating a mechanism that allowed the state to rid itself of parastate firms. Articles 16 and 32 allowed the line ministers who presided over parastate firms to propose the alienation of decentralized organisms or majority-held firms to the SPP when the parastate firm ceased to fulfill the goals for which it was created.

The LFEP, however, contained a variety of other measures that either contradicted the broader goals of privatization or simply muddied the process. Article 39 of the LFEP, for example, granted line ministers the right to intervene in the privatization process in order to "protect the public interest, the stockholders . . . social parties and the labor rights of workers in public firms." Article 56 of the LFEP encouraged the creation of worker-management "mixed committees" to improve productivity in parastate firms. Later attempts to sell public firms would often require the elimination of these mixed committees as a precondition for the sale of the firm.

Most important from the standpoint of pushing privatizations through the bureaucracy, the LFEP failed to wrest operational control over public firms from authorities who had a vested interest in *preventing* their privatization. Formal authority over privatizations changed hands a number of times during the de la Madrid administration, and financial control over parastate firms was tightened by the LFEP. However, operational control over parastate firms remained in the hands of the various line ministries to which the firms were attached. From their positions on the executive committees and boards of directors of parastate firms, line ministers could keep a watchful eye on the efforts of would-be reformers. When the SHCP sought information on parastate firms to evaluate whether they were candidates for privatization, line ministers withheld data or presented contradictory and incorrect data, making it virtually impossible to evaluate a company, much less create a sales prospectus for potential buyers (Interviews 64, 69).

Thus, while the number of firms in the parastate sector declined dramatically during the de la Madrid presidency, most of the reduction came from the elimination of relatively insignificant firms. Privatizations during this period were primarily (1) simple sales of small companies requiring no restructuring, (2) liquidations of the assets of firms that were not viable, and (3) "extinctions," the legal elimination of *fideicomisos* and other funds that existed only on paper.[2] Furthermore, the LFEP changed the official definition of the parastate sector to exclude firms in which the state held only a minority of the shares, removing forty-eight firms from the books with a stroke of the pen.[3] All the divestitures during this period—including sales, liquidations, extinctions, transfers, and mergers—amounted to only 2.1 percent of public-sector GDP (Pichardo Pagaza 1988, 28). As shown in Table 3, employment in parastate firms held steady through the de la Madrid administration, while total federal employment actually grew.

Privatization efforts during the de la Madrid administration did, nonetheless, teach public officials important lessons. According to the guidelines for privatization, the responsibility for identifying the firms to be sold was placed on the line ministries; however, in practice these guidelines were quickly abandoned. Flowcharts from the SHCP show the initiative for privatization coming from the line minister to which the parastate firm was assigned. Officials from the Ministry of the Treasury, however, admitted that the process also operated in reverse: officials from the SHCP pressured individual ministries to identify firms that could be privatized. As one official from the UDEP put it, in the early stages of privatization, Treasury asked the ministries what they wanted to privatize. In later stages, Treasury asked the ministries what they wanted to keep and why (SHCP 1994a, 33; Interview 65).

Another official from the UDEP noted that early privatization attempts under de la Madrid had been "anarchic and disorganized" (Interview 58). With each ministry in charge of carrying out the privatizations of firms in its sector, it was impossible to establish standards for privatization or to monitor the process. Minister of the Treasury Jesús Silva Herzog attempted to impose

2. Government officials reported that the number of public firms reported in 1982 was actually an exaggeration because it included firms that existed only on paper, thus making the number of firms privatized during this initial period an exaggeration as well (Interviews 56, 64; see also Teichman 1995, 132).

3. While there is a dramatic drop in the number of minority-owned firms in the year that the LFEP passed—from sixty-nine in 1985 to seven in 1986—it is unclear why *any* minority-owned firms remain on the books. See SHCP (1994b, 9).

TABLE 3 Public-sector employment, 1980–2000

Year	Employment in Parastate Firms	Total Federal Government Employment
1980	723,200	3,151,900
1981	805,700	3,457,000
1982	885,200	3,668,600
1983	1,000,200	3,943,500
1984	1,042,000	4,186,600
1985	1,057,100	4,292,400
1986	1,028,000	4,344,400
1987	1,027,300	4,373,700
1988	1,055,538	4,652,753
1989	1,013,194	4,660,202
1990	949,588	4,683,129
1991	850,092	4,711,853
1992	626,319	4,533,410
1993	539,737	4,477,065
1994	526,719	4,557,432
1995	518,515	4,595,218
1996	517,430	4,626,535
1997	510,181	4,727,178
1998	497,908	4,804,973
1999	479,045	4,808,949
2000	476,471	4,814,116

Sources: 1980 from Carrillo Castro and García Ramírez (1983, 157); 1981–87 from Salinas (1994, 38); 1988–2000 from INEGI (2002).

some order by assigning the Subministry of Banking to handle divestitures, but without much success. When Herzog left the SHCP in the summer of 1986, Gustavo Petricioli took over and placed responsibility for privatizations with a smaller group within the SHCP, the Council of Advisors.

The structure of the Council of Advisors was compact. The head of the council oversaw the work of three advisors, each devoted to a single project. One advisor was placed in charge of restructuring the debt of Mexico's large, private conglomerates, the *grupos*. Another worked on consolidating the banking system into a smaller number of stronger banks. The third was placed in charge of privatizations. When President Salinas took office, he appointed the head of the Council of Advisors, Jacques Rogozinski, as the lead official for privatizations. Rogozinski, in turn, adopted a similar organizational structure for the unit devoted to carrying out privatizations under President Salinas.

Another important innovation that occurred at the same time that the Council of Advisors began handling privatizations toward the end of 1986 was the use of "agent banks" as financial intermediaries in support of the privatization process. An official who had been recruited from a position in one of the banks to work on privatization within the Salinas administration noted that the process was exceptionally complicated, with thousands of details that needed to be handled. By leaving the loose ends to the agent banks, the Council of Advisors was freed up to focus on selling the firms. This official noted that the use of the agent banks had important side effects, because it was in the process of serving as an agent bank that many of the financial institutions in the country began to get a grasp on financial engineering. Agent banks needed to place values on companies, and prior to 1986 none of Mexico's banks had experience in this. Legal, corporate, labor, financial, and accounting details all had to be dealt with by someone, and these became the responsibility of the agent banks (Interviews 60, 74).

Another official from the SHCP who had participated in the privatization process during the de la Madrid administration contrasted his experience with that of officials who privatized firms under President Salinas. He noted that out of approximately 150 sales concluded during his time working in the SHCP, it was never once necessary to change the board of directors or get involved in any way in the administrative aspects of the firms being sold. Later privatizations would require exactly this type of organizational transformation (Interview 74).

Despite the limitations of the privatization process during this period, the de la Madrid administration took essential steps toward downsizing the state. In addition to the learning that took place in the bureaucracy, specific sectors of the economy show significant changes. By the end of 1988 the state had withdrawn entirely from fifteen productive areas in which it had previously participated, completely abandoning parastate production of bottled drinks, textiles, cement, automobiles, and pharmaceuticals. The state also abandoned its participation in secondary petrochemicals, while redefining thirty-six categories of that sector as "secondary" and, therefore opening them up to private investment (Sánchez and Corona 1993, 103; Teichman 1995, 136, 137).

By the end of the de la Madrid *sexenio,* observers of privatization generally concluded that the state had completed the process of restructuring. Pardo (1986, 245) goes so far as to claim that "despite current attempts to reduce the size of the public sector, its influence is an irreversible fact." Casar and Peres (1988) and Villarreal (1988) similarly assumed that the transformation of the parastate sector had reached its conclusion. During the following six years, President Salinas would prove these assessments very wrong.

THE LONG *SEXENIO:* DEEPENING THE PROCESS OF PRIVATIZATION FROM 1988 TO 1994

The Salinas *sexenio* began early with regard to privatizations. Even before Carlos Salinas de Gortari took office in December 1988, the privatization process had begun to change in a way that was more characteristic of the methods employed under Salinas. The Interministerial Committee on Spending and Finance (CIGF) had authorized the sale of one of the state airlines, Mexicana de Aviación, in July 1986 but could not find a buyer. No offer had been made that met the minimum requirements established under the rules of the sale (SHCP 1994b). In 1988, the final year of the de la Madrid presidency, another round of bidding was being organized for Mexicana when, on April 12, ground workers at the other state airline, Aeroméxico, went on strike in protest over plans to reorganize the company. Three days later, the Ministry of Communications and Transportation (SCT) declared Aeroméxico bankrupt, allowing for the suspension of the collective bargaining agreement, mass layoffs, and a wholesale reorganization of the firm.

The privatization of Aeroméxico marked the beginning of a new phase of reform. The characteristics of Aeroméxico and the measures taken prior to its sale were both more typical of the privatizations under Salinas than they were of those under de la Madrid. Aeroméxico was a large firm in an important sector of the economy. In addition, unlike companies that were privatized during the de la Madrid administration, Aeroméxico needed restructuring: the firm was overstaffed, and selling it would entail confrontation with more than twelve thousand employees, who were organized into three relatively powerful national unions.

The different approach that the Salinas administration took toward privatization can be seen in the National Development Plan of 1989–1994. "Rationality" and "economic efficiency" are elevated in Salinas's National Development Plan and form the basis of public-sector administration. Whereas President de la Madrid's National Development Plan retained vestiges of the ideology of the Mexican Revolution with its central role for the state in economic development, the 1989 plan begins to separate the state and economy more clearly. The later plan insists upon not "confusing public firms properly understood with entities of institutional service . . . whose efficiency and productivity are not always measurable in terms of financial profitability."

Entities of institutional service—which include the two social security institutions, the National Lottery, and Conasupo—"realize a clear social function [that] they shall develop with the greatest efficiency, distinguishing their end from the results of their operation." Having placed the institutional entities on a different footing from the rest of the parastate sector, the National Development Plan goes on to note that public firms "should be subjected to criteria of profitability and should be governed according to the competition that the market imposes for their best operation and highest social utility" (SPP 1989, 5.3.9).

Regulations governing the implementation of the 1986 Federal Law of Parastate Entities were passed in January 1990, eliminating many of the contradictions that existed in the earlier Law. Like the National Development Plan, the 1990 regulations downplayed the social aspects of parastate firms and began to emphasize the responsibility of the governing body of these firms to "establish criteria of rationality, austerity, and discipline according to which the corresponding parastate entity should use its budget" (Article 23). In addition, Article 26 established rules for determining prices and fees for the services rendered by parastate firms that "shall be fixed according to criteria of economic efficiency and financial stabilization" and take into account prevailing prices in the international market and the costs of production. Deviations between the prices at which goods are produced and those at which they are sold need to be justified by marketing strategies and must be reported to the Ministry of the Treasury (*Diario Oficial* 1990c).

President Salinas also pushed through a variety of important legislative reforms, especially in finance and banking. Toward the end of 1989, he sent to the Congress a set of legislative reforms, which were approved in December. The financial package included reforms to the laws regulating credit institutions, insurance companies, the stock market, and other forms of financial intermediation and were focused primarily on the deregulation and liberalization of financial markets. In May 1990, Salinas announced his intention to reprivatize the country's banks, which would require amending Article 28 of the Constitution (see Banamex 1990, 323; Ortiz Martínez 1994, 57–69).

Legal reform that was carried out during the Salinas administration began to anticipate the sorts of reforms that would become still more prevalent during the following administration. Privatization of large firms operating in monopolistic or oligopolistic markets with clear social purposes would require more than simply the restructuring of the firm. These later privatizations would require that the entire regulatory apparatus be restructured. In some cases this involved the creation of new government entities to oversee the regulation of a specific sector, such as the National Insurance and Finance Commission, created under the 1989 reforms (Ortiz Martínez 1994, 62).

The bureaucracy dedicated to privatizations continued to evolve as well. One of the keys to privatization under the Salinas administration was the continued centralization of power in the hands of a small, dedicated group of public officials within the Ministry of the Treasury. When President Salinas took office in 1988 his minister of the Treasury, Pedro Aspe, requested that the head of the Council of Advisors under President de la Madrid, Jacques Rogozinski, draw up plans for a new agency that would be devoted exclusively to the sale of public firms (Interview 72). With the help of another official who had worked in one of the agent banks assisting privatizations under the de la Madrid administration, Rogozinski created a compact bureaucracy with an obscure name operating within the Treasury Ministry: the Unit for the Divestiture of Parastate Entities (UDEP).[4]

With a staff of only forty people and a director with the rank of subminister, the UDEP was, in the words of one official, a battleship: "It had very strong canons inside of a very small structure." Another official from the agency agreed, stating that the structure of the UDEP was "small, consolidated and focused." Because all the positions had relatively high rank, decisions could be made without having to filter their way through the bureaucracy for approval from higher-ranking officials (Interviews 58, 60).

The most important innovation adopted by the UDEP was the "resectorization" of parastate firms. As soon as a firm was earmarked for privatization, operational control was transferred to the Treasury minister, who would either act as the chief executive officer (CEO) of the firm or appoint a new CEO. Resectorization undercut the capacity of line ministers to resist privatization by denying them the resources of the parastate firm. It also made it easier to restructure public firms. By taking the helm of the public firm, Treasury officials could restructure the finances, invest in new equipment, and renegotiate labor contracts in order to make the firm more attractive to potential buyers (Interviews 64, 72, 74).

The authority of the UDEP was further enhanced through passive support from the executive branch. When the UDEP began the process of privatizing parastate firms, labor leaders, line ministers, and executives of

---

4. The UDEP was actually misnamed, in that divestitures included liquidations, "extinctions" of firms that existed only on paper, and mergers of firms as well as their transfer to state and local governments. The UDEP was concerned only with sales of public firms. Furthermore, the privatization of Mexico's banks was handled by another entity, the Bank Divestment Committee.

parastate firms often sought to circumvent the authority of the UDEP by appealing directly to the president. President Salinas regularly sent these supplicants back to the director of the UDEP, forcing them to negotiate with him. This process quickly solidified the authority of the UDEP within the Mexican bureaucracy (Interview 77).

The creation of the UDEP was followed by that of a parallel entity devoted exclusively to the privatization of the banks.[5] The Bank Divestment Committee (CDB) was established in September 1990 and consisted of officials from the Ministry of the Treasury, the Bank of Mexico and two representatives of the private sector (Ortiz Martínez 1994, 224). Part of the work of the CDB had already been accomplished within the Council of Advisors, which by 1990 had consolidated forty-four banks into eighteen (Ortiz Martínez 1994, 24).

A senior official within the UDEP described the problem of selling public firms in terms that indicated familiarity not only with the practice but also with the theory of privatization. This official explained that in the hands of the government, the parastate firm was like a broken watch, the value of which is zero. The watch, however, has some value in the hands of private actors who know how to fix it. The key to selling public firms, then, is finding out how much a firm could be worth and then determining how to split the potential value between the public and private sector (Interview 77).[6]

The notion that only the private sector knew how to make parastate firms viable, however, did not prevent state officials from undertaking a wide variety of organizational, regulatory, and financial restructurings prior to privatization. One of the most important steps taken prior to privatization during the Salinas administration was the restructuring of labor contracts in parastate firms. The senior official from the UDEP conceded that "collective bargaining agreements were almost always renegotiated prior to the sales of firms because the government has more ability to deal with labor than the private sector does."

Working with the assistance of agent banks, international consulting firms, structural adjustment loans, and technical support from the World

---

5. Although the legislation formally establishing the UDEP was not published in the *Diario Oficial* until October 30, 1990, the UDEP had been operating since the beginning of 1990. In addition to informants who pointed this out to me, Ortiz Martínez (1994, 218) notes that the UDEP was already functioning in the summer of 1990, when officials were still working on creating a new apparatus for the privatization of the banks.

6. An influential study from the Massachusetts Institute of Technology, *Selling Public Enterprises*, makes a similar observation, arguing that "assets are sold when buyers and sellers value them differently, thus creating a positive-sum game where both parties can gain" (Jones, Tandon, and Vogelsang 1990, 3).

Bank, reformers within the Salinas administration undertook ever larger and more complicated privatizations. Between 1990 and 1993, the UDEP managed to sell some of Mexico's biggest public firms, including two steel mills, a fertilizer plant, a diesel truck and engine plant, Telmex (the telephone monopoly), and the state-run television corporation. The UDEP also completed a number of large privatizations begun during the previous administration, including the sale of Mexicana de Aviación, and that of all the remaining sugar mills. In Salinas's first full year in office, revenues from privatization increased modestly over 1988 but still remained below $1 billion. In 1990, however, revenues jumped to in excess of $3 billion, more than doubling the combined receipts of all previous privatizations. In 1991, sales receipts from parastate firms jumped once again, to more than $10 billion (see Figure 4).

While leaving the control of large firms in the hands of Mexican investors, in many cases the Salinas administration allowed foreign investors to enter as minority shareholders. It completed the sale of Mexicana de Aviación to a group that included the Chase Manhattan Bank and sold Telmex to a group that included both Southwestern Bell and France Telecom. Although the state-owned petroleum firm, Pemex, remained in the public sector, the Salinas administration continued the process of stealth privatization begun during the de la Madrid administration by removing fifteen more products from the definition of basic petrochemicals that were reserved for state development. In 1992, Salinas went further and proposed a major restructuring of Pemex to break the petroleum giant into four decentralized divisions (Ramírez 1994, 31, 32; Teichman 1995, 137).

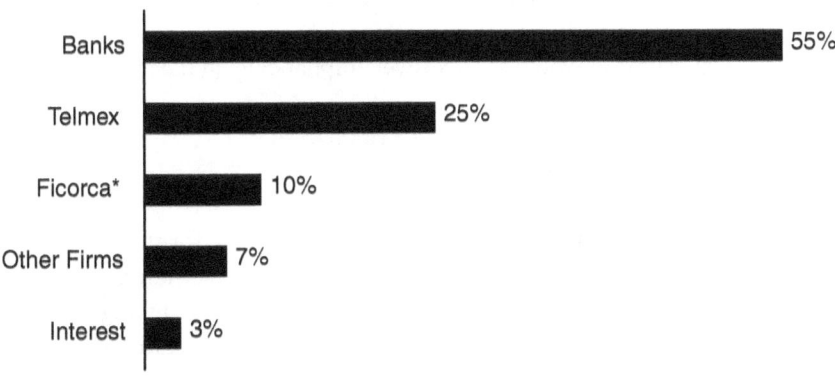

FIG. 5 Sources of revenue from privatization, 1988–1993

Source: SHCP 1994a, 64.
* Special Fund to Cover Exchange Rate Risk. See Maxfield 1989, 226.

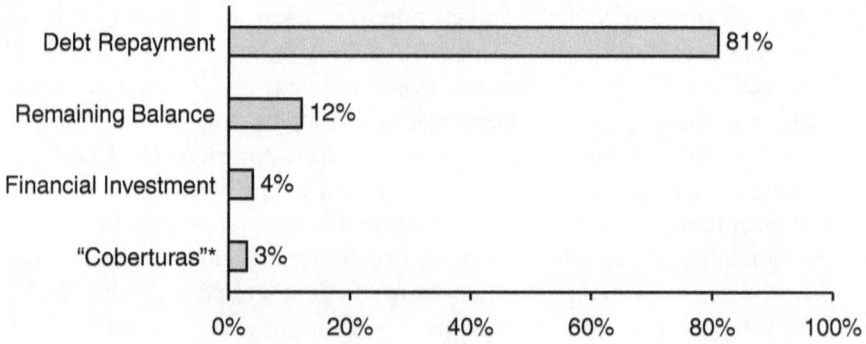

FIG. 6 Uses of revenue from privatization, 1988–1993
Source: SHCP 1994a, 64.
* Guarantees purchased to insure against a fall in the price of petroleum.

The consequences of these large privatizations may seem paradoxical in theory: while ridding itself of the firms, the state received large amounts of revenue for the public coffers. The Salinas administration, however, was quick to resolve this potential contradiction. Resources from privatizations were placed in a "contingency fund," of which 80 percent went toward amortization of the foreign debt. Figure 5 shows the sources of revenue from privatizations during the Salinas administration. Figure 6 shows the use of these revenues.

By the end of the Salinas *sexenio*, the process of privatization appeared once again to have run its course. One analysis of economic restructuring under Salinas asserted, "It is probable that the process of privatization of public firms will be completed before Salinas leaves power in 1994" (Lustig 1994, 137). On the eve of President Salinas's departure from office, the Mexican business magazine *Expansión*, asked: "Is there anything left to sell?" (1994b, 71). The Zedillo administration would soon answer emphatically in the affirmative.

PRIVATIZATION UNDER ZEDILLO: DECENTRALIZATION
AND STATE REFORM, 1994–2000

In 1997, the number of firms in the parastate sector rose for the first time in nine years, growing to 232 firms from 229 in 1996. A year later, the official count of firms in the parastate sector increased again, reaching 261 (see Fig. 3). At first glance, the growth in the number of parastate firms might suggest that the parastate sector was making a comeback or, at the very least, that its decline had bottomed out. Closer examination, however, reveals that the apparent growth in the sector was actually a prelude to further privatization.

The Zedillo administration's zeal to push for further privatization was all the more impressive because it occurred against the backdrop of the staggering failure of the recently privatized banks. The peso devaluation, in President Zedillo's first month in office, and subsequent economic collapse in 1995 left most of the private banks holding worthless debt. In an effort to shore up the banking system, the Zedillo administration created the Portfolio Purchase and Capitalization Program and then the Bank Fund for Savings Protection (Fobaproa). These new agencies acquired the bad debt of private banks, imposed more strict oversight and regulation of the banking sector, and even took control of failed banks while studiously avoiding any suggestion that the state would nationalize the banks.

Further, President Zedillo sought to shore up the failing banking system through a partial privatization of social security, channeling public retirement savings to private banks. Beginning January 1, 1997, Mexican workers were required to choose from among a number of private banks that were designated to act as Retirement Fund Administrators (Afores). Under the new law, workers, employers, and the government all pay their portion of the roughly 11 percent of each worker's salary into the private accounts established by the Afores rather than into the state's social security fund (see Pérez Sandi 1996).

At the same time, the Zedillo administration moved forward with a series of privatizations that had already been contemplated as the natural extension of the policies of the de la Madrid and Salinas administrations. These privatizations would target two sectors in particular: energy, and communications and transportation.

As with previous privatizations, legal reforms were needed before sales could take place. Some of this reform had already been undertaken during the Salinas administration, including a major reorganization of the Mexican ports and the passage of a 1993 law allowing for the creation of private concessions within individual ports (*Diario Oficial* 1993). Barely two months after entering office, President Zedillo pushed a constitutional amendment through the Mexican legislature allowing for the privatization of railroads and satellites by removing them from the Constitution's definition of "strategic" sectors. New laws governing the railroads and use of satellites and the radio spectrum were passed in 1995 (*Diario Oficial* 1995a, 1995b). At the beginning of 1998, new regulations allowing for the privatization of the airports were published.

In order to carry out these privatizations, the Zedillo administration created the Interministerial Committee for Divestment (CID) in 1995 to take over responsibility for the privatization process. The UDEP, in turn, was renamed the Unit for Investment and Divestiture of Parastate Entities and largely removed from the business of privatization (SHCP 1995a). The composition of the CID, in contrast to that of the UDEP, depended on the sector in which privatization occurred. The Ministries of the Treasury, of Commerce, of Labor, and of Control and Administrative Development all had permanent seats on the commission, with line ministers joining the commission when privatizations touched on parastate firms within their sectors.

While it is possible under the structure of the CID for all ministries to move forward with privatizations simultaneously, the sectors that have received the most attention are energy and communications and transportation. The justifications for focusing attention on energy and on communications and transportation are clear enough: these sectors provide essential inputs for all other sectors of the economy (see *El Mercado de Valores* 1995, 19). At the same time, the characteristics of these particular sectors required changing the mechanisms by which privatizations were conducted. Even proponents of privatization have commented on the need for adequate regulation of sectors that have monopolistic and oligopolistic structures (18). Where many of the privatizations under the Salinas administration required the restructuring of parastate firms, privatization under Zedillo required restructuring the ministries that regulate the activities of firms.[7]

Rather than "resectorizing" firms and bringing them under the control of the SHCP—as the UDEP had done—the Zedillo administration seeded line ministries with former officials from the SHCP. Decentralization of the privatization process under Zedillo, then, did not entail a return to the relatively anarchic period of privatizations during the early years of the de la Madrid administration. Instead, former SHCP officials who were familiar with the privatization process were placed in strategic positions within the Ministry of Energy and the SCT to implement ambitious privatization plans from within those ministries (Interviews 64, 65).

Upon entering office, President Zedillo appointed Guillermo Ortiz Martínez as the minister of communications and transportation and Carlos Ruiz Sacristán as the minister of energy. Ortiz Martínez had acted as the head of the CDB and directed the privatization of the banks under the Salinas administration. Ruiz Sacristán served with Zedillo in the Treasury Ministry, restructuring the debt of the private sector in the early 1980s (Jáquez

---

7. The need for better regulation over these sectors was anticipated in the law that established the CID. Article 2 entrusts the CID with "seeking the decentralization and regional establishment of companies to the degree that this is compatible with their efficient functioning" and "promoting production and commercialization practices that are both healthy and competitive" (SHCP 1995a, fracciones VIII, IX). In addition, the president of the recently created Federal Competition Commission has a permanent invitation to sit on the commission to ensure the creation of competitive conditions in newly privatized sectors of the economy.

1998). During the Salinas administration, he worked as the general director of public credit and served on the CDB. In addition, Ruiz Sacristán brought with him a team made up of some of the key players from the UDEP. Two of the UDEP's four general directors—Jorge Silberstein and Rosana Ingle de la Mora—accompanied Ruiz Sacristán.

The economic crisis of 1994, however, changed President Zedillo's plans slightly. In an attempt to calm the fears of foreign investors, Zedillo replaced Jaime Serra Puche, his minister of the Treasury, with Ortiz Martínez. Ruiz Sacristán and his team then moved to the SCT. Not surprisingly, privatizations in communications and transportation have advanced considerably further than in other ministries, including energy. As one senior official from the UDEP during the Salinas administration observed, "It's logical that the structure within the SCT is better because that's where all the people who have experience and 'know how' are. All of them are in the SCT, at the Treasury or in the private sector" (Interview 72).

The legislation establishing the CID suggests in vague terms the possibility of individual ministries designating a special commissioner to act as an intermediary between the line ministry and the CID (SHCP 1995b, Article 10, fracción IV; and Article 11). In fact, within the SCT, the special commissioner became the central player in the privatization process. The new structure for privatizations in the SCT relied heavily on the special commissioner, Jorge Silberstein, who was a former general director from the UDEP. Silberstein was placed in charge of the newly created Unit for the Support of Structural Change (UACE) which still operates out of the basement of the main SCT offices in Mexico City. The UACE answers directly to the minister of communications and transportation, while working closely with the CID. Despite the problems caused by recurrent economic crises, the UACE managed to generate more than $4 billion in sales from privatizations in communications and transportation (see Table 4). A brief review of one the areas of privatization will serve to illustrate the complexity of this phase of restructuring.

The primary area of growth in the number of parastate firms since 1994 is in the creation of Complete Port Administrations, known as APIs. During the 1970s, various attempts were made to coordinate the operation of ports and bring them under the central control of the federal government. In addition to the lack of coordination among states with different ports, four different national-level agencies that were organized along functional lines attempted to administer the operation of the ports.[8] During the Salinas

---

8. A National Port Coordinating Commission was created in the early 1970s; this body operated alongside a dredging service, a General Director of Maritime Work, and a General Director of Port Operation, covering twenty-two ports that were administered by the federal government.

TABLE 4 Revenue from privatizations by ministry 1995–2000 (millions of nominal dollars)

	1995	1996	1997	1998	1999	2000	Total
Ministry of Energy	0.0	0.0	0.0	295.9	0.0	0.0	295.9
Ministry of Governance	0.0	0.0	0.0	121.9	0.0	0.0	121.9
Ministry of Treasury	0.3	14.5	21.6	110.3	0.0	0.0	146.8
Ministry of Communications and Transportation	162.2	57.3	2,653.7	469.2	273.5	406.7	4,022.7
TOTAL	162.6	71.9	2,675.3	997.4	273.5	406.7	4,587.4

Source: Amozurrutia 2002.

administration, a new institution, Puertos Mexicanos, was created to merge the various activities of these four agencies into a single agency. According to one source, Puertos Mexicanos was effectively under the control of the SHCP, not the SCT (Interviews 50, 55).

The consolidation of the entire port system was necessary in order to bring the resources of the central government to bear on improving the infrastructure and equipment in the ports. One official argued that if the ports had not passed through this phase, they would never have been attractive to the private sector: the state needed to fix them up first. In addition to providing physical infrastructure, a 1993 law passed by the Salinas administration gave investment security to private investors by allowing for the creation of private concessions within individual ports (Interview 55).

Toward the end of the Salinas administration, Puertos Mexicanos was dissolved, and the SCT began creating individual port authorities, the APIs. Stock is created for an API and the government, initially, is the sole owner. The goal of establishing each port on an independent basis is twofold. First, the separation of ports from one another allowed for the elimination of cross-subsidies from profitable to unprofitable ports so that each port can be required to operate on a self-sustaining basis. Second, once the ports are established as separate, incorporated entities, the rights to operating them may be sold as concessions. The board of directors of the API includes representatives of local and state governments where the port is located as well as representatives of the private sector, even though that sector does not yet hold shares in the port (Interview 50).

During the Zedillo administration, the SCT opened twenty-three port concessions to private capital. In addition, the SCT granted concessions to private companies to operate railroads, airports, and satellites. Out of forty-two concessions opened to bidding, thirty-five succeeded in transferring control to the private sector. Out of more than $4 billion collected from privatizations between 1995 and 2000, all but about $500 million came from privatizations within the SCT (SCT 2000, 413–15; SHCP n.d.).

State Autonomy, Parastate Firms, and Political Coalitions

Contributors to the literature on economic reform in the developing world argue that the centralization of power within the state apparatus is critical to the implementation of neoliberal reforms. They also emphasize the need for the state to be insulated from political pressure (Centeno 1994, 47; Petrazzini 1995, 6). In the previous section, examining the internal workings of the

Mexican state, I indicated that the centralization of authority within the Ministry of the Treasury was in fact essential to the ability of the Mexican state to undertake increasingly large and complex privatizations.

The insulation of the state from interest-group pressures is somewhat less certain. The Mexican state was certainly capable of suppressing the demands of popular sectors, as Petrazzini and Centeno argue. Yet it achieved this suppression not through insulation but, rather, by penetrating, co-opting, and dominating organized labor. Moreover, as the Mexican state implemented policies detrimental to labor and small, domestically oriented capitalists, it tightened the links that joined it to the interests of large, internationally oriented capitalists and to international institutions supporting liberalization of the Mexican economy. Privatization not only satisfied the demands of these groups to open up large swaths of the economy to private investment, it also provided the material resources with which to reward this constituency.

In the following section I show that while the Mexican state did manage to overcome the pressure from labor and other popular-sector groups, growing independence from these sectors reflected an increased dependence on other sectors. First I examine the changing structure of relations between the state and its principal domestic interlocutors: labor and capital. I then assess the international influences on the process of privatization.

PRIVATIZATION AND LABOR CONTROL

Privatization contained both macro- and microeconomic goals. The macroeconomic goal was to use revenues from privatization to repay the foreign debt while transferring public assets into private hands. The microeconomic goal was the "modernization" of these firms to make them competitive in the international economy. Modernization entailed both investment in new plant and equipment, which the government was incapable of making under the conditions of extreme austerity that prevailed in the 1980s, and making labor contracts more "flexible," to restore the rights of owners over their property. Privatization in large firms involved the wholesale restructuring of labor contracts so that workers could no longer block management initiatives regarding the organization of production.

Because President de la Madrid generally avoided privatizations in strategic sectors of the economy, he was able to avoid direct confrontation with powerful labor unions in the public sector. Even without direct confrontation, however, the policies of austerity seriously undermined the wages of Mexican workers. By the end of de la Madrid's first year in office, the real minimum wage in Mexico City had fallen by 75 percent. By the end of his term, in 1988, the real minimum wage in Mexico City stood at barely half its 1982 value and unrest was evident in a number of important parastate firms. Flight attendants struck over wages and contract changes in 1983; workers at the state-owned diesel truck and engine manufacturer, Disel Nacional, struck over contract changes and layoffs in 1983 and 1986; Telmex workers went on strike in 1984 and again in 1987 over wage and contract issues; workers at the Mexican National Railroad protested contract changes in 1984 and struck over layoffs in 1986 (Middlebrook 1995, 215; Teichman 1995, 118, 119).

Even before Carlos Salinas assumed the presidency, the posture that he would adopt toward organized labor could be seen in the declaration of bankruptcy at Aeroméxico. Once in office, Salinas sent even clearer signals to organized labor that he would not permit workers to stand in the way of his economic reforms. As with the general thrust of the privatization program under Salinas, the attack on organized labor showed continuity with the policy of the previous administration, while its application was intensified.

Within the first four months of taking office, Salinas jailed the powerful leader of the Mexican Petroleum Workers Union, Joaquín Hernández Galicia, on charges of corruption, and replaced the heads of the national teachers union and the state workers union (Correa and Corro 1989). The dramatic gestures of Salinas met little resistance and even began to earn him a reputation for decisive leadership. That the leadership of the Petroleum Workers Union was corrupt was never in doubt. However, Salinas replaced Hernández Galicia with equally unrepresentative leaders, who acceded to a reorganization of Pemex, job cuts, and major changes to the collective bargaining agreement. Salinas's handpicked leadership of the national teachers union agreed to structural changes in the organization of public education, which weakened the union (Middlebrook 1995, 295).

Because of the great diversity among the firms being sold, however, state-labor relations varied considerably during the process of privatization. President Salinas intervened directly to broker a deal with Telmex workers—a deal that has been showcased as a model of cooperation between labor, capital, and the state. As a profitable firm operating in a monopoly market with excellent prospects for growth, Telmex was able to guarantee no layoffs to members of the Mexican Telephone Workers Union, and the state subsidized a loan to allow the union to purchase 4 percent of Telmex stock (see Chapter 5).

But positive-sum games were the exception, not the rule. When labor proved intransigent, President Salinas did not hesitate to bring the full force of the state to bear on the unions. When workers resisted privatization at a large state mining company in 1989, the firm was declared bankrupt and troops were sent to enforce the decision (Ortega Pizarro 1989). In 1991, Salinas broke the powerful labor unions at the port of Veracruz and turned over the operation of port services to private transportation companies, which had long complained of the inefficiency of the existing arrangement (Corro 1991). The restructuring of the Mexican National Railroad entailed the forced retirement and mass firing of tens of thousands of workers between 1988 and 1994 and required the iron grip of the Mexican Railroad Workers Union (see Chapter 6).

President Salinas further sought to undermine the political unity of organized labor by encouraging the formation of an alternative labor federation. However, his administration's interest in a new labor federation went beyond simply that of weakening the Mexican Confederation of Labor. Salinas wanted to encourage the organization of workers who would take a more "modern" approach to industrial relations, accepting the need for worker flexibility and a commitment to productivity. Salinas found a useful ally in the influential general secretary of the Mexican Telephone Workers Union (STRM), Francisco Hernández Juárez. Hernández Juárez adopted much of the rhetoric of productivity and modernization while attaching these concepts to the importance of maintaining a satisfied and well-trained work force (see Chapter 5 and Dubb 1992; Middlebrook 1995, 284; Zapata 1995, 130, 131).

In 1990, Hernández Juárez led a breakaway faction of state and former state-sector unions and formed the Federation of Unions of Goods and Services Workers (Fesebes). Founding members of the Fesebes included the democratic faction of the electrical workers, the airline pilots' union, and the union representing ground crews at the airlines. In these unions, workers "earn well above the average wage in Mexico and the focus of the federation has largely been to promote the signing of productivity agreements" (Dubb 1998, 17).

By the end of his term, however, President Salinas's support for the alternative labor federation diminished as the PRI began to prepare for the presidential elections and once again called on the resources of the traditional labor leadership to mobilize the vote. Although privatization paved the way for restructuring labor relations in the microeconomic environment of public firms, Salinas's attempt to pass a reform of the federal labor law that would "modernize" labor relations throughout the economy was ultimately blocked by the old guard of the CTM (Zapata 1995, 121–36).

The heterogeneity of the public sector—even among large firms in strategic sectors of the economy—has led the state to employ various tactics in order to maintain social peace. Unions representing telephone workers and airline pilots survived the process of privatization and even won important concessions for their workers. In contrast, the leadership of unions representing port and railroad workers presided over the dismantling of these unions. Once restructuring is complete, the railroads and ports will be much more decentralized in their operation, thus undermining the capacity for workers to reestablish strong, national unions.

As Table 3 shows, the number of workers employed in parastate firms declined only slightly during the first year of the Salinas administration. By 1990, however, the sales of large firms can be seen in the employment data. Between 1988 and 1994, the number of workers in parastate firms fell by almost half, declining from more than a million employees to a little more than five hundred thousand. Parastate-sector employment fell from 4.4 percent of the economically active population in 1988 to 1.9 percent in 1994. Between 1995 and 2000, the number of workers employed by parastate firms and their percentage of the total workforce continued to fall, though not as dramatically as under the Salinas administration (INEGI 2002).

The decline of parastate employment, the weakening of the PRI, and Mexico's changing relationship with the global economy have forced organized labor to rethink its alliances and strategies. Throughout 1996 and 1997, the Fesebes sponsored a series of meetings to discuss the future of Mexican labor and responses to economic restructuring. Following the death in June 1997 of Fidel Velázquez, the ninety-seven-year-old patriarch of the CTM, the Fesebes announced its departure from the official Labor Congress, which historically served as an umbrella federation in Mexico. In November 1997, the new group formed the National Workers Union (UNT), claiming a membership of more than 1.5 million workers (Dubb 1998, 19; *Trabajo y Democracia Hoy* 1997, 90–92).

The rhetoric of the new labor central suggests a new set of alliances for Mexican workers, rejecting corporatist ties between labor and the PRI and highlighting the need for cross-border solidarity among workers. The practices of the UNT leadership have been less clear. One faction within the UNT retains important links to the PRI through STRM general secretary Francisco Hernández Juárez. Hernández Juárez continues to argue for greater cooperation between labor and capital and "modernization" as the key to improving working conditions. Other factions within the UNT are less enthusiastic about the prospects of winning through cooperation with management and support the center-left opposition party, the Democratic Revolutionary Party (PRD).

Despite the imposition of pro-government leaders in many large public firms, workers in strategic sectors of the economy were often able to extract important benefits from their state employer. Workers outside the state sector, especially those affiliated with the CTM, also gained benefits, many provided by parastate firms, through their participation in the PRI. With the demise of the parastate sector it is unlikely that this dualism will change, although the mechanism for creating it has. The impersonal mechanism of economic necessity will replace the political control of organized labor. A small nucleus of relatively privileged workers will probably do well in those firms that are able to compete in the new economy. At the same time, large numbers of workers organized in the CTM will probably lose many of the benefits they gained under the PRI, especially now that the PRI has lost the presidency.

THE STATE'S CHANGING RELATIONSHIP TO PRIVATE CAPITAL

With the confidence of the private sector shattered by the bank nationalization of 1982, the de la Madrid administration moved quickly to regain the support of domestic capital. Although President-elect de la Madrid had not expressed any enthusiasm for the bank nationalization, once in office he was compelled to complete the legal formalities. President de la Madrid went beyond simply ratifying the nationalization with the necessary constitutional amendment. He also placed railroads and satellites under the control of the state and introduced into the Constitution for the first time the notion of state "rectorship" over the economy.

Although the private sector expressed alarm at the constitutional amendments, in retrospect it appears that this was actually a necessary prelude to privatization. A high-ranking official working in the UDEP during the Salinas administration claimed that the 1983 constitutional reforms were indispensable for subsequent privatizations because they established the boundaries between public and private more clearly and specified which sectors of the economy were priority and which strategic. According to the official, "[W]ithout this modification, nothing else could have been done: it was essential" (Interview 58).

And despite the concern of the private sector, the conciliatory attitude of the incoming administration was unmistakable (Bailey 1988, 130). By the end of de la Madrid's first month in office, Congress had approved new regulations of banking and credit service that allowed 34 percent of the shares in the banks to be sold back to the highest bidder (Hernández Rodríguez 1986, 251). More important, the new administration agreed to return all the shares of nonbank financial institutions to their original owners. Private insurance companies, stock brokerages, leasing firms, and warehousing companies, which made up an important part of the bankers' financial holdings, were back in business as early as 1984, with the result that "[t]hese newly reprivatized nonbank financial institutions quickly began to compete for savings and create a parallel financial market beyond state control" (Maxfield 1990, 156). Thus, even though the state controlled the banks, the banks no longer controlled money.

Although capital flight slowed slightly following the bank nationalization, the total capital leaving Mexico during the first three years of the de la Madrid administration is still estimated at more than $16 billion (153). Control over capital became still more concentrated during this period as the stock market, brokerage houses, and insurance companies became the preferred means of financing among large capitalists. While the state attempted to use banks to channel credit to priority sectors of the economy, it ended up financing the public-sector debt and investing still more in oil in order to pay off foreign obligations.

Because the large industrial conglomerates had financed their industrial firms internally through banks belonging to them, many of the industrial firms acquired through the bank nationalization were so heavily leveraged that the state could not sell them without first absorbing their debt. Although a moratorium had been placed on the creation of new public firms, the de la Madrid administration did establish one new, important *fideicomiso*, the Fund to Cover Exchange Risk (Ficorca). Ficorca socialized a large part of the losses incurred by the private sector as a result of the financial speculation and collapse, leaving the Mexican government holding "$21 billion worth of private sector debt" (Maxfield 1990, 160). Approximately 80 percent of Ficorca financing went to twenty of Mexico's largest conglomerates, including Grupos Alfa, Cemex, Visa, and Vitro (Jáquez 1998).

Despite the de la Madrid administration's attempts to reassure the private sector with both words and deeds, business groups remained suspicious and became increasingly involved in direct political opposition to the PRI. The National Action Party (PAN)—founded by conservative northern businessmen in 1938 out of concern that the Cárdenas administration was taking Mexico down the path to socialism—became the vehicle for much of this new activism. Capitalists who had long been content to influence state economic policy through consultation and external pressure were beginning to enter into politics and challenge the hegemony of the PRI (*Expansión* 1993b; Hernández Rodríguez 1986, 247).

President de la Madrid's efforts to regain the confidence of capital were also thwarted by "the heterogeneity of business interests and the lure of foreign investments" (Maxfield 1989, 216). Although Mexican capitalists had united briefly around their opposition to the bank nationalization, they were soon divided again, between large and small, internationally oriented and domestically focused. As President de la Madrid began lowering tariff barriers and allowing greater foreign investment, it soon became clear that labor would not be the only casualty of restructuring: the PRI would ultimately have to jettison the interests of its small- and medium-sized business supporters who were organized around Canacintra.

By the end of the de la Madrid administration, representatives of Canacintra were expressing their dismay at the growth of foreign competition as the state continued to cut back on its support for domestic producers (Lizárraga 1988). While the *grupos* retained access to financing through the parallel financial system, credit for small- and medium-sized firms was scarce and costly (Maxfield 1990, 159). Large firms that could compete internationally or that could use low-cost, imported inputs were benefiting from the opening and were in a position to push for still greater economic liberalization. Proponents of economic liberalization were further strengthened by the terms of the 1982 and 1986 agreements with the International Monetary Fund (IMF) that pushed for austerity and structural adjustment.

As the 1988 presidential elections approached, the Mexican economy once again began to spin out of control. GDP had fallen by one-half of one percent in 1982, then plunged by 5.2 percent in 1983. After achieving positive growth in 1984 and 1985, GDP fell again in 1986 following a collapse of world oil prices from an average price of twenty-five dollars a barrel in 1985 to twelve dollars a barrel in 1986. In 1987 and 1988 the economy recovered anemically, growing by 1.7 and 1.3 percent, respectively. Inflation was also becoming unmanageable during this period, reaching 105 percent in 1986 and 159 percent in 1987 (Aspe 1993, 69, 119; Medina Peña 1994, 242). Adding to the economic concerns, the Mexican stock market crashed in 1987. It was in this environment that President de la Madrid announced the PRI's candidate to succeed him in office: the minister of planning and budget, Carlos Salinas de Gortari.

As the architect of much of the economic restructuring undertaken by President de la Madrid, Salinas promised continuity in the programs he had promoted at the SPP. However, while Salinas had a broad network of connections within the public bureaucracy, he had virtually no base of support among the traditional popular sectors of the PRI (Centeno 1994). The head of the petroleum workers union—whom Salinas would later jail on charges of corruption—actively opposed the nomination of Salinas and was even reported to have provided financial support to the opposition candidacy of Cuauhtémoc Cárdenas (Middlebrook 1995, 293; Waterbury 1993, 251). As Salinas began to voice his opinions on deepening the reforms initiated by President de la Madrid, he alienated small domestic producers as well (M. A. García 1988).

Opposition to neoliberal reforms had also begun to shatter the internal unity of the PRI. A PRI defector—the son of former president Lázaro Cárdenas—Cuauhtémoc Cárdenas, rallied traditional PRI constituents and an amalgam of left-wing parties around an anti-austerity program. When early vote tallies showed Cárdenas in the lead, newly installed electoral computers mysteriously broke down. Two days later, Salinas was declared the winner, with the narrowest margin of victory in Mexican history. As Cárdenas rallied hundreds of thousands of protesters in street demonstrations against the electoral fraud, representatives of powerful Monterrey-based businesses met with President-elect Salinas and declared his election "the best transfer of power that we have had in the last three *sexenios*" (Mondragón 1988).

President Salinas had actually begun to tighten his alliance with the private sector even before taking office. In December 1987 the Commission to Finance and Strengthen the Endowment of the PRI was established to collect funds from the business class and channel them toward the Salinas candidacy (*El Financiero* 1997c, 49). Once in office, Salinas took advantage of the growing support for the PAN while going still further to reconfigure the class base of the PRI. Lacking the votes required in the lower house of Congress to reform the Constitution, Salinas cut deals with members of the PAN, who voted with the PRI in support of twenty-two out of twenty-five constitutional amendments and effectively isolated the PRD (Meyer 1997). In return, in 1989, Salinas allowed the PAN to win the first governorship in Mexican history to go to an opposition party in the state of Northern Baja California. He subsequently intervened on behalf of the PAN following contested elections in the states of Guanajuato and Chihuahua (MacLeod 1997).

Salinas sought to transform the structure of the PRI by reducing its dependence on organized peasants, workers, and popular sectors. In 1993, Mexican newspapers reported that President Salinas had solicited donations of $25 million dollars for the PRI from *each* of twenty-five of the richest Mexicans, who had profited enormously under his administration (Meyer 1995, 113–20; Oppenheimer 1996). While the scandal that erupted, following news reports of the plan, ultimately scuttled Salinas's efforts, it is not

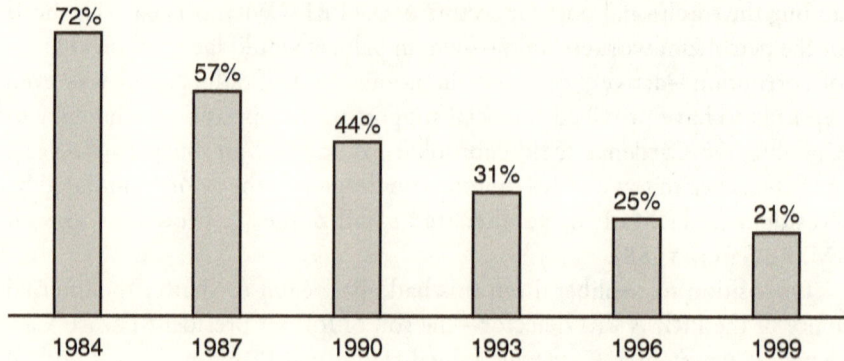

FIG. 7 Parastate firms' share of revenue among Mexico's largest fifty companies, 1984–1999

Sources: Calculated on the basis of *Expansión* 1985, 1988, 1991, 1994a, 1997, 2000.

clear that the private sector wanted to tie its fortunes to the PRI. Instead, Mexican capitalists appeared content to diversify their political holdings, using the PAN as a vehicle to contest the PRI while limiting the scope of debate surrounding economic policy (*Expansión* 1993b, 54–79).

When Salinas's handpicked successor for the presidency, Luis Donaldo Colosio, was assassinated in 1994, Ernesto Zedillo became the new heir apparent. The accidental presidency of Zedillo promised the private sector six more years of continuity in economic policy. Zedillo was the third consecutive president to ascend to the post after heading the SPP. Although it is argued that Zedillo lacked the sort of tight network of support within the public bureaucracy that Salinas cultivated, Zedillo had already established important ties to the private sector during the early 1980s. Before entering the SPP under the Salinas administration, Zedillo served as the director of

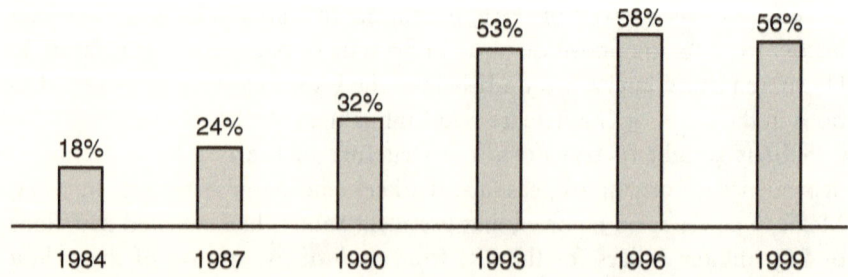

FIG. 8 Private Mexican firms' share of revenue among Mexico's fifty largest companies, 1984–1999

Sources: Calculated on the basis of *Expansión* 1985, 1988, 1991. 1994a, 1997, 2000.

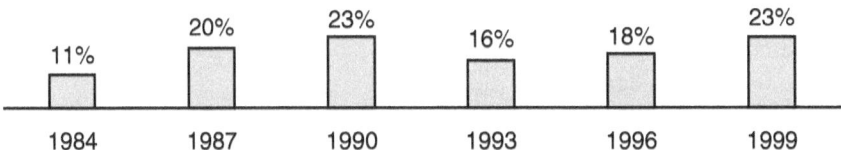

FIG. 9 Foreign firms' share of revenue among Mexico's fifty largest companies, 1984–1999

Sources: Calculated on the basis of Expansión 1985, 1988, 1991, 1994a, 1997, 2000.

Ficorca, the *fideicomiso* that absorbed billions of dollars of debt of Mexico's largest private firms (Jáquez 1998).

These same groups also benefited enormously from privatization as they began incorporating public assets into their portfolios. Figures 7, 8, and 9 show the sales revenues of Mexico's largest fifty companies from 1984 to 1999. In 1984, twelve of the fifty largest firms were state owned, with sales revenues that accounted for 71 percent of total sales. By 1999, only two state-owned companies ranked among the top fifty. Because Pemex remains in state hands, the sales revenues of these two firms still represent a significant 21.3 percent of total sales of the largest Mexican firms. But the importance of channeling this revenue stream from the state to private capital is undeniable. The share of revenue from the top fifty firms going to Mexican firms grew from 18 percent to 56 percent during this period, while foreign firms increased their share of revenue from 11 percent in 1984 to 23 percent in 1999.

With the collapse of the banking system in 1995, the Zedillo administration once again propped up private firms that had lost billions of dollars as a result of the peso devaluation, economic stagnation, and simple mismanagement. Although the bank privatization between 1991 and 1993 earned more than $12 billion, between 1995 and 1998, the state was forced to intervene in forty-nine failing financial institutions, including ten banks. The cost of rescuing the privatized banks, absorbing bad debt, and restructuring their operations is now estimated at approximately $115 billion, almost ten times the amount that the state received for their privatization (Rogozinski 1997 appendix 1; Sánchez 2001b).[9]

The struggles of large private firms to stay afloat without government intervention has not prevented a growing concentration of wealth in Mexico. In 1994, Forbes magazine's list of the richest people in the world

---

9. Osterberg (1997 n. 18) estimates the cost of the bailout at 12 percent of GDP in 1996, about $47 billion. By 1998, the estimate had grown to more than $60 billion (Preston 1998). More recent estimates (Sánchez 2001b) of the price of the bailout indicate that it has grown to $115.6 billion in 2000, or 22 percent of GDP, using figures reported in INEGI (2002).

revealed that Mexico's economic restructuring had produced twenty-four billionaires. Of these, at least seventeen participated in the privatization program, buying banks, steel mills, sugar refineries, hotels and restaurants, chemical plants, and a telecommunications firm as well as concessions to operate firms within newly privatized sectors of the economy, such as ports, private toll highways, and cellular and long distance telephony (*El Financiero* 1997b; SCT n.d.; SHCP 1994b; Zamora 1989).

Family and business links among Mexico's elite probably make the actual concentration of wealth even greater than is suggested by these figures. Three of Mexico's twenty-four billionaires belong to the Garza Sada family, and Mexico's richest man, Carlos Slim, is first cousin to Alfredo Harp Helú, number 24 on the list. Moreover, among Mexico's twenty-five largest firms in 1999, we find four firms controlled by Slim: Carso Global Telecom, Telmex, Grupo Carso, and Grupo Sanborns (*Expansión* 2000).

PRIVATIZATION AND THE INTERNATIONALIZATION OF THE ECONOMY

Throughout the 1980s the debt crisis defined Mexico's relationship to the international economy. Privatization was critical to paying down the foreign debt, with more than 80 percent of the revenues from privatization under the Salinas administration going toward debt repayment. Even before they were sold, however, public firms provided an important source of government revenue and foreign exchange. While the balance of the parastate sector stayed within a narrow range between 1.5 and negative 1.5 percent of GDP from 1965 through 1981, in 1982 the balance turned sharply positive and remained sizeable throughout the 1980s, reaching as high as 9 percent of GDP (see Fig. 2). As the parastate sector was tapped for resources to pay the debt, the inability of the Mexican state to make the investments necessary to expand services and production in important sectors such as telecommunications became part of the justification for privatization (Aspe 1993; Rogozinski 1997).

As chartered under the Bretton Woods agreement, the IMF and the World Bank operated with distinct missions. While the IMF was designed to provide short-term loans to cover balance-of-payments deficits, the bank's mission was to provide low-cost, long-term loans for large developmental projects: precisely what it was doing prior to the crisis. From 1949 through 1980 the bank provided hundreds of millions of dollars in loans to the Federal Electric Commission, regional light and power companies, the Pacific Railroad, the Mexican National Railroad, the state-owned fertilizer company, and parastate steel firms. Bank lending supported agriculture development as well as public-sector projects in tourism, ports, water supply, urban development, and industrial promotion (see Table 2; also Babai 1988; MacLeod 2001, table 3.7; Waterbury 1993).

Although World Bank loans to Mexico continued following the collapse of the economy in 1982, they could not begin to make up for the loss of private lending and the capital flight that was draining Mexico's finances. Nor could they prevent the economy from dropping into recession after Mexico negotiated a stabilization agreement with the IMF at the end of 1982 (Kaufman 1990, 100). Bank lending to Mexico only surpassed $1 billion dollars for the first time in 1986, even as capital flight exceeded $5 billion dollars annually from 1983 through 1985. While the bank continued lending for a wide range of projects within Mexico, the character of these loans changed. Instead of promoting state-owned firms, bank lending in the 1980s and 1990s began to focus on restructuring firms to prepare them for privatization.

Thus, as the debt crisis dragged on, the missions of the two multilateral financial institutions began to converge. IMF officials slowly came to realize that payment of the debt would require economic restructuring; World Bank officials began to doubt the effectiveness of public firms in promoting long-term growth and stopped supporting their development. In the 1980s, for the first time, the World Bank used structural adjustment loans to support reform of public enterprises in the developing world (Babai 1988, 273, 268).[10]

By the early 1990s the policy of the bank turned sharply against public-sector enterprises. In a survey of state ownership in the developing world that was published in 1995, the World Bank (1995, 1, 25) argues that state-owned firms hinder growth and undermine macroeconomic stability, while lamenting the fact that "the size of the state-owned enterprise sector has significantly diminished only in the former socialist economies and a few middle-income countries." The bank goes on to claim that "large and inefficient state-owned enterprise sectors are costing developing economies dearly, especially the poorest among them," and provides practical guidelines for would-be privatizers (24).

Increasing support from the bank for privatization coincides roughly with Mexico's turn to a more intensive and extensive phase of privatization under President Salinas. In 1988 the bank provided $400 million for restructuring parastate steel firms (World Bank 1988, 142). Two years later, the Ministry of Energy, Mines, and Parastate Industry (SEMIP) proposed

---

10. While pointing to the role of the World Bank in support of privatization, Babai (1988, 274) argues that "neither the Bank nor the Fund has engaged in a rush toward privatization."

the sale of the complex of mines and steel plants that had been restructured using the World Bank loan. These firms were subsequently sold in 1991 in two packages for a little more than $170 million (See Rogozinski 1997, appendix 1). In 1989, the bank provided $1.25 billion that was earmarked for different forms of restructuring: $500 million for "public enterprise reform," another $500 million for "financial sector adjustment," and another $250 million for an unspecified "industrial restructuring project" (World Bank 1989, 168).

Another way that multilateral financial institutions helped push privatization was by encouraging government officials to attend international conferences on privatization and public-sector reform. One of the lead officials in the UDEP under the Salinas administration noted that the IMF, World Bank, and International Development Bank teamed up with private, international consulting firms to sponsor and fund such conferences. During his time working with the UDEP, the official attended conferences in Bolivia, Peru, El Salvador, Chile, and Venezuela. While the multilateral financial agencies encouraged attendance at these conferences, expenses would frequently be paid for by private consulting firms seeking contracts to participate in public-sector reform (Interview 58).

These consulting firms were yet another important international influence upon Mexico's privatization program. International consulting firms worked in tandem with Mexican banks to assess the assets and prepare the sales prospectus for all of the most important privatizations. McKinsey Corporation and Booz Allen & Hamilton helped prepare the sales prospectus on ten and eleven of the eighteen bank privatizations, respectively. Goldman Sachs advised Mexico on the privatization of Telmex and on two of the remaining bank privatizations. First Boston, Bank of America, Barclays, Price Waterhouse, Salomon Brothers, Merrill Lynch, Credit Commerciale of France, Donaldson Lufkin & Jenrette, and Manufacturers Hanover all participated in the evaluation or elaboration of the sales prospectus of at least one bank (SHCP 1994a). Mercer was the principal consulting firm advising the restructuring of the Mexican National Railroad and First Boston worked as the international financial agent (Interview 62). Government officials asserted that they sought out international firms to work on projects as much for their expertise as for their name. That is, private, international accounting firms and banks provided a measure of legitimacy to the process and helped win the confidence of prospective buyers in the international financial community (Interview 62).

Mexico undertook much of its trade liberalization unilaterally. Officials within the SHCP and the SPP came to believe that overcoming inflation was linked to economic liberalization (Teichman 1995, 87). The battle between supporters of reviving economic growth through domestic investment and supporters of introducing an entirely new model that would transform Mexico's relationship with the international economy had been won by the latter group. The strength of this group is evident in Mexico's decision to negotiate a free trade agreement with the United States and Canada. Although most of the reforms of the 1980s had been driven by economic crises, by 1993 the second Mexican economic miracle was being widely proclaimed within international financial circles. New foreign investment, both direct and indirect, had recovered from the slump of the early and mid-1980s and, by 1991, had increased, to more than three times the amount of annual new investment prior to the economic crisis (Aspe 1993, 142).

Close scrutiny of specific privatizations and provisions of the North American Free Trade Agreement (NAFTA) reveals that Mexico did not open its markets indiscriminately. Large privatizations in banking, telecommunications, and airlines reserved control for Mexican nationals. Similarly, Annex 1 of NAFTA reserves a variety of newly privatized activities for Mexican nationals. Retail distribution of gasoline, diesel, lubricants, oils, or additives, as well as the transportation and storage of petroleum gas—all of which once belonged to the Pemex empire—are still reserved exclusively for Mexican nationals. Foreigners may own no more than 25 percent of a company providing commercial air service. Foreign ownership in cable television and agriculture is limited to 49 percent and only Mexican nationals may obtain concessions from the Ministry of Communications and Transportation to provide stevedoring and warehousing services at ports, or construct and operate roads for land transportation.

But, despite the goal of maintaining Mexican, if not state, control over strategic sectors of the economy, the economic crisis of the mid-1990s led to greater foreign participation in Mexico's financial sector. When the Zedillo administration imposed higher reserve requirements on the private banks to prevent another collapse of the banking system, most of the privately owned banks could not comply. The administration stepped in and absorbed much of the debt of the private firms. It also took control of insolvent banks, though it later attempted to return these to private hands as quickly as possible, wiping debt off of the balance sheets and selling the banks at a fraction of the cost to restructure their finances. But to strengthen the banking system, it was also necessary to open this sector to foreign investment on a level not seen since the nineteenth century. Mexican banks have been allowed to join with foreign financial institutions or be taken over by them. Table 5 shows the changing ownership of Mexican banks from 1990 to 2000. While

TABLE 5 Changing ownership of banks, 1990–2000

	1990	1995	2000
Mexican	19	26	14
Foreign	1	18	20
TOTAL	20	44	34

Source: Sánchez 2001b, 48. Note that total shown for 1995 in Sánchez does not equal the sum of Mexican and foreign banks. I have changed the reported total of 41 to 44.

only one of Mexico's twenty banks was foreign-owned in 1990, by 2000, twenty of thirty-four banks operating in Mexico were owned by foreign capital (Pérez-Moreno 2002; Sánchez 2001a).

Discussion

In the introductory chapter I discuss the transformation of the meaning of state autonomy and the two distinct aspects of autonomy identified in literature that examines neoliberal transformations. State autonomy refers both to the independence of the state from specific social groups and also to the capacity of the state to implement policies. Various authors have argued that the Mexican state was autonomous in both these ways and that the success of economic restructuring can be traced to these factors (Centeno 1994, 47; Haggard and Kaufman 1992; Petrazzini 1995, 105; Waterbury 1993).

In this chapter I call into question both the independence of the Mexican state from external pressures as well as its technical capacity to implement economic reforms. To begin, the Mexican regime was not insulated from interest-group pressure. Instead, the state was buffeted by recurrent economic crises that were generated, in part, by the actions of elites who engaged in speculation against the peso and capital flight. Although virtually all the literature on Mexico's transformation acknowledges these structural constraints on the state, it conclude that the Mexican state remained independent nonetheless because the state was free from the direct control of elites—instrumentally autonomous—or that it was capable of contradicting the interests of popular sectors (Centeno 1994, 191; Haggard and Kaufman 1992, 9; Waterbury 1993, 3).

In Mexico's privatization program, however, we find various instrumental links between the state and private capital. The Council of Advisors and later the UDEP drew on personnel from private consulting firms and agent banks, and the CDB granted the private sector representation for the bank divestiture. Moreover, the state relied on private-sector firms to perform many of the administrative tasks involved in privatization, a point that I return to below. While the state has shown itself to be impervious to the demands of workers and managers of parastate firms, it accomplished this not by insulating itself from these groups; rather, in the time-honored tradition of the PRI, the state overcame the interests of these groups by penetrating, co-opting, disorganizing, and, where necessary, repressing these groups.

While large capitalists were heavily dependent on the state to reimburse their losses, they nonetheless retained their independence from the state. Indeed, it was the autonomy of capital—the control that private actors enjoyed over vital resources that were independent of the state—that enabled these groups to pressure the government to carry out policies favorable to them. Recurrent economic crises caused by capital flight and disinvestment pushed the Mexican state to enact even more far-reaching reforms.

The question of state autonomy as administrative and bureaucratic capacity is more complex. Haggard and Kaufman (1992, 25) argue that the only way to take the state out of the economy is by bringing it back in: "Privatization . . . calls for technical expertise in financial restructuring, rehabilitating companies, and preparing them for divestiture, as well as procedures that guarantee equal access by potential buyers." While the evidence from Mexico's privatization program is more ambiguous here, it still fails to support the assertion of Mexican state autonomy understood as state capacity.

Without question, state involvement was necessary to the implementation of economic reforms. The centralization of power within the Ministry of the Treasury was crucial to the success of Mexico's privatization efforts. Within the UDEP, a small, focused group was able to overcome the opposition of labor unions and public-sector managers, whose interests were tied to the preservation of public firms. Yet this core of reformers within the Treasury Ministry also relied heavily on the expertise of the private sector. Rather than develop the capacity of the state, the Ministry of the Treasury chose to outsource critical aspects of the financial restructuring and rehabilitation of companies to agent banks and international consulting and accounting firms. Indeed, by drawing on the resources of the private sector, Mexico's reformers actually privatized part of the privatization process itself, contradicting Kahler's (1990, 55) assertion that the state is the "only instrument . . . to change policy in a less statist direction."

The state's capacity to enact reforms, then, cannot be fully distinguished from the state's relationship to elites or from the content of its policies. Allying itself with Mexico's most powerful capitalists, the state enhanced its capacity in two ways. First, it used the organizational and technical expertise of the private sector to assist it in carrying out the privatization program. Second, by enacting policies so clearly favorable to the elite, the state assured itself of the support of these groups. State capacity, then, appeared quite high because the reform efforts met with little or no resistance from the most powerful groups in society. If the state had attempted to enact policies that were contrary to the interests of Mexico's elite, the outcomes—and the conclusions drawn in the literature on Mexican state capacity—would have been very different.

Mexico's privatization program also calls into question some of the common assumptions of the literature in institutional economics regarding the nature of public and private ownership. Soft budget constraints are said to prevail when a loss-making firm does not exit the economy but, instead, is sustained through public subsidies of one kind or another (See Kornai 1992, 143; Waterbury 1993, 112). Thus, public-sector firms are said to operate in an environment of soft budget constraints because the state will support their activities regardless of their actual levels of profitability, while private-sector firms face the possibility of the ultimate hard budget constraint: bankruptcy.

Yet in Mexico, the concept of hard and soft budget constraints has been turned on its head. Throughout the 1980s, the Mexican state actually imposed relatively hard budget constraints on firms in the public sector. Even if bankruptcy was not possible in the years prior to the crisis of 1982, as a whole, the parastate sector generally operated within its means, as evidenced by the fiscal balance of the public sector from 1965 through 1982 (Aspe 1993, 157; Waterbury 1993, 112).[11] The financial discipline imposed by the Mexican state on its public sector can be attributed, in large part, to the hard budget constraints that faced the state itself: its dependence on private capital and the links between these groups and key actors in the Ministry of the Treasury. In the 1970s, soft budget constraints were introduced not by the state but, rather, by international financial agents seeking to recycle petrodollars. It was the market, then, and the speculative bubble of the 1970s, that temporarily reduced the hard constraints on the Mexican state and allowed for rapid expansion of the public sector during these years.[12]

---

11. The Mexican National Railroad is an important exception here; see Chapter 6.
12. This is consistent with Centeno's (1994) assessment of the effects of easy credit on the international market in the 1970s.

When crisis struck in 1982, the Mexican state showed that it could apply very hard budget constraints, bankrupting, liquidating, and selling public-sector firms. It was the most powerful *private* firms that the Mexican state could not permit to fail. Throughout the 1980s and 1990s, the Mexican state repeatedly intervened to prop up the largest industrial and financial conglomerates, frequently subsidizing the losses of private capitalists with revenues generated from the application of hard budget constraints to the parastate sector. From 1983 until the privatization of the banks between 1991 and 1993, the Mexican state recapitalized and reorganized the financial sector, absorbing tens of billions of dollars of private-sector debt and transforming the banks back into viable entities. Following the peso crisis of 1994, the state absorbed even larger amounts of bad debt from the private sector to prevent the collapse of the financial system.

For markets to allocate resources on the basis of hard budget constraints, profitability itself must be accepted as the ultimate value toward which economic activity is directed, and firms that are unable to achieve profitability must fail. Yet the state cannot impose hard budget constraints on social actors whose failure would fundamentally disrupt the economy and, by extension, undermine the state. The privatization of firms operating in essential sectors of the economy will make it still more difficult for the state to avoid stepping in should these firms find themselves in financial distress.

Even among large firms in strategic sectors of the economy, however, we find a great deal of diversity in both the process of privatization and the outcomes. State officials repeatedly emphasized that there was no single formula for privatizing a firm: each privatization needed to be tailored to take into account macro- and microeconomic conditions affecting the firm (Interviews 48, 56, 58). On the macroeconomic side, new regulatory agencies might need to be created to monitor the activities of firms that previously were controlled directly by the state. Regulation of the macroeconomic environment may even extend beyond the borders of Mexico through specific provisions of NAFTA or the World Trade Organization. Microeconomic reorganization was crucial to the sale of public firms, but took very different forms depending on the firm. Like macroeconomic restructuring, microeconomic changes might extend beyond the borders of Mexico to include the introduction of foreign investment or mergers and alliances with foreign firms.

In the following chapters I narrow the focus of inquiry to examine the details of specific privatizations. What accounts for the very different outcomes of privatization across firms? Did successful privatization occur only in firms that were already profitable, such as Telmex? Or, did macro- and microeconomic restructuring make a difference in the operation of these firms after their privatization? What role did labor play in shaping privatization outcomes? And how has private ownership of these firms influenced their operation? In Chapters 4, 5, and 6, I seek to answer these questions in three important industries in the communications and transportation sectors. These case studies span the entire period of privatizations of large firms, beginning with the first of the large privatizations, Aeroméxico, then through the highly touted success story of Telmex, and finally ending with the controversial restructuring and sale of the Mexican National Railroad.

FOUR

The Turbulent Privatizations of Aeroméxico and Mexicana de Aviación

> Communications and transportation are the infrastructure of economic and social development. In recent years, they have become a decisive factor for decentralizing national life and modernizing the economy.
> —*Carlos Salinas de Gortari, "Comunicaciones y transportes: Modernización y decentralización"*

> If a country is truly committed to its people, then why would its government maintain ownership of an airline and use billions of pesos of taxpayers' money to modernize the fleet when only 2 percent of the population has ever flown and, at the same time, better public services are needed?
> —*Pedro Aspe, El camino mexicano de la transformación económica*

"Communications, Instrument of Social Justice"

On the northern face of the buildings that house the central offices of the Ministry of Communications and Transportation (SCT) in Mexico City, a series of immense mosaics provides a graphic depiction of the role that transportation and communications were once understood to play in national life. In one panel, an enormous crane lifts the entire country, which is criss-crossed with highways and rail and phone lines. Below, in the past that Mexico is leaving behind, Spanish conquistadores supervise the work of Native Americans who are speaking in hieroglyphs. Above, in the future toward which Mexico is being carried, an industrial worker leads a woman and her child up a flight of stairs to the stars; another worker guides a peasant up the same steps while teaching him to read. Around the crane that is lifting the country toward its modern, industrial future is wrapped a banner that reads, "Communications, Instrument of Social Justice" (Comunicaciones Instrumento de la Justicia Social).

The murals on the SCT buildings were completed just as Mexico began the period of stabilizing development and they illustrate the social and political value associated with communications and transportation. Even today, despite the very different orientation of economic policy in Mexico, advocates of neoliberal reforms continue to affirm the social use value of these sectors. Writing for the World Bank, Saunders, Warford, and Wellenius speak of the high "private and social returns on the investment" in telecommunications and assert that "[t]elecommunications contribute to the development of a shared environment that reaches a country's most remote areas and can facilitate political, cultural, economic, and social integration" (1994, 16, 18).

The willingness of neoliberals to accept the social use value generated by these sectors creates a dilemma for reform-minded officials. Despite their conviction that markets are the most efficient means of allocating resources, many neoliberals also acknowledge that the private economy does not produce communications or transportation infrastructure in the quantities or locations that is needed. In every economy, the state plays at least a guiding role in providing this essential infrastructure; often it plays a much more active role. Where and how did the Mexican state intervene historically in the development of these sectors? And how did its intervention change with the privatization program? In what parts of these sectors did competition, price, and profitability determine allocations? And where did the state introduce restrictions upon market allocations in communications and transportation?

In the previous two chapters I examined the rise and demise of the parastate sector: the broad transformation of state and market structures and the relationship between the state and its major pressure groups both domestic and international. Although I emphasized the interdependence of state and economy and the absence of state autonomy, I also argued that a critical feature of capitalist economies is that they include a certain amount of "disembedded" exchange: transactions in which price, profitability, and competition may exert a dominant influence over other values and allocative mechanisms.

In the chapters that follow, I use this same theoretical framework while tightening the focus of inquiry to explore case studies of privatization: the organization of specific public firms, the process of selling these firms, and the attempts by these firms to negotiate the newly created market environment. I will concentrate in these chapters on the privatization of firms in communications and transportation because these are areas in which there is very little disagreement regarding the legitimate right of the state to intervene. Some communications and transportation networks have historically been assumed to be "natural monopolies," making the provision of services in these areas problematic through market mechanisms. As part of the "social infrastructure," the development of communications and transportation infrastructure is widely agreed to exert positive externalities on other parts of the economy. These sectors are also vital to achieving political control over the national territory, providing the state with further justification for intervention (see Fraser, Donheiser, and Miller 1972, 68–70; Mattelart 1994).

Privatizations in communications and transportation, then, should be limiting cases in which the introduction of market allocations is more difficult than in other, more easily commodified sectors of the economy. Despite the widespread acceptance of a positive role for the state to play in these sectors, in recent years the Mexican state has nonetheless attempted to extend the principles of market organization to these sectors, with varying degrees of success. The case studies focus on large firms whose sales were socially and technically complex, skipping entirely the first, easy phase of privatization and picking up the process with the second phase of privatizations, begun in 1988. The three cases that I examine in depth provide a range of experiences within the privatization program, spanning the period of large privatizations that began at the end of the de la Madrid presidency and continued into that of Ernesto Zedillo.

In the present chapter, I examine the sale of Mexico's two major airlines, Aeroméxico and Mexicana.[1] As noted in Chapter 3, the privatization of Aeroméxico signaled a departure from previous privatizations of the de la Madrid presidency in that it entailed major restructuring of the firm and confrontation with labor unions. Since the privatizations of Aeroméxico and Mexicana, civil aviation has experienced considerable upheaval. An explosion of competition in the industry led the newly privatized firms into bankruptcy even before the economic crisis of 1994–95. Although the government stepped in to prevent the complete collapse of the two firms, they were placed under the control of their creditors. As the industry consolidated internally—both firms now operate under the control of a single holding company—it also expanded internationally—developing international alliances and buying foreign firms.

In Chapter 5, I examine the privatization of the telecommunications monopoly, Telmex. The privatization of Telmex, under President Salinas,

---

1. In the process of privatization, both of Mexico's trunk carriers changed their names. Aeronaves de México became Aerovías de México, while Compañía Mexicana de Aviación changed its name to Corporación Mexicana de Aviación (Leyva, Mecalco López, and Mendoza Molina 1998, 46). For the sake of simplicity, except where I use the full name of the company, I refer to the first firm as Aeroméxico and the second as Mexicana both before and after the privatization.

has been praised as a model of benevolent modernization. The growth potential of the industry allowed for trade-offs and compromises between the state, labor, and capital. The Mexican state got a high price for the firm, workers got job security, and private investors got a high rate of return, with Mexican capital retaining control over the company. As I will show, however, Telmex's success as a private company is largely a result of the fact that the state eased the firm's transition into the market, restricting competition and limiting the effects of price on allocations in telecommunications.

Finally, in Chapter 6 I explore the difficult privatization of the Mexican National Railroad (FNM). In contrast to the turmoil that characterized the privatization of the airlines and the cooperative model of privatization that was showcased by the sale of Telmex, the sale of FNM involved the steady, inexorable repression of FNM workers and a wrenching process of reorganizing the assets of the company. One of the most troubled public firms, FNM was a perennial money loser and an incubator of labor unrest and political opposition to the ruling party. Yet even as a public firm, FNM operated in a competitive environment. From the 1940s, the railroads competed against private bus and trucking companies that were controlled by powerful Mexican capitalists who employed a workforce that was more fragmented and malleable than railroad workers. The state promoted competition against FNM through the construction of highways and the licensing of private trucking companies. The privatization of FNM highlights, once again, the importance of microeconomic restructuring for the macroeconomic changes taking place in Mexico and the reorientation of Mexico's economy away from inward-oriented national development and into a global network of production and exchange.

The range of experiences covered by these three privatizations offer important insights into the limits of Mexico's current economic development strategy and the neoliberal model more generally. In the concluding chapter, Chapter 7, I draw together the various strands of the argument and show how the case studies illustrate these limitations.

Civil Aviation and the State

The development of civil aviation in Mexico cannot be disentangled from the international development of this sector. The transfer of technology and organizational forms as well as governmental regulations and bilateral agreements with industrialized countries all impinge upon air transportation in the developing world. Therefore, I begin with a brief overview of the historical development of the world air transportation sector in order to identify key elements of the relationship between the state and the market in civil aviation. Then, following the structure of the two previous chapters, I examine the development of this sector in Mexico, highlighting the growing role of the state through 1982. After showing how and where this sector of the economy was embedded in social and political relationships, I examine the privatization of the two principal Mexican airlines. In the final section I assess the Mexican air transportation industry following privatization.

Although the focus of this chapter is the privatization of Aeroméxico and Mexicana, it is worth repeating a fundamental premise that guides this research. Economic actors—firms, states, and social classes—are social constructions, and the boundaries that define them are contested terrain. The process of privatization in Mexico has challenged established boundaries not only between state and market, but also between Mexico and the international economy, workers, and management, and between firms themselves as they struggle to place certain goods and services on the market, withdraw others from it, and adapt to the changing environment in which they operate.

It is important to reiterate this point here because of the interdependence between civil aviation and a wide range of other industries. Air transportation depends on numerous services beyond simply that of carrying passengers between two points. While the Mexican state stepped in to provide some of the infrastructure necessary for civil aviation—airports, airport services, and air traffic control in particular—there were still clear gaps in the sector that remained under the control of private capital or were subject to pressures of price, profitability, and competition. As in many of the larger and more complex privatizations of the Salinas administration, the interdependence between Mexico's airlines and these other components of civil aviation meant that privatization would have to go beyond simply selling a state-owned firm. It would also require a reorganization of an entire set of social relationships and a restructuring of the markets in which these actors operated.

In the privatization of the airlines, we not only see a scramble to disassemble the state sector and the regulatory apparatus that governed air transportation in Mexico. We also see how social actors affected by the reorganization of this industry respond and how they attempt to reassemble the various pieces of civil aviation into viable organizations. After privatization, the state opened the market to increased competition. A debilitating price war among private companies in civil aviation soon led firms to seek stability in the industry by forming alliances, acquiring competitors, and generally seeking to limit the influence of price and competition on crucial allocations.

Despite these private efforts to regulate the market, privatized air carriers in Mexico were undermined by an excess of profit maximization and price competition, leading the state right back into civil aviation barely six years after the first airline privatization. Yet while the federal government picked up the tab to keep the airlines aloft, it refrained from reentering into the operation or management of the airlines, allowing a holding company, Grupo Cintra, to reassemble the various components of civil aviation. By the end of the Zedillo administration, this company owned both of Mexico's trunk airlines, Aeroméxico and Mexicana, as well as three regional feeder companies, an express cargo service, a firm that provides ground services at Mexican airports, a training center for airline personnel, 49 percent of the reservation system that links the various airlines with travel agents, and a minority share of Turborreactores, a firm devoted to the maintenance and repair of jet engines (see Cintra 2000, 32). Instead of decentralizing the economy, privatization and liberalization in this sector had the effect of increasing the concentration of control and ownership in the airlines.

THE INTERNATIONAL AIR TRANSPORTATION INDUSTRY

The role of the state in the development of civil aviation has been so pervasive that it is impossible to imagine what civil aviation would look like without state intervention. It is unlikely that anything even remotely resembling the current worldwide network of airports, air-traffic control, passenger and freight routes, and airlines would have emerged without the heavy hand of government subsidization and regulation. According to Lewis (2000, 2), "[I]n the United States there was little air transport of anything but mail until 1925." Initially, federal government subsidies for carrying mail provided the impetus for the development of a national, and later an international, system of civil aviation.

Airlines require large capital investment and have uncertain returns on investment, which, in part, explains the heavy involvement of the government in their development. The fact that profitability depends on economies of scale also provides something of a public-goods justification for intervention.

Although the U.S. government was content to let private capital own and operate the airlines, it imposed strict and detailed regulation upon these companies. Even before the advent of the New Deal and the growth of government regulation in the 1930s, the U.S. Congress passed the Air Commerce

Act, in 1926. In the years that followed, no aspect of this industry escaped federal oversight and regulation. The Civil Aeronautics Act of 1938—later amended under the Federal Aviation Act of 1958—established the basic framework of regulation under which private capital would operate for four decades. The act treated civil aviation like a public utility and established the Civil Aeronautics Board, with broad powers "to regulate virtually every facet of the industry's structure, operations, and relationships to other industries," including routes, prices, rules of cooperation and collusion, safety, airway maintenance, and the licensing of aircraft, pilots, and mechanics (Fraser, Donheiser, and Miller 1972, 68, 71; see also Lewis 2000, 2, 3).

In most advanced industrial economies, the state not only exercised direct regulation and control over the development of air transportation, it also strongly influenced civil aviation through its investment in military technology. As early as 1931, manufacturers of military airplanes applied their technological developments to civil aviation, creating "a tightly knit family of mutually dependent government agencies, military contractors, and consumer-oriented industries" (Lewis 2000, 15). In the United States, in 1969 alone, "over half of all the total funds expended on civil aeronautic R&D originate[d] with the government" (Fraser, Donheiser, and Miller 1972, 62).

The widespread intervention of the U.S. government in this sector is typical of other advanced industrialized countries, making international civil aviation a patchwork. Domestic routes are generally served exclusively by domestic carriers, with no competition from foreign airlines. On international routes, domestic firms must compete with the national firms of the country of destination. Competition on international routes, however, is dependent on bilateral agreements forged between national governments. A multilateral airline industry association also seeks to influence policy at the international level. Even now, more than two decades since the beginning of airline deregulation in the United States, most countries' national markets remain closed to foreign competition. With the exception of the European Union (EU) where any EU airline is now allowed to compete in the market of any other EU country, most states only allow companies that are owned by their nationals to operate in the domestic market; foreigners are prohibited from owning more than 25 percent of the voting shares of a U.S. carrier (*Economist* 1993c, 69; 1994, 15).

In addition to making the distinction between the national and international segments of civil aviation, it is important also to distinguish between equipment and service providers in air transportation. The very fact that equipment and service are not vertically integrated has some historical and operational significance. In the United States, aircraft manufacturers and air carriers once had much closer ownership and operational ties. In the 1930s, the airlines used these ties to shift costs, with unregulated manufacturers charging inflated prices to the regulated airlines, which allowed the airlines to justify increases in ticket prices. In response, in 1934 the U.S. Congress passed the Black-McKellar Act, prohibiting "interlocking connections between airline operators and aircraft manufacturers" (Lewis 2000, 7).

Despite the wide range of government interventions that still influence the operations of civil aviation, deregulation and industry attempts to circumvent the regulatory apparatus have transformed the industry in recent years. In 1978, the United States began to lift restrictions on pricing and routes, leading initially to the rapid growth of low-cost carriers, followed by an industry shakeout. The immediate effect of increased competition was lower ticket prices, which, in turn, forced many firms to reorganize their operations.

The introduction of market allocations in certain segments of the industry did not lead to the immediate transformation of air travel into a uniform product in which the service provided by one carrier is indistinguishable from that of another carrier. To the contrary, competition in the airlines has been characterized by a sophisticated system of niche pricing among different classes of service. In the 1980s, carriers began to introduce variable pricing in which the amount of time prior to travel becomes the crucial variable determining how much a seat costs (*Economist* 1993c, 70).

By the late 1990s, however, low-cost competition began to undermine some of the largest carriers in the world. Even before the economic blow of the terrorist attacks of September 11, 2001, major U.S. airlines were finding it increasingly difficult to operate profitably. Increasingly, air travel was being sold as a commodity in which new technologies—in particular, online sales and the complete pricing information provided by the Internet—have driven down ticket prices. In the wake of September 11, the U.S. government provided loan guarantees and created the Air Transportation Stabilization Board with sweeping powers to decide which airlines would survive the turmoil in the industry (Hansell 2002; Labaton 2001; Zuckerman 2001).

Increased competition at one level has led to increased cooperation at another. Firms have managed to sidestep the political and organizational boundaries that limit their route systems by entering into a variety of different cooperative arrangements with other firms. These arrangements allow firms to provide "seamless service" to customers—selling tickets from one country, through another and to a third, for example—in a gray area that is neither governed nor market. Firms may create franchising arrangements with other firms, sell blocks of seats on their routes to other carriers for resale, and enter into "code-sharing" agreements that allow firms to share revenue and coordinate schedules and baggage handling (Hanlon 1996, 95).

EMBEDDING THE AIR TRANSPORTATION INDUSTRY IN MEXICAN SOCIETY

Limits of the Market in the Macroeconomy
Many of the features that characterize the development of civil aviation in the United States can also be seen in the industry in Mexico—even if they appear different through the lens of a developing country. The domination of the air transportation sector by firms in advanced industrial economies meant that property rights were established to a large degree outside Mexico. The production of aircraft as well as the development of firms to organize the transportation of passengers and cargo were both strongly influenced by the interaction between foreign firms and their governments. Upon arrival in Mexico, these organizational arrangements, technologies, and rights of control present themselves as predetermined social facts. Thus, while most developing countries jealously guarded their sovereignty over a flagship carrier, the equipment on which these services depended was still owned and controlled by actors in the industrialized world.

The Mexican state was relatively slow to take an interest in civil aviation. Although as early as the 1930s foreign carriers were required to obtain a concession from both the Ministry of Foreign Relations and the SCT (Ruiz Dueñas 1988, 165), this regulation appears to have been concerned more with restricting foreign ownership and defending Mexican sovereignty over its airspace than with any proactive interest in promoting air transportation.[2] Unlike the governments of industrial countries that subsidized the development of air transportation by purchasing planes for military use and through research and development, the Mexican government had little incentive to participate in the sector.

Civil aviation was never designated a "strategic" or "priority" area of the economy under Article 28 of Mexico's Constitution, even though other components of transportation and communication—railroads, the mail, telegraphs, and later, satellite telecommunication, for example—were (see Ruiz Dueñas 1988, 298). At the same time, the public character of air transportation was clearly acknowledged under the law. Article 39 of the 1936

---

2. Until the 1950s, the SCT was the Ministry of Communication and Public Works. For the sake of simplicity, I refer to both as the SCT.

General Law of Means of Communication and Transportation prohibits any action that interrupts public transportation (Guerrero 1988). This law also grants the state the right to intervene in labor disputes that would interfere with communications and transportation. The *requisa*, or confiscation, allows the state to nullify a strike, order workers back to their jobs, and even replace workers with temporary personnel when labor unrest threatens to disrupt public communication and transportation networks.

The principal mechanism by which the state attempted to direct and control the industry in the early years was through price regulation. In the 1930s, a General Directorate of Fares was created within the SCT to regulate prices on domestic routes. Changes in fares had to be approved by the general directorate. Airfares for domestic service did not change between 1960 and 1973 and by 1982 had fallen to 40 percent of 1970 prices. Even the nationalization of Aeroméxico in the late 1950s did not indicate any clear state policy on the development of this sector. Instead, the nationalization appears to have been an ad hoc response to specific problems at the firm (Sánchez et al. 1993b, 176; Tamayo 1988, 738).

The principal country of departure and destination for most of Mexico's international flights was and remains the United States; almost 90 percent of all Mexico's international air traffic either originates or terminates there (Tandon 1992a). As noted above, fares on international routes are determined through bilateral agreements between governments that set prices at higher levels than fares for domestic routes. Because the vast majority of Mexico's international traffic has been with the United States, the most important of these bilateral agreements is with that country. In 1960, Mexico entered into an agreement on air transportation with the United States that designated the airlines to provide services between the two countries. This agreement remained in effect until shortly after the privatization of Mexico's two carriers in 1988 (Hernández Vélez 1988b; Salmon 1988).

The influence of the United States upon civil aviation in Mexico went still further. In order to enter U.S. airspace, foreign carriers must conform to Federal Aviation Administration regulations regarding air safety. Thus, in addition to purchasing most of its fleet from U.S. firms, Mexican airlines needed to maintain and staff any plane that entered U.S. airspace in accordance with the U.S. regulatory guidelines for civil aviation.

It was not until the mid-1960s that the Mexican state began to formulate a more proactive approach to guide the development of the civil aviation sector. That approach consisted primarily of providing essential infrastructure and backward linkages upon which both the nationalized Aeroméxico and the still private Mexicana, depended. In 1965, the state created Airports and

Auxiliary Services (ASA), a parastate firm charged with managing the construction, maintenance and operations of Airports under the control of the SCT. In the early 1970s, ASA focused primarily on building airports to serve the national market.

Through ASA, the state imposed what amounted to a luxury tax upon the affluent users who could afford to travel by air. ASA even became a source of revenue for the federal government through the fees it charged the airlines for landing rights, air traffic control, arrival and departure gates, and ground service. At the same time, the state also provided subsidies to civil aviation. Through Pemex, the state supported the industry with low-cost jet fuel. In 1980 the state took over a majority of the shares of the company that provides repair and maintenance services for jet engines, Turborreactores, which included both Aeroméxico and Mexicana as minority shareholders (Casar and Peres 1988, appendix 1; Tamayo 1988b, 723, 735).

And, as Table 6 shows, air transportation received a growing percentage of federal investment in transportation through the 1960s. After that, federal investment in civil aviation fluctuated considerably until the crisis of 1982 when suddenly federal investment for air transportation jumped to almost 24 percent of total federal investment in transportation. Despite its relative disinterest in civil aviation, the Mexican state still exercised important influence on both sides of the airlines' budgets, regulating revenues through price controls and airport taxes but also providing subsidies through other parastate firms.

Restrictions on Property Rights in the Microeconomic Environment
At first glance, Mexico's two principal airlines, or "trunk" carriers, appear very different.[3] Mexicana remained in private hands and was able to earn modest profits until the economic crisis of 1982. Aeroméxico, in contrast, was nationalized in 1959 and was a chronic drain on public resources. During almost thirty years under state ownership, the company showed a profit in only three years: from 1979 to 1981 (Tandon 1992a, 2). Yet, in Mexicana's final year as a private firm, the two firms look strikingly similar in their organizational structure. The macroeconomic, political, and social environment seems to have largely determined the microeconomic organization of these two firms.

Aeroméxico was a troubled firm even before the Mexican state took it over. Founded in 1934 to fly between Acapulco and Mexico City, the airline

---

3. Trunk carriers are granted rights to operate in the entire country as opposed to regional or feeder companies that generally operate smaller equipment in a particular region.

TABLE 6 Distribution of federal investment in transportation

Year	Highways and Bridges (%)	Railroads (%)	Ports and Marine (%)	Air Transportation (%)
1952	44.80	49.60	5.70	0.00
1953	42.50	51.60	5.90	0.00
1954	41.20	49.40	9.40	0.00
1955	41.80	46.80	10.60	0.80
1956	36.50	53.00	9.80	0.70
1957	43.10	46.50	8.90	1.50
1958	40.80	49.50	8.50	1.20
1959	44.50	48.60	6.00	0.90
1960	33.40	55.50	6.70	4.40
1961	44.40	46.80	5.00	3.70
1962	41.90	44.80	6.60	6.80
1963	54.20	32.80	4.10	8.90
1964	55.30	37.50	3.70	3.40
1965	51.80	34.90	2.20	11.00
1966	67.70	23.80	4.60	3.80
1967	47.40	34.20	4.10	14.30
1968	46.80	31.90	4.20	17.10
1969	51.00	29.50	6.40	13.10
1970	50.60	32.80	13.90	2.70
1971	64.30	21.50	9.80	4.30
1972	68.50	15.70	7.20	8.60
1973	64.00	22.10	10.70	3.20
1974	50.20	28.90	10.20	10.60
1975	48.10	34.10	10.40	7.40
1976	56.20	34.80	3.10	5.90
1977	57.00	30.60	4.20	8.20
1978	60.40	27.30	4.40	7.90
1979	49.90	33.90	7.40	8.90
1980	37.30	31.30	18.50	12.90
1981	1.80	61.00	29.90	7.30
1982	0.50	46.20	29.40	23.90
1983	39.10	23.00	15.60	22.20

Source: Adapted from Islas Rivera 1990, 129.
Note: Rows may not sum to 100% because of rounding.

grew through the gradual acquisition of smaller companies. Pan American airlines purchased 40 percent of Aeroméxico in 1940, but Mexican owners retained control and, in 1957, Pan American sold its stake in the company to Mexican investors (Tandon 1992a, 2). Aeroméxico's growth strategy created unique problems for the company. With a fleet made up of a variety of different planes, Aeroméxico could not simply redeploy its assets to capture the most profitable segments of the market. Instead, the company struggled to maintain spare parts for the wide range of equipment that it operated and had to either cross-train its employees—mechanics as well as pilots—on the different equipment, or maintain separate training programs. Pilots trained to fly one type of aircraft are generally not permitted to fly another (Vázquez Talavera 1990, 118).

In July 1959, the state nationalized Aeroméxico, apparently in response to a strike by the pilots' union but also out of concern that the company was about to collapse (Tandon 1992a, 2). The exact circumstances under which Aeroméxico was nationalized are not entirely clear. The pilots' union included right-wing, anticommunist groups that made it difficult for workers to confront the company as a unified force (Waterbury 1993, 249). At the same time, given the social unrest that was sweeping Mexico in the late 1950s, allowing the firm to fail could have strengthened the hand of radical workers who were demanding far-reaching political and economic reforms (see Chapter 2).

During the almost three decades that followed its nationalization, Aeroméxico's operations went from bad to worse. Ruiz Dueñas argues that Aeroméxico consistently lost money because its obligations to fulfill political goals overwhelmed any concern about profitability (1988, 165). Politicians could fly for free and the political appointment of Aeroméxico's upper management meant that there was frequent turnover of administrative personnel (Interview 54; Vázquez Talavera 1990, 120; Corro 1988b, 30).

Mexicana, in contrast, only became a parastate firm following the economic crisis of 1982 when the Mexican state stepped in to save it from bankruptcy. Founded by U.S. nationals in 1921, Mexicana received a concession to fly between Mexico City and the Gulf Coast state of Tampico. In 1924, another group of investors from the United States purchased a majority of the company and began to fly air mail along the Mexico–Tampico route. Five years later, Pan American Airlines purchased all the shares of Mexicana and inaugurated the first international flights for the company, flying to Brownsville, Texas, and Guatemala. As the carrier designated by the U.S. government to fly Latin American routes, Pan American later acquired a concession from the U.S. Postal Service to carry mail between Mexico and the United States, flying from Los Angeles to Mexico City.[4]

In the 1960s, it appears that increased competition growing out of Mexico's bilateral agreement with the United States began to take its toll on Mexicana, leading the company to the verge of bankruptcy in 1967. Instead

4. This account is based primarily on Mexicana (2000) and Tandon (1992b). Where there are discrepancies between the two sources on dates of acquisition and the date when Pan American introduced international service, I defer to Mexicana.

of allowing the firm to fail, however, the Mexican government supported the acquisition of the firm by Mexican investors (Mexicana 2000). Despite this intervention and the company's partial dependence on government contracts, the firm remained in private hands until the economic crisis of 1982.

Both companies took advantage of boom of the 1970s and the easy credit on international markets to invest in new equipment. Between 1970 and 1982, the number of commercial planes in Mexico grew from 291 to 859. From 1979 to 1982, Mexicana's fleet grew from 34 to 45 planes. The growth of investment in new equipment was also reflected in Mexico's imports. In 1975, Mexico imported $56 million in airplanes and airplane parts. By 1981, that figure had grown to $323 million (Salinas de Gortari 1994, 123, 146, 274). With the collapse of the peso in 1982, Mexicana's foreign debt swamped the company and the Mexican state stepped in to prevent its collapse by acquiring 51 percent of the firm.

Restrictions on the rights of capital to organize production in civil aviation were primarily imposed from above and took two principal forms: direct regulation by the Mexican state and international standards and norms with which the airlines had to comply. With the exception of staffing levels, which appear to have been quite high at both Aeroméxico and Mexicana, there is little evidence that the labor unions at the airlines exercised strong influence on the organization of the industry. Apparently the state used the *requisa* on numerous occasions to keep the airlines operating, but the dearth of literature on workers in as high-profile an industry as this suggests that airline workers were not especially militant (Trejo Delarbe 1990, 300).

One reason for the apparent absence of efforts by labor to challenge the prerogatives of management in the airlines might have been the fact that labor unions in the industry are organized along functional rather than industry lines. Workers in the industry are represented by four separate national unions: one for pilots, another for flight attendants, a third representing ground workers at Mexicana, and a fourth representing ground crews for Aeroméxico and foreign airlines that fly into Mexico. Trejo Delarbe (1990, 301) notes that "although they frequently supported each other in their respective movements, these four unions remained separate."

In addition, airline pilots constitute something of an aristocracy of labor. Government regulation and international standards determined fundamental issues such as minimum hours of rest and maximum days of work, while the organization of work is strongly influenced by the technical requirements of the aircraft. Airline pilots may have seen little need to challenge management except where salaries and benefits were concerned. The average salary in aviation was more than one-third higher than the average salary among railroad workers in 1955 and by 1975 was almost three times the salary of railroad workers.[5] Lower down in the labor hierarchy in the airlines, flight attendants and ground workers may have had more grievances, but also had less clout in negotiations with management.

According to a representative of the national pilots' union, the Union Association of Pilot Aviators (ASPA), the pilots' main concern about the organization of the airlines under state ownership had to do with the failure to invest and maintain the fleet. The union representative argued that the pilots were well aware of the inefficiency of the airlines and the need to modernize and invest. He also attributed the problems at Aeroméxico to the extreme politicization of service. Claiming that Aeroméxico was run more like another ministry of state than a firm, the union representative pointed out that politicians could simply ride for free on flights and pilots had no right to deny them: "Aeroméxico had been squeezed dry, it wasn't really a company" (Interview 54).

Still, despite Aeroméxico's long history under state ownership and the private control exercised over Mexicana until 1982, the two firms shared many similarities. Waterbury (1993, 126) points to staffing levels of more than 200 employees per plane at Aeroméxico and compares the figure to the inefficient U.S. firm Eastern Airlines, which operated with 146 employees per aircraft. Yet Mexicana also had elevated levels of staffing prior to its nationalization in 1982. Table 7 shows the principal characteristics of the two firms from 1981 to 2000. Although Mexicana employed somewhat fewer workers per plane in 1981, the difference was hardly significant: 227 employees per plane at Mexicana compared with 242 at Aeroméxico.[6]

Table 7 shows other similarities between the two firms. In 1981, Mexicana's final year as a private firm, the company had slightly more planes than Aeroméxico—forty-five compared with Aeroméxico's thirty-nine—and its profits were much higher—250 million Pesos compared with 33 million Pesos. None of these differences, however, indicates strikingly different organizational structure between the private and the public firm. Instead, there appear to be two principal reasons why Aeroméxico usually lost money while Mexicana managed to make a profit.

The first reason is the different route structures of the two airlines. Aeroméxico, unlike Mexicana, was required to maintain air transportation

---

5. Estimates by Islas Rivera (1990, 27) very likely *understate* the benefits that pilots enjoyed because they represent an industry average, including lower-wage ground workers and flight attendants. Nor do the figures take into account other perquisites enjoyed by airline pilots such as company cars and travel benefits (Interview 54).

6. Waterbury (1993, 126) indicates that Aeromexico employed 284 workers per plane. Tandon (1992a, 1992b) claims that Aeroméxico never had more than 274 employees per plane in the 1980s. See Table 7.

TABLE 7 Principal characteristics of Mexico's two largest air carriers, 1981–2000

Year	Number of Employees		Size of Fleet[a]		Profits (Millions of 1992 Pesos)		Percentage of Revenue from Domestic Service		Passengers Transported Internationally (Thousands)	
	Aeroméxico	Mexicana	Aeroméxico	Mexicana	Aeroméxico	Mexicana	Aeroméxico	Mexicana	Aeroméxico	Mexicana
1981	9,448	10,212	39	45	33	250	62	39	994	1,864
1982	10,417	11,031	38	45	-378	-280	59	40	883	1,544
1983	10,463	11,910	40	43	-83	42	53	40	1,067	1,949
1984	10,791	12,158	42	44	-160	126	54	41	1,280	2,306
1985	11,010	12,980	43	46	-402	-137	56	39	1,117	2,341
1986	11,214	13,373	43	45	-416	-429	57	35	1,049	2,479
1987	11,505	13,906	43	45	-448	-337	52	33	1,396	2,976
1988	7,015	13,540	29	45	-492	104	75	37	438	3,224
1989	4,218	12,783	32	44	18	36	76	39	559	2,903
1990	5,104	11,974	39	44	19	-361	71	42	937	3,101
1991	6,004	11,289	45	52	4	-268	66	50	nd	nd
1992	7,100	9,801	nd	nd	-132	-702	nd	nd	nd	nd
1993	nd	nd	nd	nd	nd	nd	nd	nd	nd	nd
1994	5,939	6,824	56	46	nd	nd	nd	nd	1,350	2,119

TABLE 7 *(continued)* Principal characteristics of Mexico's two largest air carriers, 1981–2000

Year	Number of Employees		Size of Fleet[a]		Profits (Millions of 1992 Pesos)		Percentage of Revenue from Domestic Service		Passengers Transported Internationally (Thousands)	
	Aeroméxico	Mexicana	Aeroméxico	Mexicana	Aeroméxico	Mexicana	Aeroméxico	Mexicana	Aeroméxico	Mexicana
1995	5,355	6,499	53	44	nd	nd	nd	nd	1,194	2,201
1996[b]	5,314	6,288	53	42	nd	nd	nd	nd	1,473	2,315
1997[c]	6,000	6,800	58	51	nd	nd	nd	nd	1,715	2,792
1998[c]	19,224		63	54	nd	nd	nd	nd	1,646	2,828
1999[c]	20,028		66	56	nd	nd	nd	nd	1,924	2,936
2000[c]	7,414	6,966	64	57	nd	nd	nd	nd	2,462	3,441

Sources: 1981–92, number of employees and profits, Tandon 1996, 229, 234. All other figures for 1981–92, Tandon 1992a, figs. 17-8 and 17-12; 1992b, figs. 18-1 and 18-6. All figures for 1994–96, Cintra 1996, 37.

nd = No data

[a]Size of fleet includes planes that are owned by the company as well as leased aircraft.

[b]Figures for 1996 only cover the first three months of the year, ending March 31, 1996.

[c]Figures for employees from Cintra annual reports 1997 and 1998 are consolidated, making it impossible to distinguish between the two carriers. Consolidated employee figures include regional carriers owned by Cintra, which explains why the number increases so sharply from the sum of Aeroméxico and Mexicana from previous years. Figures reported separately for employees of Aeroméxico in 1997 is an approximation: Cintra (1997, 7) reports that Aeroméxico had "more than 6,000 employees." See Cintra 1997, 7, 9; 1998, 1, 14; 1999; 2000, 26, 30, 42. Figures for aircraft for 1997–2000 are separated in Cintra annual reports. Figures for passengers transported internationally, 1994–2000, from SCT 2002.

services to smaller Mexican cities, covering routes that did not pay for themselves (Ruiz Dueñas 1988, 166). As part of its political and social obligations, Aeroméxico flew to cities with low demand for air travel. At the same time, Mexicana operated more of the lucrative international routes. Although Aeroméxico flew to Paris and Madrid, it carried only about 10 percent of the total passenger traffic to Europe, lagging behind the foreign carriers Air France and Iberia. Mexicana flew the bulk of the flights into and out of the United States (Tandon 1996, 224).

Table 7 shows that in its final year as a private firm, Mexicana flew almost twice as many passengers on international flights as Aeroméxico, a ratio that stayed fairly constant through the first years of Mexicana's operation under government ownership. While 62 percent of Aeroméxico's revenue came from its domestic operations in 1981, only 39 percent of Mexicana's revenue came from domestic flights. Table 7 also shows that Mexicana's revenue from national flights fell as low as 33 percent in 1987, while Aeroméxico's dependence upon the domestic market grew. In 1989, Aeroméxico earned 76 percent of its revenue from the domestic market. By 1988, Mexicana earned 55 percent of its income from the international market, where it flew only 38 percent of the total number of passengers that it carried (Sánchez et al. 1993b, 176).

The second reason for Aeroméxico's consistent failure to turn a profit appears to be related to the company's growth strategy and the equipment that the company operated. Because Aeroméxico had grown through the acquisition of smaller airlines, the aircraft that it acquired in the process were a hodgepodge. Operating different types of aircraft imposes extra costs in part because of the need to maintain a more diverse stock of spare parts for the planes. Operating different aircraft also imposes a significant expense in the training of both mechanics and pilots on the equipment and limits the ease with which personnel can be transferred from one plane to another.

Where Is the Market in Civil Aviation?
Given the variety of ways in which the state intervened in civil aviation in Mexico, it would be tempting to argue that market allocations were irrelevant. The state clearly limited private control over property rights, directed the operations of private firms, regulated and controlled prices, subsidized important inputs, and restricted the forces of competition within the industry. Yet, despite the myriad restrictions upon the forces of price, profitability and competition in this sector, market forces remained important in a number of crucial areas.

To begin, consumers seeking transportation services actually enjoyed a considerable range of choices. Tandon (1992a, 8) argues that "[e]ntry [into the domestic air transportation market] was strictly controlled, both of new airlines and new routes or flight schedules to be offered by existing airlines." Islas Rivera (1990, 160) also points to the limited competition between Aeroméxico and Mexicana, noting that the two firms covered very different routes. Yet air transportation represents a very small percentage of the total amount of transportation services provided in Mexico and consumers could easily substitute air transportation for bus or rail transportation. Given the limited income of a majority of Mexicans, most did.

Indeed, despite the large size of the airlines in terms of revenue and capital relative to other firms in the transportation sector, the number of passengers that they carry is quite small. While Aeroméxico and Mexicana ranked as the largest and third-largest transportation firms in Mexico in terms of revenue in 1994, air transportation accounted for less than 3 percent of all passenger and only one-half of one percent of all cargo transportation in Mexico (*Expansión* 1995, 24).

In addition to competition from bus and rail transportation, some limited competition existed in the form of international airlines and smaller regional airlines operating within Mexico. In domestic air travel, Aeroméxico and Mexicana dominated the sector. Table 8 shows the transportation of passengers on domestic flights by carrier. Between 1975 and 1985, the two major carriers, combined, never controlled less than 81 percent of the market and in the latter years of the economic crisis controlled as much as 92 percent of all domestic passenger air transportation. Table 8 also shows that the total number of passengers flying on domestic routes stagnated at around 13.5 million a year between 1981 and 1984. After recovering slightly in 1985, the number of passengers on domestic flights declined sharply in 1986 and by 1988 had fallen by almost 35 percent of the 1984 figure, bottoming out at 8.8 million passengers.

As demand for domestic air transportation dropped off, so did the number of carriers operating in Mexico. Table 9 shows the decline at the beginning of the 1980s in the number of domestic airlines offering regular, scheduled service in Mexico. Consistent with the decline in the number of passengers traveling on domestic flights, the number of Mexican carriers operating in the domestic market hit bottom in 1986, falling by more than half from eleven in 1980 to five in 1986. The number of foreign carriers flying to Mexico also fell, from 34 in 1980 to 26 in 1984.

Internationally, however, Aeroméxico and Mexicana faced very real competition. Although the bilateral agreements under which international

TABLE 8 Passenger transportation in domestic scheduled service, 1975–2000 (thousands of passengers)

Year	Total	Aero-california[a]	Consorcio Aviacsa	S.A.R.O.	T.A.E.S.A.[b]	Other	Aeroméxico	Mexicana	Aeroméxico (%)	Mexicana (%)	Combined Aeroméxico Mexicana (%)
1975	7,305	nd	nd	nd	na	958	3,234	3,113	44.3	42.6	86.9
1976	8,642	nd	nd	nd	na	1,174	3,653	3,815	42.3	44.1	86.4
1977	9,608	nd	nd	nd	na	1,508	3,609	4,491	37.6	46.7	84.3
1978	11,204	nd	nd	nd	na	1,978	4,007	5,219	35.8	46.6	82.4
1979	12,987	nd	nd	nd	na	2,274	4,607	6,106	35.5	47.0	82.5
1980	12,330	nd	nd	nd	na	2,610	4,152	5,568	33.7	45.2	78.8
1981	13,693	nd	nd	nd	na	3,001	4,545	6,147	33.2	44.9	78.1
1982	13,307	nd	nd	nd	na	2,768	4,614	5,925	34.7	44.5	79.2
1983	13,854	nd	nd	nd	na	2,936	4,915	6,003	35.5	43.3	78.8
1984	13,540	nd	nd	nd	na	2,664	4,915	5,961	36.3	44.0	80.3
1985	14,484	nd	nd	nd	na	2,509	5,527	6,448	38.2	44.5	82.7
1986	11,563	nd	nd	nd	na	1,100	5,004	5,459	43.3	47.2	90.5
1987	10,068	nd	nd	nd	nd	900	4,407	4,761	43.8	47.3	91.1
1988	8,840	nd	nd	nd	nd	1,053	2,715	5,072	30.7	57.4	88.1
1989	10,194	380	nd	nd	nd	588	3,790	5,436	37.2	53.3	90.5
1990	11,438	369	18	nd	nd	734	4,540	5,777	39.7	50.5	90.2
1991	12,892	587	203	76	104	784	4,993	6,145	38.7	47.7	86.4
1992	14,280	699	341	302	720	1,108	5,752	5,358	40.3	37.5	77.8

TABLE 8 *(continued)* Passenger transportation in domestic scheduled service, 1975–2000 (thousands of passengers)

Year	Total	Aero-california[a]	Consorcio Aviacsa	S.A.R.O.	T.A.E.S.A.[b]	Other	Aeroméxico	Mexicana	Aeroméxico (%)	Mexicana (%)	Combined Aeroméxico Mexicana (%)
1993	14,972	768	398	465	1,267	1,374	5,953	4,747	39.8	31.7	71.5
1994	18,394	790	488	782	2,400	1,631	7,029	5,274	38.2	28.7	66.9
1995	14,857	852	498	42	1,507	1,736	5,580	4,642	37.6	31.2	68.8
1996	14,200	965	471	ob	1,057	2,004	5,409	4,294	38.1	30.2	68.3
1997	15,428	1,152	499	ob	1,114	2,187	5,820	4,656	37.7	30.2	67.9
1998	17,046	1,621	849	ob	1,550	2,181	6,169	4,676	36.2	27.4	63.6
1999	18,248	1,602	1,103	ob	1,858	2,347	6,690	4,648	36.7	25.5	62.1
2000	17,762	1,418	1,383	ob	ob	2,362	7,291	5,308	41.1	29.9	70.9

Sources: SCT (1987) for 1975 through 1979; SCT (1990a) for 1980 through 1988; SCT (1996a) for 1989 through 1995; and SCT (2002) for 1996 through 2000. Although the data series from SCT (1987) extends from 1975 through 1986, I use this source only where another source is not available because of discrepancies between this source and SCT (1990a), assuming the later source to be the more reliable.

na = Not applicable
nd = No data
ob = Out of business
[a] Founded in 1960
[b] Founded in 1987

TABLE 9 Scheduled commercial airlines operating in Mexico, 1980–2000

		Mexican Carriers			Foreign Carriers
Year	Total	Subtotal Mexican	Trunk	Regional	
1980	47	13	2	11	34
1981	40	8	2	6	32
1982	44	12	2	10	32
1983	38	10	2	8	28
1984	35	9	2	7	26
1985	35	8	2	6	27
1986	34	7	2	5	27
1987	36	8	2	6	28
1988	43	10	2	8	33
1989	46	15	2	13	31
1990	43	15	4	11	28
1991	44	17	6	11	27
1992	47	17	6	11	30
1993	50	18	6	12	32
1994	54	19	6	13	35
1995	55	19	6	13	36
1996	52	16	5	11	36
1997	51	14	5	9	37
1998	50	13	5	8	37
1999	48	13	5	8	35
2000	47	12	4	8	35

Source: For the years 1980–94, *Expansión* (1995, 27); for 1995–2000, SCT (2002). Mexican subtotal excludes "Other" category, which consists primarily of charter companies.

routes are established seek to distribute market share evenly between carriers of both countries (Tandon 1996, 224), the fact that 90 percent of European traffic flew into and out of Mexico on foreign carriers indicates that this goal was not always achieved. Table 10 shows the distribution of passengers on international flights between Mexican and foreign airlines. In the mid-1970s, when the data series begins, foreign carriers handled almost 60 percent of the traffic carried internationally. Aeroméxico and Mexicana gained market share during the first years of the economic crisis. Yet by the end of the 1980s they had once again fallen behind the international carriers, dropping to 46 percent of the market in 1988.

The fact that Mexico had only two carriers capable of operating internationally made it difficult to compete against foreign carriers. While both Aeroméxico and Mexicana flew to the United States, several U.S. carriers

TABLE 10 Passenger transportation in scheduled international service, 1976–2000

		Foreign Carriers		Mexican Carriers	
Year	Total (Thousands)	(Thousands)	Percentage of Total	(Thousands)	Percentage of Total
1976	3,570	2,108	59.1	1,462	41.0
1978	4,885	2,854	58.4	2,031	41.6
1980	6,543	3,744	57.2	2,799	42.8
1982	5,683	3,256	57.3	2,427	42.7
1984	6,733	3,147	46.7	3,586	53.3
1986	6,459	2,931	45.4	3,528	54.6
1988	7,981	4,307	54.0	3,674	46.0
1990	9,011	4,831	53.6	4,180	46.4
1992	10,142	5,791	57.1	4,351	42.9
1994	10,737	6,778	63.1	3,959	36.9
1996	12,292	7,913	64.4	4,379	35.6
1998	13,876	8,680	62.6	5,196	37.4
2000	16,212	9,799	60.4	6,413	39.6

Sources: SCT (1987) for the years 1976–86; Pérez Escobedo (1995, 521) for 1988; SCT (1996a) for 1990–94; SCT (2002) for 1995–2000. There is a discrepancy between the figures that Pérez Escobedo shows for 1990 and 1992 and SCT (1996a). I use the primary source here (SCT 1996a) and use Pérez Escobedo only for the year that is not covered by either SCT (1987) or SCT (1996A).

competed in this market (Tandon 1996, 224; see also SCT 1987). In addition, the size of the U.S. market permitted U.S. airlines to undercut Mexican airlines because they could transfer profits earned in the domestic market to cross-subsidize international operations (Leyva, Mecalco López, and Mendoza Molina 1998, 48). The deregulation of the U.S. market and increased competition among domestic carriers within the United States after 1978 may also have increased the incentives for these carriers to expand their international operations.

Another important area where market forces influenced the development of civil aviation in Mexico was in the acquisition of airplanes. Although the Mexican government identified the goal of promoting the domestic manufacture of airplanes in its six-year plan of 1941–46, it never succeeded in establishing this backward linkage for the industry (Islas Rivera 1990, 109). Mexico's airlines, therefore, had to pay international prices to acquire planes. The international market for jets presented a significant budget constraint on both of Mexico's major carriers. Aeroméxico and Mexicana both began investing to replace old planes with new equipment during the economic boom of the late 1970s. With the crash of 1982, both were left holding a large amount of dollar-denominated debt, which ultimately led to the acquisition of a majority of Mexicana by the state (Leyva, Mecalco López, and Mendoza Molina 1998, 47; Salinas de Gortari 1994, 146).

Finally, even while Mexicana's ability to operate profitably may not have been the consequence of a savvy business plan or the entrepreneurial spirit, profits still mattered to the private firm. The allocation of more lucrative international routes to Mexicana clearly benefited the company and enabled it to maintain a moderately profitable operation. At Aeroméxico, there was never any illusion that the company was supposed to operate on a profitable basis. It fulfilled a different social function: as a public firm, it was expected to lose money.

Privatization and Microeconomic Restructuring

As I argue in Chapter 3, the privatization of Aeroméxico marked the beginning of a new phase of privatizations. While novel in many of the particulars, the process of privatizing Aeroméxico provided valuable lessons for the state. To begin, bankrupting Aeroméxico allowed the government to centralize control over the firm. Bankruptcy achieved informally what would later be institutionalized through the "resectorization" of firms under the control of the Ministry of the Treasury, placing the firm under the direct control of banks and subordinating the interests of state actors connected to the firm to those connected to finance.

In addition, the bankruptcy of Aeroméxico marked the beginning of a more direct offensive against public-sector workers and showed that a major firm could be privatized with only muted protest from workers and public administrators. The meager resistance that opponents of privatization mustered was more than compensated by the vocal support from members of the private sector. The ideology of privatization espoused during the Aeroméxico privatization—no more public support for insolvent firms—would soon appear in President Salinas's National Development Plan.

AEROMÉXICO: EARLY LESSONS IN THE ART OF PRIVATIZATION

In early 1988, the press reported on Aeroméxico's shaky financial condition and suggested that Aeroméxico and Mexicana might be merged. Aeroméxico director, Rogelio Gasca Neri, however, denied the existence of any plans for selling the company. The director of Aeroméxico affirmed the importance of the firm to the public sector stating that Aeroméxico "has been fundamental to the modernization of the country for 53 years" (*Expansión* 1988, 39; Valencia 1988). Aeroméxico's administration did, however, attempt to restructure the company and announced its plans in early March. Gasca Neri introduced a plan to retire and possibly sell thirteen of Aeroméxico's forty-three planes. The thirteen planes were all older-model DC-8s and DC-9-15s and Gasca Neri argued that these planes alone accounted for 40 percent of all cancellations and delays and that they required 50 percent of Aeroméxico's human resources in maintenance (Cardoso 1988). Restructuring would also take the form of canceling Aeroméxico service on its least profitable routes.

The plans for restructuring the firm immediately drew the fire of Aeroméxico workers and even from some private-sector representatives in cities to which Aeroméxico would stop flying. Aeroméxico workers attacked Gasca Neri, insisting that the firm could be profitable if not for the succession of political appointments of Aeroméxico directors. In the de la Madrid *sexenio* alone Aeroméxico had three different directors (Corro 1988b, 30). The labor unions representing Aeroméxico's pilots and ground workers—ASPA and the National Union of Aeroméxico Technicians and Workers (SNTTAM), respectively—offered to buy the planes that were slated for retirement and even to buy the firm outright (Calleja 1988). Business leaders in the city of Chetumal, Quintana Roo, protested the announcement that service to the southern capital would be canceled, arguing that Chetumal would be set back forty years in terms of communication (Aguileta 1988).

As the restructuring went forward, ground workers from the SNTTAM and the flight attendants' union, the Union Association of Flight Attendants (ASSA), set a strike date for April 12, accusing the company of violating the collective bargaining agreement (Corro 1988b, 30).[7] In an attempt to avert the strike, the minister of labor met with ASSA leaders and convinced the flight attendants to delay the strike (*Expansión* 1988, 38). The minister of labor also alerted pilots that a strike would lead to the bankruptcy of the company (Interview 54). Ground workers, however, voted to go forward with the strike and on April 12, Aeroméxico's 223 daily flights were suspended. Three days later, the SCT declared Aeroméxico bankrupt (Corro 1988b, 30).[8]

7. Mexican law requires labor unions to file a formal petition (*emplazamiento*) stating their plans to strike and the period during which the strike will take place prior to taking any such action (see Middlebrook 1995, 69).

8. It is very possible that the state sought a pretext for bankrupting Aeroméxico and encouraged ground workers to strike. The leader of the ground workers' union, Melchor Montalvo, sat on the board of directors of Aeroméxico and was reported to have agreed in early

The declaration of bankruptcy at Aeroméxico suddenly swept aside all the difficulties inherent in restructuring the firm. Bankruptcy released the firm from labor obligations under its various collective bargaining agreements and allowed management to fire workers. Bankruptcy also centralized control over the company during the process of restructuring. A federal judge ratified the decision of the SCT to bankrupt the firm even while stating that he was uncertain exactly how much debt Aeroméxico had. The same judge placed the firm in receivership, naming the National Bank of Public Works and Services (Banobras) as the trustee of Aeroméxico and Nacional Financiera as the "provisional auditor." Although Aeroméxico actually owed part of its debt to Nacional Financiera, the judge determined that money earned from the sale of Aeroméxico's assets would be used first to pay indemnization to the company's workers (Guerrero and Viale 1988).

Tandon attempts to calculate the welfare costs and benefits to workers as a consequence of Aeroméxico's privatization and argues that the loss in welfare to workers was "negligible because of the sizeable severance payments that workers received." Tandon also argues that workers stood a good chance of finding work following the bankruptcy and that "the welfare loss to the worker is not the full wage loss because he or she has additional leisure" (1992a, 32 n. 21).

It is important, however, to distinguish between the different groups of workers at Aeroméxico. As Table 7 shows, between 1987 and 1989, more than 60 percent of all Aeroméxico workers lost their jobs, with employment at the firm falling from 11,505 workers to 4,218. Pilots, however, apparently did not fare too badly under the reorganization. According to one official from the pilots' union, all Aeroméxico pilots were given severance pay from the company even though about half of the eight hundred pilots were rehired almost immediately with the new company. Within three years, all the pilots were working again (Interview 61).

With fewer technical and safety-related restrictions on the use of gate agents, ground crews, ticket agents, and other ground personnel, fewer of these workers were rehired. Where Aeroméxico had previously used ground crews of sixty workers, under the receivership of Banobras it began operating with thirteen (Tandon 1996, 227). Moreover, the pilots' group had enough economic clout to buy itself a seat at the negotiating table with Banobras and began bargaining with Banobras to purchase a stake in a

---

1988 to join forces with Aeroméxico management to improve services. Once the firm was declared bankrupt, Montalvo proposed the elimination of 60 percent of Aeroméxico's groundworkers (see Martínez Vargas 1988, 2).

restructured company. Neither the ground workers nor the flight attendants had enough resources to purchase part of the firm (Interview 54).

Unlike most of the sales of public firms discussed so far, the sale of Aeroméxico was a liquidation of the company's assets and reorganization under a new corporate identity, not simply a sale. The process of reorganizing Aeroméxico through bankruptcy, like bankruptcy in other countries, allowed the company to fend off creditors, suspending some of its payments while the restructuring took place. Within eleven days of the announcement of bankruptcy, the company formerly known as Aeronaves de México was already flying again (Martínez Varga 1988).

Banobras quickly liquidated all of Aeroméxico's contracts except for its aircraft rental agreements and began to rebuild the company from the ground up. The company resumed operations with five aircraft and began flying to thirteen cities for which Aeroméxico had been the sole air carrier. On May 17, Banobras took out full-page advertisements in Mexico City daily newspapers to announce that it would expand service still further and reported that the limited and restructured operations of the company were profitable (Sindicatura 1988). Slowly expanding its operations, the company increased the number of aircraft to eleven and the number of routes it flew to twenty-one by June. By October, Banobras had rehired twenty-four hundred workers, was operating twenty-nine planes, and had plans to hire an additional seventeen hundred workers. More important, Banobras was able to report profits of 5 billion pesos in the four months that it had operated the company (Vieyra 1988).

The final step in the makeover of the company was to rename it. Simply restoring the company under its previous name would have also restored the union contract. Therefore, on September 7, Aeronaves de México became Aerovías de México. The new Aeroméxico started up with a capital stock of about forty-four thousand dollars: 35 percent from the pilots' union and the remainder from Banobras.[9] It also began its life with significantly more flexible labor contracts. Pilots took a 10 percent pay cut and agreed to work longer hours. Flight attendants signed a contract described as a "similar efficiency-enhancing agreement" (Tandon 1996, 227). The union for ground workers was dissolved entirely and a new union, the National Union of Workers of Airlines and Related Companies, took its place.

9. Banobras loaned ASPA part of the capital necessary to purchase a share of the airline. The pilots' union then raised money to pay the loan through mandatory deductions from workers' paychecks, optional purchase of stock by individual workers, and union funds (Interview 54).

With its obligations to both creditors and workers restructured, the company now operated with fewer than half the workers that it previously employed per aircraft. Almost overnight, Aeroméxico was transformed into an exceptionally attractive candidate for privatization. The state wasted little time in putting the new, lean company up for sale. Less than six months after declaring the old Aeroméxico bankrupt, on September 11, the state announced the offer to sell the new Aeroméxico. Minimum prices were set and bidders could either purchase all the assets of the firm for $337 million, or buy the firm's assets except for its aircraft and engines for a minimum price of $162 million (Tandon 1996, 228).

Only two bids were offered for Aeroméxico, one of which failed to meet the minimum price and requested favorable tax treatment. The winning bid was made by a group of investors that included Mexico's old and new elite as well as capitalists with experience in the area of transportation. Grupo Dictum, the corporate name of the winning bidders, included Miguel Alemán Velasco—owner of the regional air carriers Aeromar, Aeromorelos, Aerocalifornia, and Aerocaribe—José Serrano Segovia of the shipping and trucking empire Transportación Marítima Mexicana, and Enrique Rojas Guadarrama, described by the daily newspaper *Uno Más Uno* as "an entrepreneur with a renowned trajectory in the transportation sector" (*Uno Más Uno* 1988a, 1988c). Gerardo de Prevoisin Legorreta, the owner of Mexico's largest reinsurance company, was also part of the group (Tandon 1996, 228).

Interestingly, Grupo Dictum also included the still-state-owned bank Bancomer. Why a state-owned bank was allowed to participate in the "privatization" of Aeroméxico is not entirely clear. Grupo Dictum also insisted on taking 75 percent of the company, leaving pilots with only 25 percent of the shares. The Grupo reserved the right to inspect and reject any assets that it deemed unnecessary and subsequently did, adjusting the sale price down to 655.2 billion pesos, about $285 million. With proceeds from the sale going to liquidate the liabilities of the old Aeroméxico, the state received no revenue from the sale (Tandon 1996, 229).

THE SLOW DESCENT OF MEXICANA

The speed with which Aeroméxico was bankrupted, restructured, and privatized is all the more impressive when compared with the prolonged process of privatizing Mexicana de Aviación. Almost two years before the bankruptcy of Aeroméxico, in the spring of 1986, the SCT proposed to the Intersecretarial Commission on Spending and Finance (CIGF) that Mexicana de Aviación be privatized. After the SPP approved the proposal, the

SHCP designated Banamex as the agent bank to handle the sale of the firm. Banamex prepared a prospectus and a technical-financial evaluation of the firm, solicited bids in the summer of 1987, and received two, both of which failed to meet the minimum requirements established by the government (Tandon 1992b, 3).

The request for bids was declared "deserted" and the CIGF authorized Banamex to sell 7 percent of the state's holdings in Mexicana on the stock market in an attempt to improve the company's financial situation. In a shareholders' meeting held on April 25, 1988, Banamex announced that it had sold 7.8 percent of Mexicana stock, leaving the state holding a slim majority of the firm: 50.2 percent (Ortega Pizarro 1988, 10).

At first glance, Mexicana appeared to present fewer problems for privatization than Aeroméxico. Given its long history as a private firm, Mexicana was not expected to serve public ends and, therefore, did not have a built-in constituency opposed to its privatization. In addition, Mexicana had a track record of profitable operation and had historically dominated the most lucrative routes. Nonetheless, Mexicana presented its own unique problems for public officials who had already divested more than three hundred public firms.

To begin with, the airline was larger than most companies privatized by the government to date. More important, the demand for air transportation had fallen dramatically in the mid-1980s. In 1987, Mexicana lost money for the third consecutive year (see Table 7). Although Mexicana showed a profit of 104 million pesos in 1988, the sudden improvement in the company's bottom line is probably an aberration caused by the bankruptcy of Aeroméxico. Attempts by the government to stabilize the macroeconomy and maintain support among workers also had negative effects on Mexicana. The state-owned petroleum company Pemex increased the price of fuel 230 percent during 1987 and salaries of Mexicana workers increased 116 percent. And the government's attempt to promote exports by devaluing the peso made it more difficult for Mexicana to pay off dollar-denominated debt worth $291 million.

Mexicana attempted to cushion the effect of these blows, taking out loans to cover the cost of fuel and postponing payment on its foreign debt. Mexicana's director, Manuel Sosa de la Vega, insisted that Mexicana was far from bankrupt and that it would be transferred to the private sector (Segura 1988).

The government planned to solicit a second set of bids for Mexicana in early 1988, an announcement that was warmly greeted by representatives of the major business associations in Mexico (Novedades 1988). The presidents of the Confederation of Mexican Employers (Coparmex) and the National Confederation of Chambers of Commerce (Concanaco) praised the government's decision to sell Mexicana. At the same time—as if dictating the policy that the next administration would follow—these private-sector representatives warned the government that financial and labor problems at parastate firms would need to be dealt with before the firm could be privatized. The message was clear: the government should rid itself of inefficient enterprises, but only after they had ceased to be inefficient (*Uno Más Uno* 1988b).

Bidders for Mexicana also made specific requests for reorganization before they would be willing to purchase the company. First, they wanted the company's labor agreement restructured and requested that as many as three thousand workers be fired. The cost of reducing the workforce was estimated at $45 million in severance pay that the government would have to absorb. Second, the potential investors wanted the government to restructure Mexicana's debt. One bidder even suggested bankrupting Mexicana in order to allow a debt restructuring (Tandon 1992b, 4).

The sudden declaration of bankruptcy at Aeroméxico placed the reorganization and sale of Mexicana on a back burner. Mexicana shares were withdrawn from the stock market to prevent speculation and the second round of bidding for Mexicana was postponed. The turmoil caused by the Aeroméxico bankruptcy, however, may have actually prevented a more fundamental restructuring at Mexicana because Mexicana received a temporary infusion of business on routes that it had previously shared with Aeroméxico. Barely four days after the declaration of bankruptcy at Aeroméxico, Mexicana announced that its flights were overbooked through June and that demand on routes covered by both airlines had increased by 90 to 100 percent (Ley and Velasco 1988).

By August, with Aeroméxico's restructuring well under way, Mexicana was once again put up for sale. On October 28, the government received a bid that appeared to meet all the requirements established for the sale of the firm. Grupo Serbo offered $200 million for the company; submitted a plan to restructure Mexicana financially, increasing the firm's capitalization by $60 million; promised to modernize and increase the size of the Mexicana fleet with the assistance of Air France, Lufthansa, and Mercedes Benz's aerospace division, Dornier; and proposed restructuring of the work process through negotiations with Mexicana workers. All these proposals for acquiring Mexicana were accompanied by a deposit of 10 billion pesos—approximately $6.5 million—required by the government to participate in the bidding (Campa and Corro 1988, 12).

Unfortunately for the public officials who were attempting to complete the privatization of Mexicana before the Salinas administration took office in December, Grupo Serbo's president, Sergio Bolaños, was a close associate—and, according to some accounts, a front, or *prestanombre*—of the leader of the petroleum workers' union, Joaquín Hernández Galicia (Waterbury 1993, 252). Hernández Galicia had already offended incoming president Salinas by throwing his support behind Cuauhtémoc Cárdenas in the presidential race—very likely part of the reason he was jailed on charges of corruption within the first month's of Salinas's presidency (see Chapter 3). The government would not permit Hernández Galicia to purchase the company and, once again, declared the bidding deserted.

Upon taking office in December, the new administration reassessed its strategy for selling Mexicana. In February 1989, the CIGF appointed a new agent bank—Banco Internacional—to handle the sales prospectus and organize the sale, and devised a new sales strategy. A controller was created to hold the stock of Mexicana during the sales period and a new call for bids was issued in the spring of 1989. In August 1989, a consortium of Mexican and foreign capitalists, Grupo Falcón, was named the winner of the bidding.[10]

Like the sale of Aeroméxico to Grupo Dictum, the sale of Mexicana earned the government zero dollars. Grupo Falcón promised to invest $140 million in Mexicana in exchange for the controlling share of the company, or 25.1 percent of the stock. Grupo Falcón also promised to invest $3 billion over the following ten years, $1.1 billion of which would be spent in Mexico and would create 21,500 new jobs (Tandon 1996, 226). And while the government retained slightly more than 40 percent of Mexicana's stock immediately following the privatization, the Ministry of the Treasury established a trust in Banco Internacional in which 25 percent of the shares were held to ensure Grupo Falcón's control over the firm. Falcón retained the right to purchase the remaining shares at the original price through 1992, which, in August 1992, it did (Tandon 1992b, 5; SHCP 1994a, 87).

Grupo Falcón was made up of a loose alliance led by Mexican investors with the support of foreign capitalists, including Chase Manhattan Bank, Drexel Burnham Lambert, and Sir James Goldsmith. The 25 percent, controlling share of Mexicana was divided almost in half with the Mexican investors owning a slight majority of the shares—12.75 percent. The foreign investors played a largely passive role in the subsequent operation of

---

10. This account of the sales process is based largely upon the official account contained in SHCP (1994a, 84–87).

the firm. Of the Mexican group, Pablo and Israel Brener owned the plurality of the Mexican stake in the new company. Their interest in Mexicana appeared to be linked to their ownership of the Las Hadas resort in Manzanillo and the Camino Real hotel chain, both of which they had acquired from the government earlier in the 1980s (Tandon 1996, 225). By combining Mexicana with these tourist hotels, the Brener family hoped to tap into the growing tourist market by offering packages to consumers that included both hotel accommodations and air transportation.

Although Grupo Falcón was not able to rid itself of Mexicana employees as rapidly as Aeroméxico had, Table 7 shows that between 1991 and 1994 the new owners cut the workforce by almost 40 percent, from 11,289 workers to 6,824. As with Aeroméxico, it appears that ground workers and flight attendants were hit the hardest by these cutbacks. In a meeting with the general secretaries of the three major Mexicana unions following the privatization in 1989, Minister of the Treasury Pedro Aspe is quoted as saying "Sirs, either you make Mexicana profitable, or we make Mexicana profitable" (Leyva, Mecalco López, and Mendoza Molina 1998, 63).

Aftermath: The Decentralization and Recentralization of Civil Aviation

In theory, simply transferring the ownership of firms from public to private hands should alter their operation. As studies of privatization have concluded, even in monopoly industries, "ownership matters" (Ramamurti 1996, 38). At the same time, the discussion of the historical development of civil aviation in Mexico above indicated that the macroeconomic environment exercised a decisive impact on the microeconomics of Mexico's airlines. Without macroeconomic reform, it is unclear how much privatization alone would alter Mexico's civil aviation sector.

Deregulation of the domestic market and the reduction of barriers to trade with external markets are essential macroeconomic reforms associated with neoliberal policies. Both deregulation and international liberalization should create increased competition among firms, allowing price to exercise a stronger influence on allocations and making profitability the central focus of corporate organization. The following section looks first at changes in the international market within which the privatizations of Aeroméxico and Mexicana occurred, followed by an examination of changes in the domestic market. Deregulation did increase price-based competition within civil aviation. However, the price wars that erupted following deregulation created chaos in the industry and a subsequent consolidation of Mexico's major carriers into larger corporate structures and international alliances.

UNCERTAINTY AND COMPETITION IN THE INTERNATIONAL MARKET

Important changes in the international airlines industry were already well under way by the late 1980s.[11] Even apart from policy changes governing the relationship between Mexico and the international economy, there were other developments in the international economy that had clear repercussions for civil aviation in Mexico. The timing of Iraq's invasion of Kuwait in 1990 followed by the Gulf War in 1991 was particularly damaging for the recently privatized airlines, driving up the costs of fuel and reducing the number of tourists traveling on international flights.

In late January 1988, the SCT along with Mexico's Ministries of Foreign Relations and Tourism renegotiated the Bilateral Air Convention with the Departments of State and Transportation in the United States (Salmon 1988; Hernández Vélez 1988a, 1988b). The convention allowed U.S. airlines greater freedom of operation in Mexico. Although Mexican airlines received reciprocal rights to operate in the United States, neither Aeroméxico nor Mexicana was in a position to compete with the major U.S. carriers. Through the early 1990s, Mexico entered into bilateral agreements with a growing number of other countries. In 1988, Mexico had bilateral agreements in effect with eighteen countries. By 1994, that number had grown to thirty, and by 2000 Mexico was a party to thirty-six bilateral agreements on civil aviation (Pérez Escobedo 1995, 520; SCT 2002).

Although chapter 12 of NAFTA excludes civil aviation from the trade agreement, the growing number of bilateral agreements appears to have undermined Mexico's position in the international market. Table 9 shows the number of carriers, foreign and domestic, operating regular, scheduled service in the Mexican market between 1980 and 2000. During the first half of the decade of the 1980s, the number of carriers, both foreign and domestic, declined. While thirty-four foreign airlines flew to Mexico in 1980, by 1984, only twenty-six foreign carriers operated in the market. Ten years later, in 1994, the number of foreign carriers in the Mexican market grew

11. Although Tandon (1992a, 8) asserts that "deregulation of the airline market took place in July 1991," well after the privatizations, he overlooks important changes that had already begun to transform the structure of the market: in particular the renegotiation of the bilateral agreement with the United States in 1988; the elimination of exclusivity on routes with sufficient demand, in the same year; and the growth of charter air transportation.

back and surpassed its 1980 level slightly, with thirty-five carriers flying to Mexico, the same number it had in 2000.[12]

As noted above, the United States is the single most important market for international air travel to and from Mexico. Beginning in 1990, U.S. carriers have taken a growing share of the total market for air transportation to and from Latin America including a growing share of the Mexican market. In 1990, U.S. airlines carried just less than 50 percent of passengers to and from Latin American countries; six years later, they carried 57.9 percent of the passenger traffic. In Mexico, U.S. carriers increased their share of the market to and from Mexico from 53.4 percent in 1990 to 61.2 percent in 1996 (Booth and Garvett 1998, 53).

The loss of international market share by Mexican carriers can also be seen in the broader figures showing the percentage of all international passenger transportation carried by Mexican and foreign airlines. Table 10 shows that even in the highly regulated environment of the mid-1970s, Mexican airlines only carried slightly more than 40 percent of all international traffic into and out of the country. Although this figure rose during the crisis years to 54 percent by 1986, the percentage declined again beginning with the year of Aeroméxico's privatization and the renegotiation of the U.S.-Mexico Bilateral Air Convention. By 1994, the international air traffic carried by Mexican airlines fell to a low of 35.6 percent before recovering to 39.6 percent by 2000.

Charter flights also began undercutting Aeroméxico and Mexicana in the international segment of the market. Although not reported in statistics on regular scheduled international service, international charter flights represented a growing percentage of the total international market in the 1990s. From fewer than a million passengers transported in 1990—about 11 percent of the market for scheduled international service—international charter flights grew to 3.8 million passengers by 2000, or 23.8 percent of the market for scheduled service (SCT 1996a; 2002).

GROWING COMPETITION IN THE NATIONAL MARKET

In addition to the expansion of competition from international carriers, "National Air Transportation System Guidelines published in 1988 eliminated exclusivity on routes where there was sufficient demand to justify service by

---

12. Mexico was hardly alone in its privatization efforts. Argentina, Bolivia, Chile, Ecuador, Guatemala, Nicaragua, Paraguay, Peru, Uruguay, Venezuela, and the Brazilian state of São Paulo all privatized a state-owned airline in the late 1980s and early 1990s. Booth and Garvett (1998, 55) note that "Latin America is the only region in the world, other than the United States, that has a 100 percent privately controlled commercial airline industry."

two or more airlines" (Sánchez and Corona 1993, 176). Although the General Directorate of Fares still governed pricing in the sector, these regulations were loosened to permit airlines to give discounts within certain limits, depending on market conditions.

In late May 1988, with Aeroméxico still in receivership, the SCT issued a communiqué in which it outlined the goals of "structural change" that would lead "Mexican society toward modernity in its air transportation" (Vera 1988, 14). Citing the international trend toward privatization and deregulation of the airline industry, the SCT emphasized the need to decentralize the industry: "[T]he accelerated growth of the trunk airlines Aeroméxico and Mexicana, which exclusively dominated national markets, inhibited the development of other airlines, particularly at the regional and local level." In its Governing Project for the National Air Transportation System, the SCT argued for the need to promote three types of airlines: trunk, regional and feeders (Vera 1988, 14). The Mexican government began opening up concessions for the private sector to operate regional carriers in the wake of the Aeroméxico bankruptcy.

Aerocalifornia—a regional carrier founded in 1960 and operating in the northwestern region of Mexico—was able to capture a growing share of the market, upgrading its fleet of single- and twin-engine light aircraft to include some small jets (Cintra 1996; Vera 1988, 14). TAESA, a regional carrier created in 1987, offered service on lucrative routes that Aeroméxico and Mexicana flew, breaking into the Mexico City–Monterrey and Mexico City–Tijuana routes and driving down prices. Price wars eventually led TAESA to offer a Tijuana–Mexico City round trip of ninety-nine new pesos, approximately thirty-two dollars at the time (*Expansión* 1993a, 90). Although Tandon (1996, 223) shows that air fares were artificially depressed by heavy regulation prior to the deregulation of the market, nothing in the predivestiture period came close to thirty-two dollars for a round-trip flight of more than a thousand miles.

Table 8 shows that in the aftermath of privatization and deregulation, the number of passengers traveling on domestic flights began to grow rapidly in 1989, more than doubling from the depths of 1988 to 18 million passengers by 1994. With the peso crisis of 1995, the number of passengers dropped once again. Table 9 shows the growing number of firms operating in the domestic market in the immediate postprivatization period. After peaking at nineteen firms offering scheduled service in 1995, the number of firms drops back to twelve by the year 2000.

Despite the spike in demand for domestic air travel at the beginning of the 1990s, the increased competition prevented the recently privatized airlines from earning profits. Mexicana actually continued to lose money in the first three years of the decade, while Aeroméxico registered anemic profits in 1989, 1990, and 1991 and then lost money in 1992 (see Table 7). Part of the growth in the number of travelers on domestic flights may have been artificially induced, an outgrowth of price wars that broke out on Mexico's most important routes.

Nonetheless, Aeroméxico and Mexicana clearly began to lose their dominant position in the domestic market following privatization. In the late 1970s and early 1980s, Aeroméxico and Mexicana combined never carried less than 78 percent of the total number of passengers flying on domestic routes. As smaller airlines collapsed during the crisis of the mid-1980s, Aeroméxico and especially Mexicana captured a growing proportion of the market. Between 1986 and 1990, the two airlines combined never carried less than 88 percent of all passengers flying on domestic routes. As Table 8 shows, however, new airlines and even established, regional carriers began cutting into the market share of the two major airlines following privatization. Aerocalifornia increased its share of the domestic market in the late 1980s and early 1990s. New companies, especially Saro, TAESA, and Aviacsa all grew as Aeroméxico and Mexicana dropped from carrying a combined total of 90.5 percent of all domestic passengers in 1989 to carrying only 68.8 percent in 1995.

Finally, although the smaller airlines were supposed to serve as feeders for Aeroméxico and Mexicana so that the trunk carriers could focus on the more profitable segments of the market, the smaller companies also acted as feeders for international airlines that were increasing their presence in Mexico (Hernández Vélez 1988a).

ESCAPING THE MARKET THROUGH CONSOLIDATION AND ALLIANCES

The burst of competition that followed the privatization of the two major carriers was suddenly halted by a new economic crisis in 1995. In that year, the number of passengers on domestic flights fell by 19.2 percent from the previous year and GDP fell by more than 5 percent in three consecutive quarters. Severe as the crisis of 1995 was, however, Mexico's newly privatized airlines were already in trouble before the more generalized economic crack. The business magazine *Expansión* (1995, 28), reported that by the end of 1994 both Aeroméxico and Mexicana were technically bankrupt. Competition in both the international and domestic markets had taken its toll, and in 1995 the government stepped in once again to rescue the struggling firms.

Yet even before the Mexican state stepped in to rescue the recently privatized firms, Aeroméxico and Mexicana had taken steps to defend themselves from the harsh effects of market allocations. In 1990, Mexicana participated in privatization on the buyers' side, integrating its operations vertically by purchasing the jet engine maintenance company Turborreactores from the state for $18.8 million (SHCP 1994a). In 1993, still loaded down with debt—Mexicana owed $149.5 million to Mexican banks and another $400 million to the companies from which it was renting aircraft—Mexicana managed to soften these particular budget constraints through renegotiations with its creditors, restructuring its debt over seven years with a two-year grace period; Mexicana employees also agreed to relieve the airline of its contractual obligations, accepting a wage freeze that allowed Mexicana to increase its capital by $75 million (*Expansión* 1993a, 91, 92).

In the spring of 1993, Aeroméxico made a dramatic move to limit competition by acquiring 80 percent of Mexicana. To comply with regulatory requirements that at least two airlines operate on all routes, the two companies were not actually merged and were supposed to operate as separate carriers. By the end of 1993, however, Mexicana and Aeroméxico announced a number of measures designed to limit the debilitating competition. Operating under the same ownership, the two companies began eliminating discounts on routes that they served exclusively, and also began sharing maintenance and training facilities and fusing cargo operations (*Expansión* 1993a, 91; Tandon 1996, 234).[13]

Still, the losses at Mexico's recently privatized firms continued to mount. The final blow to the attempt to carve a market out of Mexico's air transportation sector came in the fall of 1994. Gerardo de Prevoisin Legorreta, chairman of the board of directors of the two firms and chief executive officer of Aeroméxico, resigned in September 1994 under accusations that he misappropriated $50 million of Aeroméxico's assets and transferred an additional $220,000 to a company wholly owned by his attorney (see Cintra 1996, 11, 47). Aeroméxico filed a civil suit against de Prevoisin in an attempt to recover the money, but the legal wrangling and the specter of malfeasance by the company's top management soon led to additional legal proceedings from Aeroméxico's suppliers and creditors. A supplier of leased aircraft went to court, demanding $14 million from Aeroméxico and another $12.8 million from Mexicana for alleged defaults on its lease agreements. A civil suit was also filed against Mexicana in Geneva, Switzerland, for breach of a purchase agreement for a flight simulator (47, 48).

13. Aeroméxico also attempted to expand into foreign markets and other niches of the Mexican market. The company operated an express delivery service, Aeromexpress, and in early 1993 acquired 47 percent of the equity capital of Aeroperu, Peru's flagship airline (Cintra 1996, 16).

With the civil aviation sector in disarray, the government stepped in, as it had when it nationalized Aeroméxico in 1959 and Mexicana in 1982. This time, however, the state bailed out the firms financially while keeping its distance from the operational concerns of the two airlines. The federal government encouraged the creation of a holding company for the two airlines that would maintain their control in private hands. In 1995, Grupo Cintra was incorporated to own the stock of the two companies and their subsidiaries while directing the restructuring of the companies' operations and debt. Creditors to Aeroméxico and Mexicana who controlled Grupo Cintra quickly turned over their bad debt to the federal government's bank restructuring agency, Fobaproa. Yet even with the government owning 70 percent of Cintra shares in 1999, the company is effectively under the control of the banks and creditors from whom Aeroméxico and Mexicana borrowed (*Expansión* 1999a, 38; Wills 1999, C4).

Most of the pieces of the civil aviation sector in Mexico—spun off as independent, profit-making entities in the late 1980s and early 1990s—were now reassembled under the control of Grupo Cintra. By the end of 2000, Grupo Cintra owned Aeroméxico, Mexicana, and three regional airlines—Aerocaribe, Aerocozumel, and Aerolitoral. In addition, Cintra owned the express cargo delivery service Aeromexpress; a baggage-handling and aircraft dispatch company, Ground Support Services; a training center for airline employees; and the Sabre reservation system, which links the company to fifteen hundred travel agencies in 109 cities throughout the country; Cintra also holds 29 percent of the stock of the jet engine maintenance company Turborreactores (Cintra 2000).

Emphasizing competition on a global scale, between Mexico and foreign competitors, officials from Mexico's major airlines now speak of the dangers of too much competition in the domestic market. Mexicana's director of planning argues that important synergies exist between Mexicana and Aeroméxico, permitting the rationalization of routes and freeing up planes for the development of new routes. According to the director, the competition is now "more ordered" (*Expansión* 1999c, 45). Seeking to insert Mexico's airlines into the international market more successfully, both Aeroméxico and Mexicana have entered into alliances with international firms. Mexicana is now part of the Star Alliance, a multinational group of some of the world's largest carriers who have agreed to coordinate their operations and provide "seamless service" to consumers (*Expansión* 1999b, 30). In addition, Aeroméxico is attempting to create international alliances with Delta Air Lines and Air France, and has sought membership in a global alliance that competes with the Star Alliance (*Expansión* 1999a, 40).

Smaller regional airlines, travel agents, and representatives of the hotel and restaurant industry in some of Mexico's tourist destinations, however, are less enthusiastic about the operations of Grupo Cintra. They argue that the market is now too controlled. The government agency charged with preventing monopolistic practices in the economy, the Federal Competition Commission (CFC) sued Cintra in 1999, arguing that the holding company manages the two companies as one, has raised prices, and has reduced service on routes that are not served by another carrier, damaging the tourist industry in resort towns such as Huatulco, Ixtapa, and Cozumel (Wills 1999, C4).

Discussion

Introducing market reforms into civil aviation in a developing country such as Mexico poses a dilemma for policy makers. For a variety of reasons the sector does not appear to lend itself to allocating resources through the mechanisms of price, profitability, and competition. The limited size of the market, the strength of international carriers, a tendency toward internecine struggles within the market, and the inability to contain utility-maximizing strategies of capitalists within the bounds of the law—much less the developmental needs of society—all contributed to market failure in this sector. Even with domestic routes reserved for national carriers, the privatized airlines found it difficult to turn a profit.

According to Booth and Garvett (1998, 56) new owners of recently privatized airlines in Latin America "have managed to instill a new, different, profit-oriented, marketing-driven culture in their airlines." Yet the problem of linking the production of socially useful goods and services to private gain and the problem of containing individual profit-maximizing strategies within the "rules of the game" are clearly illustrated by the case of civil aviation in Mexico. Even Tandon, a strong supporter of privatization and liberalization efforts, notes that "it is unclear whether free competition is sustainable in such a market" (1996, 237).

Nor is there much support in this case study for the contention that state autonomy enabled the Mexican government to rid itself of inefficient public firms. Mexico's major business organizations were vocal in their support of privatization. Although proponents of privatization have argued that the restructuring of inefficient state firms is based on a purely technical cost-benefit calculation, the restructuring appears to point more to a shift in the state's base of support away from workers and public-sector officials and toward financial elites. After absorbing the losses of the two firms and effectively giving away both Aeroméxico and Mexicana, the state once again stepped into the breach to bail out the firms and their creditors when they ran into trouble.

The motivation behind the government's willingness to intervene in civil aviation can probably be traced to the repercussions that bankruptcy would have had in other sectors of the economy. The most significant consequences of a bankruptcy in the airlines would have been for the same political coalition that had encouraged the privatization in the first place: Mexico's major financial groups. In its 1996 public offering, Grupo Cintra lists the holders of Aeroméxico and Mexicana debt. It reads like a Who's Who of Mexico's recently privatized banking system: Banco Mexicano, Banco Serfin, Bancomer, Banamex, Banco Inverlat, Banco del Centro, Bital, Banpaís, Multibanco Mercantil Probursa, BanCrecer, Banco Obrero, Arrendadora Sofimex, and Banco Union. International creditors included Chase Manhattan Bank, Internationale Nederlanden, Capital Corporation, the Bank of New York, Royal Bank of Canada, and Swiss Bank Corporation. The airlines also still owed money to the Mexican state (Cintra 1996, 10).

In 1988, the privatization of the airlines had drawn a warm reception from the most powerful factions of Mexico's business class. In response to the declaration of bankruptcy at Aeroméxico, the president of Concanaco, José Chapa Salazar, stated that "the declaration of bankruptcy was necessary and should set the standard for the treatment of all parastate firms" (*El Día* 1988b). Although the Mexican state has allowed some of the smaller, regional carriers to fail or be absorbed by Grupo Cintra, it has generally been more successful in bankrupting public firms than it has in allowing private firms to fail.

Finally, it is worth noting how important the definition of the market and the unit to which competition should apply can be in debates over the appropriate role for the state. Opponents of Grupo Cintra have argued that its monopoly position in the market impedes competition. The response of Cintra assumes competition at a different level altogether. Brushing aside the importance of competition within the national market, Cintra officials point to the importance of "strengthening Mexican aviation" (*Expansión* 1999c, 45). In order to compete in the international market, it is argued that the Mexican airlines need the continued protection of a captured, domestic market.

FIVE

Positive-Sum Games and the Sale of Telmex

In many ways, the telephone call is a commodity just like a bushel of corn.
—*Peter F. Cowhey and Jonathan D. Aronson, "Trade in Services and Changes in the World Telecommunications System"*

Long distance companies don't want to make it [a long-distance call] a commodity: they want to create a service.
—*Senior telecommunications executive, Mexico City, May 1997, Interview 15*

Government intervention in telecommunications, or telecom, was more intensive and proactive than in civil aviation. As early as the 1880s, the Mexican state placed restrictions on the operation of foreign telephone carriers in Mexico and, at the same time, became the single most important consumer of telephone equipment. Telephone service was considered too important to both national security and economic development to leave to the fate of the market. Private owners, foreign and national, also failed to invest as much in the network as government leaders wanted. And struggles between telecommunications workers and management disrupted other areas of the economy while threatening to undermine the steady expansion of the telecom network that the state envisioned.

The sale of Mexico's monopoly telecom provider, Telmex, in 1989 is one of the great successes of the privatization program. I argue in this chapter, however, that the success of the Telmex privatization was not the result of unfettered liberalization of the telecom market. To the contrary, this sector has been opened very gradually: Telmex's position in various segments of the market was strengthened prior to privatization, the company retained its monopoly over long-distance service for seven years following privatization, and it continues to operate as a de facto monopoly in local service. Although market forces have been unleashed and reinforced in important segments of this market, a number of critical restrictions on the market remain, while new, nonmarket mechanisms have been developed to govern the allocation of resources in this sector.

Following the structure of the previous chapter, I begin with a brief overview of the general characteristics of telecommunications followed by a review of the historic development of this sector in Mexico. In this section I pay special attention to the progressive embedding of telecom services in nonmarket relationships through the late 1970s. At the same time, I will show that increased state intervention in this sector did not entirely eliminate the pressures of the market on allocations: telecom in Mexico remained connected to market forces in a number of crucial areas. I then examine the sale of Telmex and its transformation from a public to a private firm. I conclude with an assessment of the attempt to create a market for telecom service in Mexico, looking once again at where the forces of price, profitability, and competition influence allocations, and where these pressures are restricted.

Telecommunications and the State

AN OVERVIEW OF THE GLOBAL TELECOMMUNICATIONS INDUSTRY

Like the air transportation market, the telecom industry can be divided into two major segments: equipment and service. In the production and development of telecom equipment, private actors have historically dominated the sector. States have influenced the telecommunications equipment market primarily as consumers of large amounts of equipment for public networks and government use.

On the service side, states have played a much more central role in this industry, not only regulating but also owning and operating telegraph and, later, telephone, satellite, and digital-communications networks. The centrality of the private sector on the equipment side and the state on the service side made telecommunications a mixed market of public and private interests from its very inception. Even where the state abstained from outright ownership of telecommunications networks, the natural monopoly that this service represented provided clear justification for heavy state regulation (see Aronson and Cowhey 1988; Mattelart 1994).

In addition to the mix of public and private interests that shaped this industry, domestic and international forces strongly influenced the development of telecom. Telecom services were intimately connected to the project of nation building: they were a strategic asset in times of war, a tool for integrating the national economy, and a source of revenue with which to subsidize other areas of the public sector, usually the postal service. The development of telecommunications service was far too important to leave to the mercy of the market, much less under the control of foreign companies that dominated the industry in the developing world.

The need for coordinating connections between many different national telegraph providers further restricted the extent to which the exchange of telecom service could be organized along market principles. Private and state actors who might otherwise be engaged in economic and political competition with one another were forced to agree, at the very least, on basic technical issues with regard to the connection of different national telecom networks and how the cost of transferring calls from one network to another would be distributed. The need to create rules governing the exchange of international telegraph signals led to the creation, in 1865, of one of the first international organizations, the International Telegraph Union, today the International Telecommunications Union (ITU). International cooperation in this industry was also institutionalized in the creation of the International Telecommunications Satellite (Intelsat).

Thus, even while states sought to shield domestic telecom service from market forces, an international market for telecom equipment developed in which the principal consumers were states themselves. Early industry leaders in advanced industrial countries quickly dominated and diffused the new technology around the globe.

This combination of public and private, domestic and international, and market and nonmarket forces still strongly influences the telecom industry today. Telecom service remains heavily regulated within national economies. There is little dispute that the state still has a right to intervene in the development of this sector, if not to participate directly in its provision. Even where telecom services have been privatized and competition is introduced in domestic markets, the need for interconnection between competing networks forces states to remain involved in the regulation of the sector and requires that firms engage in a high level of coordination among themselves.

The telecom equipment sector remains dominated by private actors who, in recent years, have transformed both equipment and service by linking phone lines to computers. The combination of the telephone with the computer has revolutionized the sector. While telecommunications were once considered a public utility providing a standardized product, digital technology, along with high-speed connections, the growth of data transmission, and the Internet, have all enabled telecom service providers to offer differentiated products and value-added services to go along with basic telephony (see Crandall and Flamm 1989; Petrazzini 1996).

EMBEDDING THE TELECOM INDUSTRY IN MEXICAN SOCIETY

In many ways, Mexico's telecommunications industry exemplifies the historical development of the industry at the global level. While state intervention in telecom services increased steadily throughout the post–World War II period, the industry remained linked to and affected by the international market in equipment. The increase in state intervention in the delivery of telecom services was largely an outgrowth of the failure of private capitalists to develop this vital infrastructure in a way that was compatible with state policy on national development.

Yet even while the state sought to rectify market failure in this sector, it remained extremely deferential to the interests of capital. Even after taking over a majority stake in Telmex, the state continued to accept the leadership of private shareholders and guaranteed high rates of return to shareholders in the company. The state's decision to encourage the consolidation of the industry into a single monopoly carrier, however, led to the creation of a single national union of telecom workers, who had their own interests in restricting the rights of capital over this industry.

Macroeconomic Regulation and the Growth of State Intervention
There was never any question that the state had a right to regulate the provision of telephone service in Mexico and channel allocations toward social use rather than exchange value. Even under the authoritarian laissez-faire policies of Porfirio Díaz toward the end of the nineteenth century, the state played a central role in promoting the development of telecommunications service in Mexico. The state required foreign telephone companies to obtain a concession to provide service; rate hikes were subject to governmental approval; and the state was the most important consumer of the new technology, purchasing phone lines to link the Ministry of War with military offices and linking police stations with the governor of Mexico City and the Interior Ministry in the late 1870s (Telmex 1991, 40).

Within this heavily regulated industry, from the early 1900s through the late 1940s the market for telecommunications service was a foreign-dominated duopoly. Ericsson, a subsidiary of the Swedish company L. M. Ericsson, obtained a concession in 1905 to compete against the already established Compañía Telefónica Mexicana—later known as the Compañía

Telefónica y Telegráfica Mexicana (CTTM)—a subsidiary of ITT. The duopoly resulted in a duplication of service in the lucrative Mexico City market but very little development outside the industrializing, urban center. The competing companies also used different numbering systems and incompatible technologies that prevented subscribers of one company from connecting calls to subscribers of the other company.

The failure of this arrangement to promote the development of the industry—much less the integration of the national territory—led the state to begin increasing its regulatory oversight in the 1930s. Unlike the nationalization of foreign holdings in petroleum, however, the Mexican state gradually assumed a role in developing the telecommunications infrastructure: first supporting the consolidation of a single, monopoly, service provider, then promoting the "Mexicanization" of the company, and finally, acquiring a majority share of the telecommunications company in 1972.

Initially, the state encouraged organizational change in the industry without direct intervention in the planning or administration of telecommunications. In 1936 the Cárdenas administration ordered the two existing companies to interconnect their systems and passed the General Law of Means of Communication and Transportation (Telmex 1991, 96). In 1947 the state encouraged the creation of Teléfonos de México (Telmex), jointly owned by Ericsson and Corporación Contintental and with a very small amount of stock held by Mexican nationals. In 1950, Telmex was allowed to acquire its only competitor, CTTM, and was granted a monopoly on local and long-distance services (Székely and Palacios 1995, 46; Telmex 1991, 110).

During the 1950s, the Mexican government began intervening more directly in this industry, becoming more involved in fundamental issues of planning and administration. In 1953, the Ministry of Communications and the Ministry of the Treasury formulated the first five-year plan for the development of the telecommunications network, in which they sought to increase the number of telephone subscribers by twenty-five thousand a year. To assist Telmex in meeting these targets, the state imposed a telephone-service tax of 5 percent on local service and 10 percent on long distance, with revenues targeted for the improvement and expansion of telephone service (Telmex 1991, 114).

The state also sought to link Telmex more tightly to the national market and shield the company from currency devaluations by promoting the domestic manufacture of telecommunications equipment. Telmex had been hurt by devaluations of the peso in the past because it depended entirely on imported equipment. The state encouraged ITT and Ericsson to jointly form a company for the domestic manufacture of telecommunications equipment, Industria de Telecomunicación (Indetel) (Telmex 1991, 118, 119).[1]

To help Telmex finance expansion of the network the government provided long-term, low-interest loans to Telmex and encouraged the sale of shares of Telmex to the public (Telmex 1991, 117, 120). With this support, Telmex easily surpassed the goals of the first five-year plan. The number of telephones in service grew by more than 30,000 a year during the 1950s—almost tripling the total number of phones in service by the end of the decade. During the 1960s, almost 100,000 new phones were installed each year and during the 1970s more than 340,000 annually, increasing the size of the network from 1.4 million in 1970 to 4.9 million in 1980. Even during the "lost decade" of the 1980s, the size of Telmex's network doubled once again, growing by more than 500,000 phones annually and reaching 10.1 million phones in service by 1990 (see Table 11).

The advance of state intervention in telecommunications culminated on August 16, 1972, when the state purchased a majority of Telmex shares and placed a legal limit of 49 percent on the holdings of the private sector (Xelhuantzi López 1988, 25, 26). The government's decision to extend its control over the firm appears to have grown out of the deepening distrust between the private sector and the administration of president Luis Echeverría, discussed in Chapter 2.

Yet the acquisition of a majority stake in Telmex had little impact on the day-to-day operations of the firm or the developmental implications of its activities. According to one company officer, even after the state acquired a majority of shares in the company, "nothing changed in the way the company operated" (cited in Ramamurti 1996, 74; see also Székely and Palacio 1995, 47, 55). So despite the rapid growth of the network and the persistent intervention of the state, the gap between the number of phones per capita in rural as opposed to metropolitan areas remained enormous. While Mexico City had a density of 22.2 phones per one hundred inhabitants in 1992, the southern state of Oaxaca had a density of 1.9 (SCT 1992: 89).

The role of the minister of communications and transportation was limited primarily to establishing technical standards and granting concessions and permits (Pérez Escamilla 1989, 107, 108, 110). At the same time, the most conservative wings of the federal government with the closest ties to the private sector—the Ministry of the Treasury (SHCP) and the Ministry of

---

1. In 1974 the state became a minority shareholder in Indetel (Casar and Peres 1988, appendix 1).

TABLE 11 Telephones in service, 1880–2000

Year	Telephones in Service	Population	Telephones per 100 Inhabitants
1880	100	9,908,000	<1
1890	1000	11,642,700	<1
1900	3065	13,607,000	<1
1910	16,735	15,160,000	<1
1920	29,005	14,000,000	<1
1930	84,862	16,000,000	<1
1940	92,787	19,000,000	<1
1950	182,173	25,791,017	<1
1960	492,476	34,923,129	1.4
1970	1,459,276	48,225,238	3
1980	4,903,073	66,846,833	7.3
1990	10,103,360	81,249,645	12.4
2000[a]	12,068,993	97,500,000	12.4

Sources: Fox 2002, 3; Meyer and Sherman 1987, 466, 648; Telmex annual reports, 1991, 2000; Universiteit Utrecht n.d.; Zedillo 1997, 243.
[a] In its annual report for 2000, Telmex reports 24 million "Equivalent Lines in Service." This figure includes not only the more traditional measure of wired lines in service, but also wireless, data communication, and Internet access. The figure I use is only the wired service.

Planning and Budget (SPP)—participated in the rate-setting process through Telmex's Expenditures and Finance Commission. Pricing rules developed by the Ministry of the Treasury and approved by Telmex's mixed public-private board of directors were designed to guarantee a 12 percent annual return on capital (Pérez Escamilla 1989, 110; Ramamurti 1996, 75).[2]

Labor Relations and Organizational Change at Telmex
The official chronicle of the Mexican Telephone Workers Union (STRM), also suggests that the nationalization of Telmex in 1972 changed very little in the telecommunications sector. This history of the Telmex workers goes on to argue that the real barrier to directing production toward social use in this sector stemmed from Mexico's dependence on foreign technology. María Xelhuantzi López, the author of the chronicle and an advisor to the general secretary of the STRM, maintains that "[n]ationalism and the new state rectorship were not possible in the telephone sector due to the fact that in structural, administrative, and legal terms, transnational private capital continued to be the decisive factor and the dominant project" (1988, 26).

2. See Pérez de Mendoza (1989, 91) for an overview of the organization of Telmex under state control.

It is not clear, however, that telephone workers were themselves any more interested in the generation of social use in the industry than was Telmex management. Without a doubt, telephone workers wanted a greater share of the profits that were generated in the telecom industry. But the state would prove to be a more daunting obstacle to achieving this than private capital. State policy placed the burden on workers to sacrifice their self-interest for the national interest to ensure the continued expansion of the telecom network regardless of whether it was privately or publicly owned. Insisting on the importance of the sector for national development over private profits could cut both ways in the union's struggles against Telmex management.

But if the growing presence of the state in this sector did not radically transform the operation of the firm, it did have an important impact on workers. To begin with, the consolidation of telecommunications into a single monopoly carrier in 1950 facilitated the unification of workers into a single union. Moreover, the rapid expansion of the telecom network that the state encouraged could not be accomplished without hiring more workers. At the same time, by emphasizing the strategic nature of telecom and the public-goods qualities of this sector, successive governments had a rationale for breaking strikes and limiting the gains of workers.

Thus, the state defended the sector from external threats of competition and price—channeling capital through stockholding in the firm and tax subsidies—while simultaneously protecting firms from the internal threat of workers engaging in profit-maximizing behavior. While the General Law of Means of Communication and Transportation obliged competing telephone companies to interconnect their systems in the 1930s, it also gave the government the right to break strikes in public services if they threatened the national economy (Trejo Delarbe 1990, 328). As in the airlines, the *requisa* allowed the state to use force to remove striking workers from the premises of the firm and maintain operations with replacement workers (see Chapter 4).

As with other national unions, the government sought to create labor peace by manipulating the leadership of the telecommunications labor unions. Although the consolidation of the industry under the auspices of Telmex facilitated the merger of labor unions into a single national industry union, the STRM, a faction with close ties to both management and the state immediately took control of the new union (Telmex 1991, 113). Under this leadership, STRM workers quickly lost pay increases that were tied to increases in phone rates—a benefit that Ericsson workers had previously enjoyed—and the company gained the right to subcontract with companies that were not represented by the STRM (Xelhuantzi López 1988, 12).

When the social conflict of the late 1950s spilled over into telecommunications, an insurgent movement within the STRM gained control of the union for a brief period in 1958. The electoral victory of the Union Democracy Restoration Movement would prove short-lived. The democratic faction led STRM workers in work stoppages rather than strikes in late 1958 in an attempt to avoid the *requisa,* which had been used to break strikes in 1944 and 1952 (Telmex 1991, 104; Xelhuantzi López 1988, 13). But after months of wrangling, the government responded in a manner that was consistent with its broader national development strategy. Using soldiers and police, the state brought in strikebreakers to restore telephone service; the Federal Conciliation and Arbitration Board declared the strike illegal, and; the Confederation of Mexican Workers (CTM) condemned the strike as subversive, counterrevolutionary, and anti-union (Telmex 1991, 121).

The struggle for control of the STRM continued through the 1960s and 1970s. Constant pressure from the democratic faction within the STRM forced even the pro-management union leaders to make demands on the company on behalf of Telmex workers to legitimate their leadership (Xelhuantzi López 1988, 25, 26). In strictly material terms, the pro-government, pro-management leaders of the STRM generally delivered the goods. In the contract revisions of 1974, under pressure from the democratic faction, the secretary general of the STRM, Salustio Salgado, won important concessions from the company. Telmex management also attempted to build worker morale by providing scholarships for children of workers, profit sharing, company stores and sporting events for workers.

As Salgado made more demands on Telmex, however, the company sought to replace him with a more malleable leader. During the attempt to do so, workers from the Democracy Restoration Movement stepped into the breach and wrested control of the union from the pro-management forces. Led by Francisco Hernández Juárez and the more skilled workers who maintained the switching equipment, the democratic faction took control of the STRM with strong support among operators.

Hernández Juárez remains the president of the STRM to this day and his leadership has been the subject of considerable dispute. Dubb (1992) claims that Hernández Juárez has not utilized the antidemocratic measures that are typical of pro-government union leaders but that he has found other means to overcome internal opposition in the STRM. Dubb is particularly critical of Hernández Juárez's willingness to work with Telmex management to increase productivity. Trejo Delarbe (1990, 327), in contrast, points to five strikes that the STRM led between 1976 and 1987 and argues that "no other national union has sustained as many strikes as the telephone workers in so brief a period of time." Although Hernández Juárez was not always in favor of these strikes, he respected the will of the majority of the union workers when they approved them.

Regardless of whether it was due to the leadership of Hernández Juárez or the slow accretion of rights through decades of resistance and struggle at Telmex, telephone workers did win important gains in restricting the property rights of Telmex management. By the mid-1980s, the collective bargaining agreement placed limits upon the rights of Telmex management to subcontract work or use "temporary" workers in place of STRM workers. The agreement also restricted the rights of Telmex management to move workers from one job category to another or from one Telmex facility to another. And the agreement transferred a certain amount of authority over supervision of workers from management to the union (See de Buen 1989; Vázquez Rubio 1989, 59, 60).

In 1986, management and STRM leaders formally addressed the issue of new technology when they met in contract negotiations. Rather than reject Telmex's calls for increased productivity, the STRM leadership agreed to work to improve the quality of phone service while seeking to redefine the concept of productivity so that it would include the quality of work life. At the same time, Telmex made important concessions to the STRM. Workers gained representation on a committee to assess the impact of new technology on safety and health and Telmex agreed not to reduce the number of workers as a result of new technology (Daza, Sandoval, and de la Garza 1988, 67).

Where Is the Market in an Embedded Telecommunications Sector?
Given the various means by which property rights were restricted and budget constraints softened in Mexico's telecommunications sector, it would be tempting to suggest that the market was largely overwhelmed by these social facts: that the embeddedness of the industry made market allocations irrelevant. Clearly in many areas, the organization of production subordinated basic principles of market exchange to state policy, social demands, and technological necessity.

Yet despite the restrictions that prevented the allocation of resources on strict principles of price, profitability, and competition, market allocations still mattered in essential areas of this sector. Certain allocations were simply beyond the control of the state or Telmex. Although the state could encourage domestic production of telecom equipment, Mexico remained dependent on foreign technology that was privately owned. Internationally, hard budget constraints and property rights still mattered. Private, international equipment providers competed with one another to sell equipment to Mexico's telecom companies on the basis of profitability and price. The international market also influenced allocations in the capitalization of Telmex, which was already trading on international exchanges in the early 1960s (Telmex 1991, 122).

The Mexican state also sought to impose a market structure on certain allocations. Telmex retained two equipment suppliers—Ericsson and the French company, Alcatel, which had replaced ITT in the early 1980s—so that it would not be entirely dependent on a single manufacturer. The state also structured domestic-service provision to create marketlike exchanges. However artificial the process of rate setting may have been, Telmex still sold service to customers. Customers paid for service and Telmex attempted to reconcile the revenues that it received with the costs it incurred, along with all the other calculations that accompanied the provision of telephone service.

Telephone rates are generally broken into five different categories:

1. a fixed cost for connection, or "access";
2. a monthly fee for service that usually includes a set number of local calls;
3. a fee for "measured service," local calls that exceed the monthly allowance;
4. an individual charge for each national long-distance call;
5. an individual charge for each international long-distance call.

Between 1970 and 1982, inflation eroded the value of rates in all five of these areas. Rates declined in real terms between 15 percent for international long distance and 73 percent for local measured service during this period (Pérez Escamilla 1989, 112).

The division of telecom services into these different categories is highly artificial. All long-distance calls begin and end on the local network, but the very definition of *long distance* is arbitrary, depending more on domestic and international political boundaries than on any sort of technical standard. The distinction between local and long distance has become still more problematic with the introduction of new technologies such as cellular phones, which operate in local areas and regions but are not considered local service.

Distinguishing between the different categories of services, however, does allow for telecom firms to establish more clearly where revenues are coming from and costs are going. Making these distinctions also makes it clear when the firm chooses not to balance revenues and costs of each service but, instead, spreads costs over the entire firm through cross-subsidization. State planners attempted to increase the total number of subscribers and maximize the access of Mexican consumers to telephones by raising the prices on long-distance service and business subscribers while subsidizing local and residential service.

Thus, while only one percent of Telmex's subscribers used long-distance service in the late 1980s, they generated 30 percent of the company's revenue. In contrast, local phone service accounted for 52 percent of Telmex's operating costs during the same period but made up only 15 percent of revenue (Pérez Escamilla 1989, 113). International long distance was—and remains—even more important for Telmex. Mexico and the United States constitute the second largest international market for long-distance traffic in the world. In 1986, income from international long-distance calls made up 58 percent of Telmex's revenue (see Table 12).

Prices for international long distance are determined through negotiated settlements between governments, a process that has become increasingly complicated as more governments deregulate their national telecommunications networks. Calls sent from one country to another are subject to a "settlement" rate, a charge to the network from which the call was sent for completing the call on the receiving end. If two countries established the same settlement rate and sent calls of the exact same number and length to one another, then there would be no need for any settlement because the amounts due would cancel each other.

However, Mexico has long maintained a higher settlement rate than the United States and the telephone traffic to Mexico from the United States far surpasses the traffic from Mexico to the United States, leaving Mexico with a healthy trade surplus. Cowhey and Aronson (1989, 8) estimate that in 1989 Telmex generated one percent of government revenue. Indeed, in contrast to the argument that public firms place a financial burden on the state, in Mexico the state was a financial drain upon Telmex.

Yet even as a profitable firm, Telmex could not secure outside financing for expansion of the network following 1982, because Telmex could not be assessed independent of the Mexican state. According to one senior-level official who worked in the SCT during the de la Madrid administration, a private company with the exact same books, debt levels, and assets would have been able to secure external financing to upgrade the network. But as long as the Mexican state was in crisis, so was Telmex (Interview 18).

The perception that Telmex's connection to the Mexican state made it a less attractive investment was supported by the fact that Telmex had by the

TABLE 12 Sources of Telmex revenue, 1976–2000 (percentage)

Year	Local Service	National Long Distance	International Long Distance	Other	Interconnection
1976	25	nd	nd	nd	na
1977	24	46	27	3	na
1978	24	46	26	4	na
1979	23	39	33	5	na
1980	20	46	29	5	na
1981	20	44	32	4	na
1982	17	36	44	3	na
1983	14	33	50	3	na
1984	20	34	43	3	na
1985	18	30	48	4	na
1986	13	27	58	2	na
1987	15	30	54	2	na
1988	21	30	47	2	na
1989	21	33	43	3	na
1990	32	37	30	2	na
1991	36	38	24	2	na
1992	40	36	22	2	na
1993	43	34	20	4	na
1994	45	31	20	4	na
1995	42	27	27	3	na
1996	41	29	29	3	na
1997	50.4	25.1	18.9	4.5	1.1
1998	57.1	23	11.1	6	2.8
1999	49.9	25	13.5	5.3	6.3
2000	46.3	25.1	11.1	5.2	12.3

Sources: Pérez Escamilla 1989, 117, table 6; de la Garza and Herrera 1997, 331, table 9.3; Telmex annual reports, various years.
Note: Rows may not sum to 100 because of rounding
nd = No data
na = Not applicable

mid-1980s increasingly become a cash cow for the state. Ramamurti (1996, 75) notes that by 1988, Telmex paid as much as 60 percent of its income in taxes. Yet while the telephone service tax was originally intended to finance increased investment in the telecommunications sector, throughout the 1980s these fees were siphoned off for use by the general treasury, most likely for debt repayment. Because of the large trade surplus that Telmex normally runs with the United States, the firm became an important source of foreign exchange.

Table 12 shows the evolution of Telmex's revenue stream over more than two decades between 1976 and 2000. By 1986, local service accounted for less than 13 percent of Telmex revenue while international long distance made up 58 percent. In anticipation of privatization, the state began raising rates for local use. Following privatization, the steady adjustment of local rates upward has continued, while long-distance rates have begun falling as a percentage of the revenue collected from these different sources.

Microeconomic Restructuring: Goals and Challenges of Privatizing Telmex

Many of the factors that made Telmex attractive to prospective buyers provided equally strong arguments against the sale of the firm. In 1988, Telmex was ranked as the fourth-largest firm in Mexico in sales, behind the state oil company, Pemex, and the subsidiaries of Chrysler and General Motors. Telmex had assets that were valued at more than $4 billion with annual revenues in excess of $1 billion in 1987 (Pérez de Mendoza, 1989, 94). Indeed, the absence of competition within the Mexican telecom market made the firm all the more attractive to potential buyers.

As a profitable, monopoly enterprise, however, Telmex could not be sold using the same justifications that were used to liquidate money-losing firms such as the airlines or railroads. And despite the growing concentration of wealth in Mexico, no single buyer was in a position to purchase so large a firm. Furthermore, although Telmex was the fourteenth-largest telecommunications company in the world, measured by the number of lines installed, the low telephone density in Mexico—on average approximately ten phones per hundred inhabitants in 1988—suggested that there was considerable room to grow in the markets for both basic and new value-added phone services (see Pérez de Mendoza 1989, 93; Ramamurti 1996, 77).

Finally, there were also clear indications from all sides that telecommunications is not a simple commodity. Commentators from the media, the universities, the private sector, and even within the government were all in agreement that the modernization of the economy depended on the development of the national telecommunications system. Thus, proponents of privatization have had to argue that unleashing market forces, allowing individuals to pursue their self-interest, will ultimately improve the general welfare (see Salinas de Gortari 1987; Saunders, Warford, and Wellenius 1994; SCT 1993).

In the run-up to the 1988 elections, Carlos Salinas de Gortari placed special emphasis on the need to "modernize and decentralize" the country's communication and transportation infrastructure. In order to improve the quality of service and meet the demand for accelerated growth at Telmex,

Salinas suggested the possibility of allowing the participation of the private sector (Salinas de Gortari 1987, 13). Yet throughout the presidential campaign of 1988, close collaborators of Salinas denied that Telmex would be sold and insisted that rumors to that effect were pure speculation (*Uno Más Uno* 1988d).

Mexican business organizations appeared divided on the subject. Representatives of Canacintra went on record in opposition to the privatization of Telmex, arguing that it would violate the Constitution (*El Día* 1988a). In contrast, the leadership of Coparmex argued that an accelerated privatization process would contribute to private-sector confidence as well as cleaning up the finances of the Mexican state. Breaking ranks with more traditional, nationalist policies, Coparmex statements supported privatizations even where Mexican capitalists would not be able to buy the firms to be divested (Valero 1988).

Less than a year into his presidency, Salinas announced that Telmex would be sold to private investors. Significantly, Salinas made the announcement in a speech before the annual convention of the STRM. In September 1989, President Salinas listed six preconditions for the privatization of Telmex:

> 1) to guarantee the rectorship of the state in telecommunications; 2) to radically improve telephone service; 3) to ensure the rights of workers and improve their situation, giving them economic participation in the company; 4) to commit to a sustained expansion of the telephone system; 5) to undertake scientific and technological research to strengthen the sovereignty of the country in the area of telecommunications; and 6) to keep the parastate firm under majority control of Mexicans. (*La Jornada* 1989a)

President Salinas chose to privatize Telmex early in his administration for a number of reasons. Because Telmex shares were already traded on Mexican and U.S. stock exchanges, the firm would be easier to price than those without any established value. Selling Telmex quickly would also maximize the benefits to the Salinas administration, allowing it to apply revenues from the sale to debt repayment and political patronage. Moreover, there was symbolic value to selling Telmex as soon as possible: "Salinas needed to send a clear signal to private investors at home and abroad that the country was embarking on a truly new course . . . he believed that privatizing Telmex would send just that kind of signal" (Ramamurti 1996, 77).

The message that Salinas sent with the sale of Telmex went further still. The privatization of Telmex illustrated that the Mexican government could still offer carrots to go along with the sticks that Salinas used so well. In the delicate balance of inducements and constraints maintained by the PRI, the sale of a profitable firm like Telmex weighed heavily on the side of inducements. It provided an opportunity to cement new alliances with the private sector, with labor, and even to a certain extent with dissatisfied consumers of phone services.

As it turned out, not all the goals of the Salinas administration were compatible. Salinas wanted to attract foreign investment, but he also wanted to leave Telmex under the control of Mexican nationals; he wanted to introduce market competition but without undermining the position of Telmex in the market or creating ruinous competition; and he sought to create a regulatory apparatus that would somehow channel the resources generated through private accumulation into social uses.

Privatization, of course, could take many different forms, and it is not identical to liberalization. Indeed, privatization alone would have merely transferred a public monopoly into private hands. Whatever technical expertise the state bureaucracy had acquired during the course of previous privatizations, the goals established by Salinas for the development of Telmex created special complications. The Salinas administration would need to restructure both the firm and the regulatory framework governing telecommunications. The president intervened repeatedly and directly to help push forward the sale of Telmex.

RESTRUCTURING INDUSTRIAL RELATIONS

STRM leaders expressed their opposition to privatization in the early months of the Salinas administration. However, General Secretary Hernández Juárez had already begun to tighten his connections to the ruling party. Although the STRM had abandoned the official labor central, the CTM, in 1976, Hernández Juárez reached an agreement with CTM leader Fidel Velázquez in the mid-1980s that allowed him to become the president of the Labor Congress (CT). In 1987, Hernández Juárez became a member of the PRI (Trejo Delarbe 1990, 336). Whether Hernández Juárez feared a confrontation with Salinas or sought the rewards of membership in the ruling party is unclear. However, given Salinas's demonstrated willingness to use his power against labor, STRM leadership faced a real dilemma as it became increasingly clear that Telmex would be privatized.

In March 1989 the STRM appeared headed toward confrontation with the state. The union had announced its intentions to strike if it did not receive salary increases above the 8 to 10 percent cap that the government had imposed on parastate firms. At the same time, the government had given the STRM two weeks to come up with a plan for "industrial reconversion" at Telmex. The official position of the STRM at the time was complete opposition to privatization and three fundamental demands for any reconversion: no layoffs, respect for the gains in the collective bargaining agreement, and survival for the union (*El Financiero* 1989b).

The conclusion of negotiations between the STRM, Telmex, the SCT and SPP, was ambiguous. Telmex and the STRM signed an agreement to improve the quality of service at Telmex. However, while claiming to "respect the rights of the workers," the agreement was primarily focused on job flexibility at Telmex. It eliminated 57 Departmental Agreements that governed different job categories and replaced them with "job descriptions" that abolished all department-level negotiations and consolidated all discussions around wages and working conditions into the national-level, company-wide contract negotiations (Dubb 1992, 20).[3] The agreement also eliminated the rights that the STRM had won fewer than two year before to consult on the introduction of new technology (*El Financiero* 1989a).

In return for the STRM's acquiescence on these issues, President Salinas promised that the sale of Telmex would not entail any layoffs. In addition, he supported a government-backed loan to help the STRM purchase a 4.4 percent stake in Telmex (de la Garza 1989; de la Garza and Herrera 1997). Beyond Telmex, however, Salinas's offensive against labor opposition continued. In August 1989, he ordered troops to occupy the premises and enforce the declaration of bankruptcy of the state-owned mine, Cananea. Official pronouncements on the bankruptcy held the labor union responsible for low levels of productivity and an inability to compete internationally (see Chapter 3).

By September 1989, Hernández Juárez was calling the privatization of Telmex "inevitable" (*Excelsior* 1989b). The statement came on the eve of the STRM's annual convention at which Salinas would announce the decision to privatize Telmex. By December, in paid advertisements the STRM was promoting the official line: Mexican development required advanced telecommunications service and Telmex required more investment than the

---

3. For the most critical evaluation of this process See Dubb (1992). See also Vázquez Rubio (1990).

government could provide. Salinas's six-point plan had been adopted by the STRM leadership in its entirety (*Excelsior* 1989a).

In addition to showing that the ruling party could still reward its allies, Salinas appeared to be grooming Hernández Juárez to lead a new generation of "modern" labor unions that would concentrate more on productivity and cooperation with capital and less on class confrontation (Zapata 1995, 130, 131). Hernández Juárez played an important role in this plan through his participation in the founding of a new labor federation, the Federation of Unions of Goods and Services Workers (Fesebes) in February 1990. From within the Fesebes, the STRM provided an example of the type of union organization that Salinas claimed Mexico needed: one that would work with management in the creation of wealth rather than emphasizing outdated concepts of class struggle and redistribution.[4]

TRANSFORMING TELMEX'S ORGANIZATIONAL STRUCTURE

One way of privatizing Telmex would have been to break the firm into a number of smaller companies divided along regional or functional lines. This was how the AT&T monopoly was broken up in the United States and might have been used as a model for restructuring Telmex. Instead, Telmex was not only left intact, its position in the sector was consolidated and strengthened prior to privatization. Telmex was sold along with nineteen subsidiaries, which included Northeast Telephone (Teléfonos del Noroeste), an independently operated firm covering Mexico's northwestern region. The sale included telephone directory and publishing services, a cable-laying company, real estate development firms, and a cellular telephone company. To allay the fears of potential bidders that the state itself might continue to compete in the telecommunications market, the federal microwave network was separated from the Ministry of Communications and Transportation and sold to Telmex a few days before the Telmex sale (SHCP 1994a, 128).

To maximize the price of Telmex, the Salinas administration chose to limit the influence of the market on a whole range of telecommunications services. By far the most important means by which the state enhanced the price of Telmex was by limiting competition. Telmex was granted exclusive rights to operate in the profitable long-distance market until 1997. And

---

4. The official statement on the new unionism is spelled out in *El sindicalismo en la reforma del estado*, by STRM General Secretary Francisco Hernández Juárez and María Xelhuantzi López (Hernández Juárez and Xelhuantzi López 1993).

although the government granted cellular phone concessions in each of the nine cellular regions into which Mexico was divided, Telmex was the only company that was granted a license to operate cellular services in all the regions. While cellular traffic accounted for less than 5 percent of Telmex's revenues, it was the fastest-growing segment of the market. Competition was legally permitted in the local segment of the market, but local service is widely considered to be unprofitable, leaving Telmex with a de facto monopoly (Székely and Palacio 1995, 55; Ramamurti 1996, 84).

The title of concession governing Telmex's operation as a private firm was drawn up during the summer of 1990 in negotiations between the SCT, the SPP, the Ministry of Labor and Social Provision, and the leadership of the STRM and contains a wide range of nonmarket requirements that Telmex must fulfill. The concession requires that Telmex improve the quality of service and sets targets for the maximum number of lines in repair as well as maximum response time for meeting customer needs. The concession also mandates that Telmex increase the size of the network by 12 percent each year through 1994 (SCT 1990b, chap. 3, 3.2). These regulatory requirements were, in part, the price exacted by the state for granting the new owners of Telmex a long-distance monopoly for almost six years after the company's sale.

Yet while the title of concession restricts the force of the market following privatization, it also begins to establish the framework within which Telmex would eventually begin to operate on principles of price and competition. Chapter 6 of the concession deals with "rate regulation and financial equilibrium." Rates for each service "should permit the recovery of at least the 'long-term incremental cost' in such a way that 'cross-subsidies' between services are eliminated." Long-term incremental costs are defined as "the sum of all costs incurred by Telmex to provide one additional unit of capacity of the corresponding service." Cross-subsidies are defined as providing a service "with a rate insufficient to cover the average long-term incremental costs and simultaneously providing another service with a rate above the average long-term incremental costs" (*Diario Oficial,* 1990b, 6-2).

The elimination of cross-subsidization requires Telmex to (1) separate out as clearly as possible the costs incurred in providing service in each of the five categories identified above and (2) bring the prices of each of these services in line with the costs. Although cross-subsidies between these aspects of telecommunications were purely an internal matter at the time that the title of concession was written, their elimination would be essential to the creation of a long-distance market. Without forcing Telmex to account separately for different services, the firm could always use profits from one segment of the market to compete in another segment.[5]

Many of the problems of regulation were created by the fact that the state had chosen not to break Telmex up into various smaller firms. Creating numerous small firms that competed with one another might have created other regulatory issues. But the presence of a single, dominant firm in the sector made it necessary to regulate the internal activities of the firm. Telmex was kept intact because Telmex was worth more whole than in pieces. In addition, the government feared that numerous small firms created out of Telmex would be unable to compete when the market was opened to foreign competition (Interview 77).

FINANCIAL RESTRUCTURING, PRICING, AND SELLING THE FIRM

The decision to leave Telmex intact created still further difficulties for the Salinas administration. While Salinas had promised that Mexican nationals would retain control over the newly privatized telecommunications monopoly, no single Mexican company was large enough to buy the state's entire majority share of Telmex. Officials in the Mexican government decided to resolve this problem by reorganizing Telmex's stock. At a special meeting of the board of directors of Telmex on June 15, 1990, officials agreed to create three different classes of stock and a trust fund that would allow Mexican nationals to retain control over the firm even without owning a majority of Telmex stock.

The key to controlling Telmex would be through ownership of the 21.4 percent of stock designated "AA." The AA stock conferred voting rights on its owners and could only be held by Mexican nationals. In fact, Mexican nationals did not even need to own all the AA shares, simply a majority of them, in order to control the firm. The Mexican conglomerate Grupo Carso ultimately bought 51 percent of the AA shares—only slightly more than 10 percent of the total stock of Telmex—gaining control of the firm along with its partners, Southwestern Bell and France Telecom. More than half of the state's 55.85 percent of the company—31.05 percent of the total stock—was converted into "L" shares, to be placed on international stock exchanges; and an additional 11 percent of Telmex stock that was owned by the government was later converted into L shares so that the limited-rights stock constituted 60 percent of total Telmex stock. Thus, while sales

---

5. In Table 12, Telmex's revenue stream is consolidated into the most important categories: local, national long distance, and international long distance. Connection fees, monthly service, and local measured service are all collapsed into the "local service" category.

of the L series of Telmex stock would take place from December 1990 through May 1994, the critical sale was the first, consisting of the AA shares and determining who would buy the rights to control the firm.

Because of the size and complexity of Telmex and the need to seek foreign investment, the UDEP selected two agent banks, Banco Internacional and Goldman Sachs, to manage the sale. The UDEP also contracted McKinsey and Company to provide evaluations of Telmex and present an overview of the firm and the market to prospective buyers during their tours of the Telmex facilities (Telmex n.d.b, 24; Ramamurti 1996, 105 n. 34). The price that bidders would be willing to pay for the firm depended, in part, on the prices that Telmex could charge for its services and the rights that owners would have to reorganize the process of production. As noted above, by the late 1980s, inflation had eaten into real prices for telephone service, and regulatory constraints prevented the firm from adjusting rates upward. In order to attract investors to the firm, the state made a number of important changes in the rules governing the establishment of prices for telephone service.

The de la Madrid administration had already begun to allow increases in local use rates in the mid-1980s (see Table 12). Soon after giving notice that Telmex would be privatized, the Salinas administration announced that rates for communications and transportation services would be indexed to inflation (*La Jornada* 1989b). This was an especially important concession to prospective buyers because the Mexican state had been attempting to control inflation through "pacts" negotiated with labor and capital to maintain price stability. Ramamurti notes, "The price of Telmex shares rose more steeply in the Mexican stock exchange in response to the tariff revision than it did either when privatization was announced by Salinas or after control of the firm was actually turned over to new owners" (1996, 84).

The state required that companies interested in buying Telmex submit a detailed technical prospectus and reveal their plans for development of the telecommunications system. Firms interested in buying Telmex were also required to provide proof of experience in telecommunications and access to financing. Telmex's major equipment providers, however, were prohibited from participating in the auction, to prevent them from using Telmex as an outlet for their products (Ramamurti 1996).

Telmex's director of planning and corporate development, Alfredo Pérez de Mendoza, was assigned to coordinate the visits to Telmex by companies interested in participating in the auction (Telmex n.d.b, 19, 20). Between August 22 and October 26, 1990, thirteen companies participated in two- and three-day visits to Telmex in which they were provided with information authorized by the SHCP, the SCT, Banco Internacional, Goldman Sachs, and McKinsey Corporation. These tours of the Telmex facilities included presentations on the history of Telmex, regulatory matters, and Telmex's organizational structure as well as dinner with the top administration of the company. All the members of Telmex's top management participated in these promotional meetings. According to reports provided by Telmex, each of the interested parties was given an identical package containing 239 documents with financial, technical, labor, and administrative information on the operation of Telmex. Further questions were referred to the Ministries of the Treasury and of Communications and Transportation and to Banco Internacional (Telmex n.d.b, 23–27).

In late October, the Ministry of the Treasury announced that sixteen companies—four Mexican, twelve foreign—met the technical and financial requirements established at the outset of the process to participate in the auction. Ultimately, only three groups would make bids on Telmex: (1) an association between a major Mexican stock brokerage, GTE Telephone Corporation, and the Spanish Compañía Telefónica de España; (2) an association between a major Mexican conglomerate and businessmen associated with Singapore Telecommunications; and (3) an association between another major Mexican conglomerate Grupo Carso, Southwestern Bell Corporation, and France Telecom (see Telmex n.d.b, 30). The winning bid was submitted by the final group, which now controls Telmex.

Macroeconomic Restructuring and the Creation of a Long-Distance Market

The most important piece of legislation governing the operation of Telmex as a private firm is the title of concession. Any firm seeking to enter the telecommunications market in Mexico must obtain a title of concession from the SCT. As the sole provider of both local and long-distance service at the time of its privatization, Telmex was subject to especially detailed regulation in which the state stipulated precisely what expectations it had for the company beyond the simple question of profitability.

As noted above, the title of concession is an attempt to impose standards of use value that the firm must meet, requiring the firm to meet targets for quality improvement and expansion of the telephone network and restricting the influence of competition and price on telecommunications. At the same time, other parts of the title of concession elevate the importance of exchange value, in particular by eliminating cross-subsidization among different types of service. The continued monopoly of Telmex over local service for the foreseeable future created the danger that cross-subsidies would begin to flow in the opposite direction from their historical pattern. State-owned monopoly telecommunications firms intentionally distorted prices and used cross-subsidization as a means of equalizing access to telephone service: prices of long-distance, urban, and business service were elevated in order to reduce the costs of local, rural, and residential service (Aronson and Cowhey 1988, 31; Cairncross 1997). Following privatization, the concern was that Telmex would ignore price signals, not to provide universal service, but to capture market share, undercutting long-distance competitors with profits earned in the local market.

In preparation for privatization, the government began the process of eliminating these cross-subsidies. Following privatization, the shift in Telmex's revenue stream became still more pronounced. Between 1976 and 1990, income from local phone service averaged slightly more than 20 percent of total revenue for Telmex, while international long distance accounted for more than 40 percent and national long distance for 36.5 percent. Following privatization, local rates jumped, averaging more than 41 percent of total revenue between 1991 and 1996. While the percentage of revenue from national long distance fell modestly following privatization—averaging 32 percent of total revenue—the percentage of revenue from international long-distance rates fell dramatically, accounting for an average of only 23.6 percent of total revenue since privatization (see Table 12).

Reorganization of the income stream was essential to the preparation for restructuring the market within which Telmex would operate after 1997. Without raising local rates and eliminating the cross-subsidies from long distance, Telmex would have been hard-pressed to lower long-distance rates in order to compete against companies operating only in the more lucrative long-distance market. The Mexican government granted Telmex complete freedom to raise rates to stay even with inflation through 1996. Beginning in 1997, Telmex's pricing policy was required to take into account productivity increases by limiting rate increases to three percentage points below the rate of inflation. After 1998, regulation would allow Telmex to earn "a fair return on capital" (Ramamurti 1996, 86).

CREATING A MARKET BY IMPOSING TRANSACTION COSTS:
COMPETITION WITHIN LIMITS

The opening of Mexico's long-distance market to competition in 1997 was strongly influenced by the opening of the Chilean long-distance market two years earlier. The Chilean government managed to construct a long-distance market that closely approximated the theoretical ideal of neoclassical economics. Regulations requiring telecommunications firms to sell excess capacity to resellers allowed a large number of competitors to enter the market for long-distance services, and consumers were able to use access-code dialing, which allows each call to be placed with a different carrier. Newspaper advertisements provided information on daily changes in prices, and consumers made decisions largely upon the basis of price. This market operated with virtually no friction: consumers could move effortlessly from one long-distance carrier to another simply by dialing a code prior to placing the call.

The results of the opening of the Chilean long-distance market were disastrous. Price wars broke out almost immediately. Money was wasted in advertising to lure consumers away from other companies, while prices were driven down, preventing firms from making sufficient profits to stay afloat, let alone make productive investments. Consumers with outstanding bills to one company could simply switch to another company. Thus, while the volume of long-distance calling increased considerably following the opening of the long-distance market, so did the amount of bad debt held by phone companies (Interview 7).

While writing the legislation to govern the long-distance market in Mexico, state regulators communicated with their counterparts in Chile to avoid what newspapers referred to as the "Chileanization" of the Mexican market (Interview 8). Furthermore, a number of the firms entering in the Mexican market had invested in Chile and were keen to avoid allowing the same conditions that undermined the Chilean market to take hold in Mexico. Mexican regulators created a number of institutions that were designed to provide social structure to the market and prevent the frictionless exchange that took place in Chile. The Federal Telecommunications Law passed in June of 1995 emphasizes the social character of telecommunications and requires firms to present plans for investment, coverage, and quality of service in order to obtain a concession.

The implementing legislation, or *reglas,* governing the Telecommunications Law was passed a year later and contained a number of safeguards against the Chileanization of the market. The *reglas* require the long-distance companies to form a Committee of Operators to resolve disputes between competitors in the market and to act as an intermediary between telecom firms and the state. In addition to working out technical issues—the most important of which is interconnection between different networks—the committee was required to create a system for identifying customers who had outstanding debt to another company, and to create various forms of third-party mediation for resolving disputes and coordinating the market (SCT 1996c, chap. 6).

Also, rather than open the entire Mexican market at once, as happened in Chile, the *reglas* provided for a six-month phased opening in Mexico's sixty largest cities. The Committee of Operators was required to hire an independent company to assist in the process of redistributing customers (SCT 1996c, Regla 29). In order to tie consumers to firms, the long-distance carriers agreed to a process of "voting," in which customers would be asked to choose their long-distance carrier. Rather than create the form of atomization that emerged in Chile, long-distance companies were willing to spend six hundred thousand dollars apiece to contract the international data-management firm NCS International to organize a balloting process (Barbosa 1997, 14). By August 1997, NCS International had mailed more than 12 million ballots to almost seven million phone subscribers in the sixty cities slated to open to competition.[6]

Far from seeking to create a market of frictionless exchange, the process of presubscription was clearly designed to increase the amount of friction in the market by making the unit of exchange long-distance *service* rather than long-distance *calls*. The market could have been opened up much more effectively—not to mention more cheaply—simply by introducing access-code dialing. Yet while the *reglas* establish the possibility of access-code dialing, the major players in the market have managed to delay its implementation. Although the state has been blamed for the delay, there is no reason why the major competitors in the market would want the government to make such a ruling and no legal requirements like those in Chile or the United States for long-distance carriers to sell excess capacity on their networks (Peters 1998, A15). By mutual agreement, the principal carriers in the market have indefinitely delayed the introduction of access-code dialing. As an executive from one of the major competitors in the market noted, "The industry wanted a market of customers, not a market of calls" (Interview 15).

All these measures provide structure to the long-distance market. Coordination, rather than competition, characterizes much of the activity of companies that are operating in the market. In part the need for cooperation among competitors in telecommunications can be attributed to the technological characteristics of telecommunications. Interconnection between competing networks requires the sharing of information on phone numbers;

---

6. Summaries of the results of the first phase of the presubscription process can be found in *El Financiero* 1997 and González and Bueno 1997.

technical standards need to be agreed on; databases need to be created and maintained to coordinate the transfer of calls and match phone numbers and bills to customers.

Much of the structure being created in the long-distance market, however, is nontechnical in nature. The Mexican state actively sought to create an institutional structure to link competitors to one another and firms to consumers; it sought to limit the areas where prices or competition would determine allocations. Concessions were granted not simply on the basis of who could pay for them but in order to attract productive investment into Mexico and expand the long-distance network. Competition would be limited to sixty cities, opened up gradually, and adjudicated by an independent company. Bad debts with one company would be treated as a problem for all companies. The introduction of access-code dialing would be delayed in order to link long-distance customers and companies through an expensive process of balloting.

ORGANIZATIONAL STRUCTURE AND THE REGULATORY ENVIRONMENT

Limitations on foreign investment in telecommunications were decisive in shaping the organizational form of Telmex and its competitors in the long-distance market. Foreign firms are prohibited from entering the market except as minority partners of Mexican firms.[7] Thus, in the years leading up to the opening of the long-distance market, major international telecommunications companies negotiated agreements with Mexican firms that were interested in entering this segment of the market. The U.S. long-distance company MCI entered into an alliance with the banking and stockbrokerage firm Banamex-Accival to create Avantel. Bell Atlantic signed an agreement with the industrial conglomerate Grupo Iusa to create Iusacell. And AT&T teamed up with the Monterrey-based Grupo Alfa to compete as Alestra.

All long-distance phone calls, however, begin and end on a local network. And since Telmex is the only local telephone provider in all of Mexico, the rules governing interconnection to Telmex's local network would be crucial for the operation of a long-distance market. Chapter 5 of Telmex's concession is devoted entirely to interconnection. Under the terms of the concession, Telmex was not obliged to interconnect any competitors until January 1, 1997. However, the concession requires that Telmex reach

---

7. The 1995 Federal Telecommunications Law specifies that majority foreign ownership is allowed only in cellular phone service (*Diario Oficial* 1995a, Article 12).

individual interconnection agreements with every telecommunications operator that formally requests to interconnect to the local network. Telmex is also obliged to install interconnection circuits at the expense of the company that is requesting interconnection and to act fairly and without discrimination against other companies (SCT 1990b, chap. 5).

Telmex's concession also attempted to elicit the participation of the private sector in the construction of the rules by requiring that Telmex publish a proposal by January 1, 1994, detailing how interconnection would be achieved. Further, it gave interested parties the right to present their objections, and the SCT retains the authority to impose an agreement if the companies could not reach one on their own (SCT 1990b, chap. 5).

The attempt to involve the private sector in the drafting of the regulations governing interconnection, however, failed to take into account the fact that the private sector was barely even established. New companies entering the long-distance market were still in the process of forming. Avantel, the earliest entrant into the market, was not even created until October 1994, three months after the "Resolution Plan on Long Distance Public Network Interconnection" was published (Avantel n.d.). The joint venture between AT&T and Grupo Alfa was not announced until early November 1994 (*El Financiero* 1994).

Subsequent legislation passed by the SCT continued to emphasize both public regulation and private ordering. In June 1995, a new Federal Telecommunications law was passed to replace the 1930s General Law of Means of Communication and Transportation. The new law takes a number of the provisions of Telmex's title of concession and generalizes them to all participants in the market. Concessionaires must reach agreements on interconnection or have one imposed upon them; firms must provide nondiscriminatory tariffs; they must act on a basis of reciprocity in interconnection; and they must provide separate accounting that allows for the disaggregation of rates for different services (Articles 43, 44). Yet in negotiations between Telmex and its competitors, no agreement could be reached on interconnection rates, forcing the state to set the rate itself.

Telmex asked for interconnection rates of almost $.15 per minute, while its competitors requested that the rate be set at $.03 a minute. The SCT set the rate closer to that requested by Telmex's competitors: $.0532 for 1997, falling to $.0469 in 1998, followed by a renegotiation in 1999 at a rate not to exceed $.0315 (DePalma 1996a, 1996b). In the generally pro forma justification attached to such decisions, a new "objective" of telecommunications regulation suddenly appears: "To support tariff levels in telecommunications services similar to those that exist in the countries with which

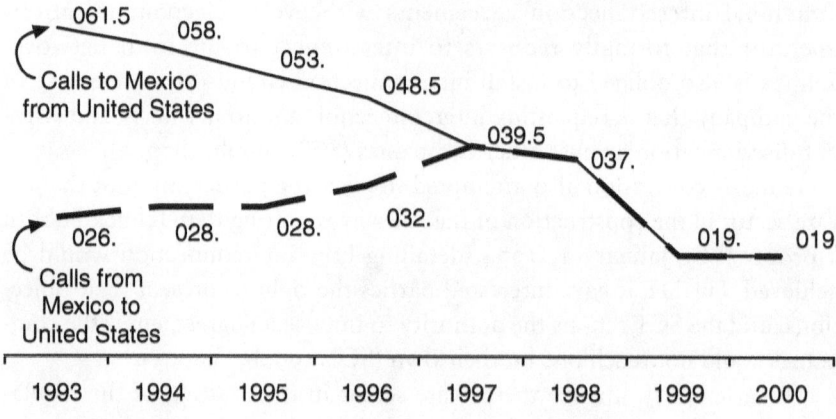

Fig. 10 International settlement rate, 1993–2000 (U.S. cents per minute)

Source: FCC 2002.
Note: The settlement rate is the price paid by one company to another for completing an international long-distance call between Mexico and the United States.

Mexico carries out a great deal of commercial exchange" (*Diario Oficial* 1996).

Nor are interconnection fees the only, or even the most important, component of prices determined outside the market. The price for completing a call on a foreign network—the settlement rate—is established through negotiations between governments. Because the principal foreign market for Mexican long-distance service is the United States, the two governments must negotiate on behalf of their respective corporations to arrive at mutually acceptable fees. As Figure 10 shows, in 1993, U.S. telecommunications companies paid Telmex more than twice as much per minute to terminate a call in Mexico than Telmex paid to them to complete a call in the United States. By the time the long-distance market was opened in 1997, that price had been equalized, with the settlement rate paid to U.S. companies rising from $.26 a minute to $.395 cents and the price paid to Telmex (and future participants in the long-distance market) falling from $.615 to $.395 a minute.

CARVING A MARKET OUT OF LONG-DISTANCE SERVICE

Within the cooperation and coordination established for preventing ruinous competition in the long-distance market, another set of rules and regulations are designed to prevent actors from working together too much: to erect legal barriers between them, encourage competition, and force them to act in a "nondiscriminatory" manner that allows for calculations to be made on the basis of price and profitability. Competition requires the functional separation not only of firms from one another but of the regulatory functions of the state from the operational functions of the firm.

Selling Telmex, of course, was the first step in this direction. The state's new regulatory function was refined still further with the creation of a Federal Telecommunications Commission (Cofetel) prior to the opening of the long-distance market. Third-party mediation has also been an important component of regulating the market, creating private forms of regulation that structure competition between firms.

Like government regulation, however, third-party mediation requires that the regulators maintain independence from individual companies that are competing in the market. And similar to the regulation governing Telmex in its title of concession, the 1995 Telecommunications Law requires each firm to disaggregate different services and set rates that allow for the recovery of "at least the average long-term, incremental cost" (SCT 1995b, chap. 5). In addition, the International Telecommunications Union and the Federal Communications Commission (FCC) in the United States have pushed the Mexican government to provide a competitive environment and to encourage firms to move prices toward costs.

Separating Regulation from Operation
Despite the difficulties of separating the interests of firms from one another or from those of the state, a number of measures have been taken to separate regulation from operation in telecom. The 1995 Telecommunications Law calls for Cofetel to have "technical and operational autonomy [and] ... have the organization and the power necessary to regulate and promote the efficient development of telecommunications in the country" (SCT 1995b, transitory 11). Cofetel was not created until August 1996, barely four months before the opening of the long-distance market, and implementing legislation governing the operation of Cofetel was not passed until the following December.

Representatives of Telmex's competitors complain that Cofetel as well as the SCT regularly favor Telmex because of the political impact that the failure of Telmex would have: Telmex represents a large percentage of the Mexican Stock Exchange and employs more than forty thousand workers (Interviews 18, 25). The fact that all five members of the governing body of the Cofetel are appointed by the president through the minister of communications and transportation does not bode well for its independence (SCT 1996b: Article 3; see also Ramamurti 1996, 96).

The separation of regulation from operation has been somewhat more successful through private, third-party mediation—though there have been problems here as well. The Committee of Operators is charged with contracting private companies to adjudicate specific aspects of the long-distance market. The law attempts to ensure the independence of these companies by stating that these firms "may not be controlled by ownership by the long-distance operators represented on the Committee nor by the principal affiliates of them" (SCT 1996d, Regla 29). In the case of disputes between companies, a committee of experts may be assembled from a list of experts named by the companies to mediate disputes. Experts who make up the committee, however, must be "independent," are subject to veto if three or more companies oppose their inclusion on the list, and can be replaced by the SCT if they fail to show "impartiality" (SCT 1996d, Regla 27).

Where these efforts to regulate the market independently have failed, there is always recourse to the courts. One of Telmex's principal competitors, Avantel, sought legal protection against paying reimbursement to Telmex for the costs that Telmex claimed to have incurred in installing the equipment necessary for interconnection. Although an independent company had been contracted to adjudicate the dispute, Avantel executives accused the firm of succumbing to pressures from Telmex (Matus 1997c, 1A). According to one of Avantel's senior executives, the regulatory process is currently subject to considerable insider pressure. Representatives of all the companies talk to the regulatory agencies, but how, when, and why a decision might be handed down is subject to informal, behind-the-scenes bargaining. Although the carriers were able to work out many of the details of the market among themselves prior to the opening of competition, one industry official argued that the current problems in the market have to do with discriminatory practices that require stronger, more independent adjudication (Interview 15).

Dividing and Combining Firms: New Alliances and New Boundaries
As noted above, Telmex's title of concession requires that it disaggregate prices in order to eliminate cross-subsidization. The principle of prohibiting a monopoly in one sector from using that monopoly power to compete in other markets is fairly well established—until AT&T was broken up, it was prohibited from entering into the computer market (Aronson and Cowhey 1988, 28; Coll 1986). However, the need to prohibit cross-subsidization in Mexico was also the result of the previous decision not to break Telmex into different companies. If different local and long-distance companies had been created when Telmex was privatized, each separate company could be

expected to generate revenues that exceed costs in each market segment. Given the current structure of the market and the presumption that Telmex's local and long-distance divisions would collude to undermine competition, Telmex local is legally prohibited from discrimination against other long-distance companies. That is, Telmex is required by law to behave as if its local and long-distance companies were distinct entities.

The promotion of competition, then, requires the creation of formal, legal equality. The language of telecommunications regulation is peppered with words such as *nondiscrimination, impartiality, reciprocity,* and *neutrality.* Interconnection must be provided on a nondiscriminatory basis to all competitors in the market. Rates cannot be discriminatory and firms must act toward their affiliates, subsidiaries, customers, and others "upon a basis of nondiscrimination" (SCT 1995b, Artículos 41 to 44). The implementing legislation requires that the ballot distributed to long-distance customers include an "impartial and neutral description" of each company and that the information be presented in "nondiscriminatory terms" (SCT 1996d, Regla 21).

The principles of nondiscrimination, reciprocity, and equality also appear repeatedly in NAFTA's chapter on telecommunications. Under chapter 13 of NAFTA, Mexico may organize its telecommunications market largely as it pleases, with one caveat: the regulatory framework must be applied equally to Mexican and foreign firms. In other words, the Mexican government is required to extend to foreign firms the juridical equality necessary for participation in the market. Under Article 1305 of NAFTA, "anticompetitive conduct" by monopolies will be prevented through

a. accounting requirements;
b. requirements for structural separation;
c. rules to ensure that the monopoly accords its competitors access to and use of its public telecommunications transport networks or services on terms and conditions no less favorable than those it accords to itself or its affiliates.

These same principles are at the center of the debate over the liberalization of telecommunications within the General Agreement on Trade in Services (GATS). Petrazzini (1996, 59) emphasizes the need for regulatory autonomy—from private firms and from the state—and notes that "structural and accounting separations are aimed at avoiding cross-subsidies within the company and other anti-competitive behavior" (64).

The creation of a single international long-distance market, however, is not yet compatible with the construction of Mexico's long-distance market.

The international long-distance market in Mexico is dominated by traffic to Mexico from the United States, which, in turn, is dominated by AT&T. If AT&T were allowed simply to enter the Mexican market as AT&T, it would constitute a single network linking Mexico and the United States. Traffic between Mexico and the United States would leave the Mexican market and be absorbed into the international organizational structure of AT&T.

Promoting a long-distance market in Mexico, then, has required that the state prevent the creation of an international market between the United States and Mexico. Just as regulations prevent Telmex from coordinating activities between its long-distance and local divisions, foreign subsidiaries in Mexico cannot legally receive preferential treatment from their parent company abroad—primarily in the United States. Calls entering Mexico from foreign countries are monitored at the international gateway by an independent accounting firm and redistributed to the long-distance companies in Mexico in the same proportion that these companies send outgoing international long-distance calls. Proportional return prevents AT&T from swallowing up the lucrative international long-distance market in one gulp and redistributes calls, guaranteeing that companies operating in the Mexican market receive the same percentage of long-distance calls—and revenue for completing the call—to Mexico as they send from Mexico.

In addition to the macroeconomic framework that separates firms and functions in the market, the opening of the long-distance market has led individual companies to separate different aspects of their internal organization beyond those required by law. Although the state chose not to divide Telmex into a number of smaller firms, Telmex executives have broken the firm apart organizationally, internalizing competition by creating ten regional divisions and one long-distance division, each of which has a measure of autonomy in its operation but is required to show a profit (Interview 21).

The opening of Mexico's long-distance market also provided the impetus for AT&T to split itself into three different firms in 1995. Following the privatization of Telmex, AT&T became an important equipment provider to Telmex, obtaining, among its other contracts, one for $130 million to install 60 percent of Telmex's fiber-optic backbone network and another $17 million to develop a direct fiber-optic network to businesses (Aguilar 1994, 17). Yet when AT&T announced in late 1994 that it would be entering into an alliance with a Mexican firm to compete against Telmex in the long-distance market, Telmex canceled its contracts with AT&T (Interview 27). Commenting on AT&T's breakup, the head of a consulting firm in Washington, D.C., noted, "There are definitely companies that chose not to buy AT&T equipment because they were competitors on the service side" (Andrews 1995, D1). AT&T's vertical structure had become a liability not because it failed to overcome transaction costs but because the service division now competed against potential customers of the equipment division.

The domination of the Mexican economy by large conglomerates that include the new telecommunications carriers, however, will continue to undermine attempts to create market allocations in this sector. Large firms, especially in financial services, are some of the most important consumers of telecom services. Each of Mexico's major competitors in the long-distance market includes as part of its corporate structure a large financial firm along with the telecommunications carrier. How prices will be established between the telecom firm and other firms that belong to the same corporate parent is impossible to determine. However, Telmex officials acknowledged that it is unlikely that other large companies that are owned by Grupo Carso will be contracting with Telmex's competitors to provide telecom service (Interview 7).

Competition in Long Distance

The decision by the major carriers and Mexico's regulatory agencies not to introduce access-code dialing prevented the creation of a market for individual calls. Instead, the goal was to create a market of customers. Fierce competition within and outside these parameters soon erupted. Because of the balloting system chosen by the long-distance carriers, competition for customers took on all the trappings of a political campaign, including misrepresentation, vote fraud, and the stuffing of ballot boxes. After eight months of campaigning in 1997, the seven major competitors in the long-distance market had spent more than $420 million in advertising (*Reforma* 1997, 4A).

The incentive for Telmex's competitors to get out the vote was especially strong because the agreement worked out with the SCT and the Committee of Operators assigned nonrespondents automatically to Telmex. Alestra offered kilos of ham to retailers to collect ballots. Avantel hired hundreds of messengers to collect ballots (Bueno 1997c, 8). Alestra was subsequently accused of attempting to corrupt postal employees, offering cash and food to workers in exchange for their filling out ballots in favor of Alestra (Cardoso 1997, 42).

Cofetel and NCS International attempted to bring the balloting under control in various ways. Many phone subscribers never received ballots, so voting rules were changed to allow customers to select a long-distance company without an official ballot. Other phone subscribers simply received their ballots late, making it difficult to respond within the time period allotted for registering the vote. Three and a half months into the balloting process, NCS International's vice president for Latin America complained that the company had never witnessed as much irregularity as it found in Mexico: "In the 300 similar processes that we have done in the world never have we encountered a situation where the mail was interfered with" (Matus 1997b, 35A).

Competition for customers moved to the courts in May 1997 when Avantel threatened to submit a complaint to the Cofetel against Telmex for failing to provide it with the interconnection circuits it needed to meet its customers' needs. Avantel also accused Telmex of charging it more for the circuits than it charged corporate clients who stayed with Telmex. The complaints against Telmex were also taken beyond Mexico, to the FCC in the United States. The FCC grants permission to foreign carriers to enter the U.S. market and may only permit the entrance into the U.S. market of companies from countries where U.S. firms receive fair treatment. Both MCI and AT&T sent reports to the FCC detailing their unfavorable opinion of Telmex and the market environment in Mexico (Matus 1997a, 1997d).

The stakes in this competition are not hard to uncover. Telmex's annual report shows that income from national and international long distance in 1996—when Telmex still enjoyed a monopoly on long-distance service— approached $4 billion. However much social and institutional structures have shaped the form that competition takes, there has been little need to create incentives to attract competitors apart from the opportunity to snatch part of this lucrative market from Telmex.

THE STRUCTURE OF THE LONG-DISTANCE MARKET

Given both the constraints that were placed on the operation of the long-distance market and the intense competition that occurred within it, at the end of the first phase of opening it appears that Mexican regulators successfully avoided the Chileanization of the market. An identifiable structure had emerged from the wrangling. Prices had fallen, but not too much. Telmex lowered its long-distance rates gradually in anticipation of the opening of the market, international rates falling 31 percent between 1992 and 1997, national rates 38 percent (Table 13). These price reductions were offset by increased prices for local service. This trend continued through the first year of competition, during which prices in national and international long distance were scheduled to fall another 12 percent, while all local fees

TABLE 13 Telmex long-distance rates, 1991–2000 (1996 pesos)

Year	Average Price per Minute National Long Distance	Average Price per Minute International Long Distance
1991	3.15	5.94
1992	3.01	4.92
1993	2.91	4.54
1994	2.57	4.4
1995	2.01	4.74
1996	1.97	4.09
1997	2.19	3.61
1998	1.98	2.64
1999	2.32	3.13
2000	2.17	2.15

Source: Telmex annual reports, 1991–96, 1998, 2001. Calculated on the basis of total revenue from each segment of the market divided by total minutes of traffic.

except installation charges continued to rise (Telmex 1997). Locked into fixed interconnection and settlement rates, Telmex's competitors have not been able to undercut Telmex's prices.

The structure of this market is reinforced by the limited number of competitors. Although seven firms entered the market to compete with Telmex at the beginning of 1997, by the end of the presubscription process, three firms—including Telmex—captured more than 99 percent of the long-distance market. Out of more than 4 million votes cast in the balloting process, Telmex received 55.7 percent, Alestra 24.9 percent, and Avantel 18.6 percent (González and Bueno 1997, 4).

The government succeeded in obtaining investments from the new owners of Telmex, and the telecommunications infrastructure expanded rapidly during the first four years of Telmex's operation as a private company. Telephone density increased substantially in some of the states with the lowest penetration of telephone service. However, the high rates of growth in telephone density in most of these states can be attributed to the low base level of telephone penetration. In absolute terms, the inequality of telephone penetration between regions in Mexico remains acute. In the southern states of Oaxaca and Chiapas, telephone density is 3 and 2.74 per hundred inhabitants, respectively, while in the northern border states and Mexico's Federal District, telephone density ranges from 10.62 to 25 per hundred inhabitants (SCT 1992: 89; Telmex n.d.a).

Investments by Telmex's competitors are not likely to reduce these inequalities. As Figure 11 shows, the fiber-optic networks of Telmex's two principal competitors, Alestra and Avantel, both connect the highly industrialized cities of Monterrey, Guadalajara, and Mexico City and link these cities to the United States without reaching into southern Mexico, where telephone density is the lowest. The new long-distance companies also failed to make the promised $2 billion of investment in infrastructure during 1996 and 1997. By mid-1997, one newspaper estimated that only $800 million of the promised investment had actually been made (Bueno 1997b, 8).

The deal that Telmex workers made with President Salinas appears to have held. In addition to acquiring a 4.4 percent stake in the newly privatized firm, the STRM has avoided mass firings. There are no reliable figures on the changes in the number of workers in different areas of the firm, though we can surmise that the number of operators has fallen dramatically. However, while dissidents speak of "stealth" firings and transfers that force workers to choose between home and job, the size of the workforce at Telmex has actually grown slightly since privatization.

With Telmex facing competition, however, the STRM finds itself in the difficult position of needing to support the company's competitive position in the market lest it allow other firms to win market share. One STRM official argues that the biggest challenge facing the STRM is transforming itself into a true "industry" union. Because Telmex has been synonymous with the industry since the inception of the STRM, it has always been more like a company union (Interview 29). Where the STRM should go next, or whom the union should try to unionize, is not clear. The new competitors in long-distance all use nonunion workers. In addition, equipment suppliers can perform certain types of maintenance that Telmex's unionized workers have historically performed. Establishing the boundaries of union cooperation and confrontation with Telmex will be a major challenge in years to come.

One important area in which the STRM has cooperated with Telmex has been in managing the retraining of workers who are displaced as a result of technological change. According to one union official, the STRM's agreement with President Salinas forced Telmex to invest heavily in retraining. Thus, Telmex and the union created and maintain Intelmex, a retraining center for workers that is run by the union. This official claimed that at any one time, 15 percent of the workforce is in the process of retraining (Interview 35).

The STRM has also pioneered international labor solidarity to advance its cause. In anticipation of the signing of NAFTA, the STRM signed a letter of understanding with the U.S.-based Communications Workers of America (CWA). Once NAFTA became law, the STRM filed the first Mexican complaint under NAFTA's labor side agreement in support of telecom workers

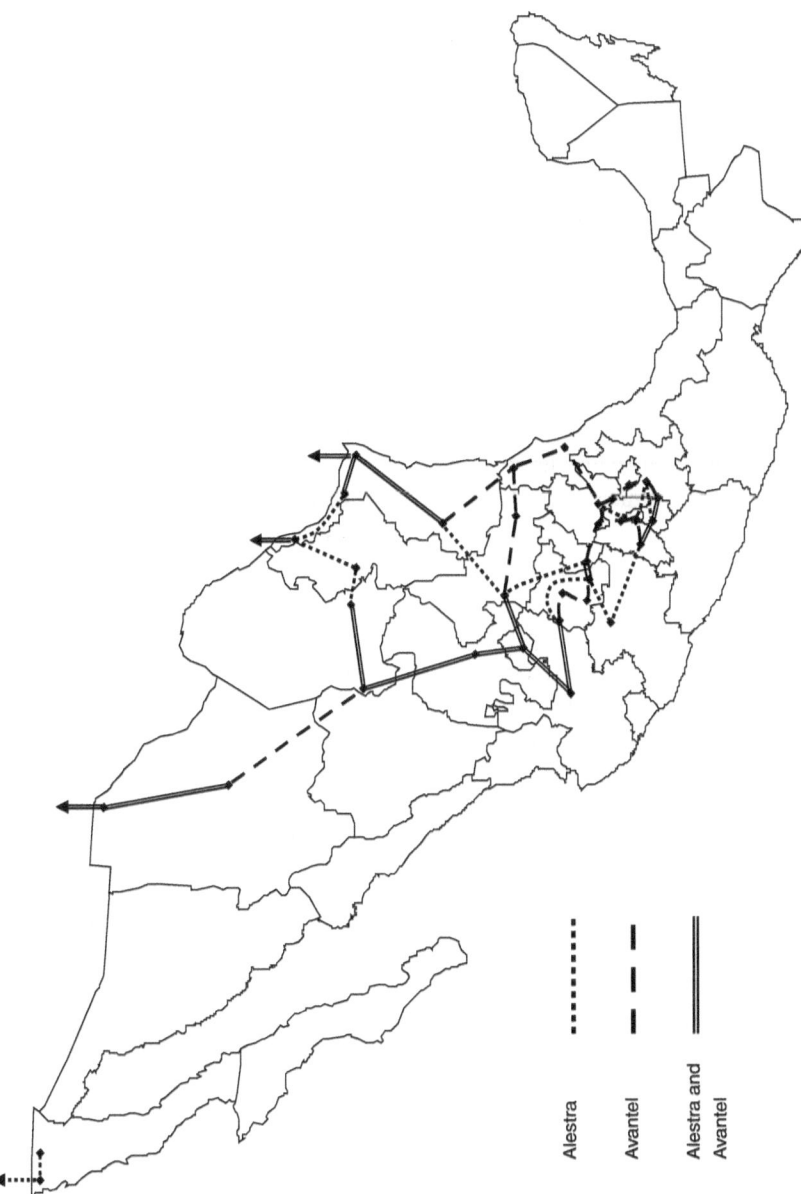

FIG. 11 Fiber-optic network of Telmex's principal competitors
Source: Alestra n.d.; Avantel 1996.

in the United States who had been fired by Sprint precisely as Sprint was engaged in negotiations to create an alliance with Telmex.

Discussion

One of the most common arguments against public ownership is that public firms operate inefficiently because political pressures override market pressures (Kornai 1992, 489; Rogozinski 1997, 27–29). The privatization of public firms and restriction of the government's role to that of regulator, then, are said to result in the following benefits to society. First, the public sector is freed from the obligations of supporting inefficient firms in nonessential sectors of the economy, allowing it to focus on basic needs such as health, education, and welfare (see Aspe 1993, 161). Second, the introduction of competition in the provision of goods and services that had previously been produced by the state encourages private actors to produce these same goods more efficiently.

In the telecommunications sector, the Mexican state showed itself to be relatively immune from the sorts of political pressures that in theory should have prevented it from responding to market pressures, especially during the Salinas presidency. The state took the first steps toward rebalancing local and long-distance rates prior to the privatization of Telmex. Furthermore, much of the discourse of the inefficient state sector simply did not apply to Telmex. Despite the cross-subsidization discussed above, Telmex was always a profitable firm and a valuable source of foreign exchange for the Mexican state because of the surplus of international calls received by Mexico when compared with calls sent (Pérez de Mendoza 1989).

The rhetoric of privatization is instructive, nonetheless, because it directs our attention to the fact that competition in the market assumes the existence not only of discrete entities competing with one another but also of a discrete—and disinterested—entity to arbitrate the process of competition. This is an especially tall order within the neoliberal cosmology because the state, which is presumed to be incapable of resisting political pressures, is entrusted with the responsibility for creating a level playing field and adjudicating disputes impartially. The need to provide strong, independent regulation of Telmex also came into conflict with other goals of the privatization program, specifically the attempt to make the firm as attractive as possible to private investors (Ramamurti 1996, 96).

In a sector such as telecommunications, there is an especially delicate balance between the cooperation and coordination necessary to maintain service provision and the competition necessary to create a market. Privatization and the creation of markets in the telecom sector may depend even more upon rules than on alternative forms of economic organization. With multiple entities now dividing functions that were previously folded into the state, privatization actually increased the need for rules to govern this part of the economy. Even with a relatively small number of competitors operating in the market, the amount of regulation—public and private as well as formal and informal—grew substantially in response to the entrance of new players in this segment of the telecommunications market (95).

Cowhey and Aronson (1989, 8) assert that a phone call is "a commodity just like a bushel of corn." Yet the technological requirements of separating a phone call into a commodity—delivering it to customers, measuring its dimensions, establishing its price and creating competition—all make the construction of a long-distance market much more complex than selling a bushel of corn. The single most essential requirement for competition in this market is the ability to transfer calls to and from competitors' networks. This process requires an exceptionally high degree of coordination among firms. Agreements need to be reached to allocate numbers, computers need to be programmed to reroute calls, contracts for interconnection need to be signed, and rates for interconnection need to be established.

At the same time that all manner of coordination must be established between firms in the market, the regulatory framework must clearly establish where and how actors can operate on principles of price, profitability, and competition. Actors may not treat one another in a "discriminatory" manner across certain market boundaries; indeed, they must treat divisions of their own company as if they were equivalent to companies against which they compete in the market. In order for the telecommunications market to function properly, this set of market and nonmarket transactions must be coordinated, which involves creating barriers to some exchanges while facilitating other exchanges.

Despite the differences between the privatization of the airlines and that of Telmex, certain similarities stand out. Increased government involvement in both these sectors through the 1970s and early 1980s failed to completely insulate Mexico from the forces of the international market. And in both telecommunications and air transportation, privatization and the subsequent deregulation of the market have pushed some workers into closer alliances with the firm.

Unlike civil aviation, however, Telmex remained profitable throughout its life as a majority state-owned enterprise and contributed significantly to the state's revenues. And, although the tripartite cooperation between labor, capital, and the state had symbolic value as a model of Salinas's vision of a modernizing Mexico, Telmex was a unique case. The profitability of Telmex, the potential for growth in the industry, the form in which Telmex was privatized, and the regulatory environment all facilitated these trade-offs among actors.

SIX

The Transformation and Sale of the Mexican National Railroad

[T]he Chihuahua-Pacific Railroad has a transcendental significance. It is not only a means of communication and of national integration; not only a factor which must stimulate the economy of a vast region of the nation. It is something more; it is a symbol of what our people is capable of accomplishing through its effort, its desire to improve, and its encompassing patriotism.
—*President Adolfo López Mateos, 1961, cited in David H. Shelton,*
"The Banking System: Money and the Goal of Growth"

With its foundation in the new legal framework, the development of a new, secure, competitive, and efficient rail system will be promoted. The key to achieving this will be attracting private capital, national as well as foreign, to the sector through transparent and stable rules and an efficient process of privatization.
—*SHCP, Plan Nacional de Desarrollo, 1995–2000*

The roots of state involvement in the railroads reach even deeper than in the previous two cases examined here. Although railroads were excluded from the list of priority and strategic industries in the Constitution of 1917, their importance to the national economy led to a variety of interventions by the state into their operations even before the revolution. The virtual monopoly that the railroads held over the transportation of export commodities and domestic goods also provided workers with significant leverage in their negotiations with owners, foreign and domestic as well as private and public. This leverage helped labor unions negotiate generous benefits for workers and placed a wide range of restrictions on managerial prerogatives over the organization of the railroads. The union's strength and its involvement in opposition politics also led the government to intervene in the operation of the railroad workers' union more systematically than in any other industry.

Although the government began the preparations necessary for the privatization of the railroads in the early 1980s, attempts to consolidate and restructure the railroads financially and organizationally had been under way since 1908, when the state first created the Mexican National Railroad

(FNM) to administer failing private firms. The restructuring of the 1980s, then, looks like the continuation of a decades-long process of bringing the entire network of national and regional rail companies under the control of the largest company in the system. Only after it combined the entire rail system into a single system was the state able to assess the viability of dividing the railroads into smaller units that might each, independently, manage to operate profitably. The economic crisis of late 1994 served as the catalyst for the final steps that were taken to place control of the railroads in private hands.

In the following section I provide a brief overview of the principal characteristics of rail transportation in general before turning to the historic development of the railroads in Mexico and the embedding of this sector in the country's political and social institutions during the years of state-led development. In the section after that I examine the process of restructuring the railroads and the subsequent sale of concessions to operate the newly created main lines. Finally, although the railroads have been in private hands for fewer years than either the airlines or Telmex, I conclude with some preliminary observations regarding the operation of this sector under private ownership.

Railroads and National Development

RAILROADS, ECONOMIC DEVELOPMENT, AND THE STATE

Like the other industries examined here, private and public interests were tightly intertwined in the development of the railroads from their earliest days. Capitalists in Europe and the United States also diffused this technology to their colonies and independent developing countries. Around the world, railroads were either developed and owned by public entities or operated as concessions granted by the state. In Europe, "governments awarded concessions to the railway companies in order to extend the rail network to all parts of the country, both for economic and for strategic reasons" (Violland 1996, 34). In the United States, with few exceptions, railroads were privately owned, but their construction was promoted through land grants to private capitalists and financed by government bonds (Chandler 1977, 82; Josephson 1934, 78).

The precise contribution of the railroads to economic development is a point of some debate among historians and economists. Alfred Chandler claims that railroads were essential to the industrial revolution in the United

States for a number of reasons. Along with the telegraph, railroads "provided the fast, regular, and dependable transportation and communication so essential to high-volume production and distribution" (1977, 81). Railroads also provided a business model for an industrial economy, as they were "the first to require a large number of full-time managers to coordinate, control, and evaluate the activities of a number of widely scattered operating units" (81). Moreover, Chandler points to the way in which the development of the railroads transformed U.S. and Western European economies through backward linkages, especially in the production of steel for rails and through the demand generated for construction and finance (1977, 90; 1990, 282).

Other economic historians have challenged the centrality of the railroads to industrial development. This literature, sometimes referred to as new economic history, has attempted to document more rigorously the contribution of the railroads to economic growth during the nineteenth century. Although this research tends to be cautious in its estimates of the amount of economic growth that can be attributed to railroads, it has challenged traditional interpretations of the railroads as a "leading sector" of or indispensable to early industrialization. One summary of the application of the new economic history to the railroads concludes that the principal contribution of the new economic history is that "historians can no longer exaggerate the contribution of railways to economic development" (O'Brien 1977, 100).

Despite the controversy surrounding the exact role of railroads in economic development, a number of points can be made about the physical and technical characteristics of railroads and the way in which these have influenced the social organization of rail transportation. To begin, the construction of railroads required enormous investments of capital that would have to be amortized over a long period of time. Railroads also depended upon public rights-of-ways. And although railroads overcame natural barriers through the construction of tunnels and bridges and could operate under severe weather conditions, they also created their own rigidities by committing large amounts of sunk capital to specific destinations, routes, and facilities (Chandler 1977, 86; Violland 1996, 34).

Governments played little role in the development of technology for the railroads. Instead, as in the telecommunications sector and the early development of the telegraph, governments stimulated the evolution of railroads more as consumers of this expensive infrastructure and through the subsidization of rail construction. Governments did, however, create detailed regulatory oversight of the operations of private carriers. This intervention was justified on a number of grounds. In addition to the fact that railroads operated on public rights-of-way, the quasi-monopolistic characteristics of the industry led to increased government intervention. Although alternative means of transportation have always been available, railroads created a unique form of dependence on their services. Owners of factories and mines along a particular route often found themselves at the mercy of the railroads for their transportation needs and often supported strong government regulation to prevent price gouging by rail companies (Chandler 1977, 82; O'Brien 1977, 21).

Since the middle of the twentieth century, railroads have faced growing competition from highway transportation. The cost of hauling heavy cargo over long distances remains cheaper for railroads than for trucking. Railroads also have clear advantages in terms of safety and environmental impact (Violland 1996, 35). However, railroads are less flexible than trucking in a number of ways. Railroads cannot climb a grade steeper than .05 and thus require either more tunneling or longer routes, making rail infrastructure more expensive than highways over steep terrain (Interview 51). Another disadvantage of railroads is that they cannot always provide door-to-door service. Instead, they require that shippers contract for the transportation to and from rail terminals. Railroads also face complex problems in managing the organization of freight hauling when trains carry more than one type of cargo, as rail cars need to be linked in a particular order and coupled and decoupled at each stop, slowing the transportation of cargo (Dávila Capalleja 1999, 213, 214).

With the growth of highway transportation, the volume of cargo shipped on railroads has fallen and government subsidies to rail companies have grown. Governments in Germany, Italy, Japan, the Netherlands, Sweden, the United Kingdom, and the United States have all undertaken major reorganizations of their rail sectors (Violland 1996, 35). In the United States, the federal government rescued the bankrupt Penn Central Railroad and five other lines that were in trouble in 1976 and created the Consolidated Rail Corporation (Conrail) (Salpukas 1999). In 1980, the Staggers Rail Act initiated a process of deregulation in an attempt to balance the interests of railroads and their users. In 1987, the federal government sold its stake in Conrail (Phillips 1996; Salpukas 1999; Tye 1990, 10). It is estimated that between 1982 and 1993, 42 percent of workers in the industry lost their jobs (*Economist* 1993b, 65).

In recent years, private companies have also begun a process of reorganizing the operation of transportation services. Increasingly, transportation firms are providing multimodal service in which railroads are only one part of a larger transportation network that includes trucking and marine shipping

(*Economist* 1993b; TMM 1996b). This trend has been accompanied by a growth in containerized shipping, which allows for faster transfer of cargo from one mode of transportation to another. By combining different modes of transportation under the umbrella of a single firm, transportation companies are able to take advantage of the benefits of each type of transportation.

The growth of multimodal transportation has coincided with the consolidation of railroads through mergers and acquisitions. In the United States, Union Pacific bought Southern Pacific for $5.4 billion in 1997 and Norfolk Southern and CSX split Conrail in a deal worth $10.2 billion in 1999. Later in the same year, Burlington Northern Santa Fe and the Canadian National Railway announced their intent to enter into a $6-billion deal that would create the largest railroad in North America (Bagli 1999, C2)

As I discuss below, the growth in merger activity and the increased coordination of different modes of transportation within the corporate structure of a single firm has not escaped Mexico. With the privatization of FNM, three major firms, all allied with North American transportation companies, now compete in the market for cargo traffic to and from Mexico.

THE MEXICAN RAILROADS

National Integration and Macroeconomic Regulation
State intervention in the operation of the railroads in Mexico predates not only the state-led development policies of the twentieth century but even the liberal dictatorship of Porfirio Díaz (1876–1910). The Mexican government granted the first concession to private capitalists to build a railway between the gulf port city of Veracruz and Mexico City in 1837. Over the following three decades, the government granted another thirty-two concessions for rail lines throughout the country, not one of which was completed (SCT 1994c, 1). Civil war and foreign invasions prevented the completion of the link between Veracruz and Mexico City until 1873 (Coatsworth 1990, 180).

Shortly after the completion of the first rail line in Mexico, Porfirio Díaz took power and began the process of pacifying and stabilizing the country. In need of foreign investment, and still deeply indebted to England, France, and Spain, Díaz relied on U.S. capitalists as the principal source of foreign investment in the 1880s until Mexico renegotiated its debts with the European powers. When European companies joined their U.S. counterparts in the 1890s, they contributed to a major construction boom in the industry. Between 1879 and 1910, the Mexican rail system grew more than twentyfold, from 893 kilometers to 19,205 (Coatsworth 1990, 180).

Concessions, however, were not the only incentive for private capital to expand the rail network. In addition, the Díaz regime found it necessary to grant a variety of incentives to foreign capitalists to build rail links. The government subsidized between one-third and one-half of the cost of construction to foreign companies and provided tax exemptions, land grants, and import permits for rolling stock and other capital goods. These incentives were all provided without any restrictions being placed on the companies' freedom to choose the location, length, and characteristics of the railways. After 99 years, the concessions would revert to government hands (Coatsworth 1990, 180; López Pardo 1997, 20, 21).

In exchange for the concessions, the state made three principal demands on private capital. First, to stimulate exports—primarily minerals—the government placed limits on prices that the railroads could charge for shipping goods to foreign destinations. The prices for shipping exports were set at one-half the rate for domestic transportation. Second, government goods and personnel, especially military troops, would receive reduced fares and mail would be carried for free. Third, the rail companies were required to lay a telegraph line for the government alongside the lines they installed for themselves.

The primary beneficiaries of these incentives for rail construction, apart from the rail companies themselves, were foreign investors in mining and the extraction of other raw materials. In 1907–8, mining and metals alone accounted for more than half the total cargo carried on Mexico's principal lines. At the same time, with the fares for exports fixed below the prices that could be charged on domestic routes, rail companies sought to recover the costs of shipping exports through higher prices for domestic transportation. Mexican farmers, industrialists, and businessmen who shipped cargo on domestic lines were forced to pay high tariffs to compensate for the losses that accrued from exporting minerals (López Pardo 1997, 22–24).

Unlike other Latin American governments, however, the Mexican government never agreed to subsidize the costs of operation or guarantee private firms a specific profit margin. When the price of silver fell on international markets and the Mexican government devalued the peso at the turn of the century, the rickety economic condition of the railroads was revealed. In 1902, with the principal rail lines in the country in danger of bankruptcy, President Díaz's minister of finance, José Yves Limantour, warned the Mexican Congress against letting the railroads fall into the hands of foreign speculators. Over the course of the following six years, the government began taking over failing firms and by 1908 owned or controlled two-thirds of the rail industry in Mexico (Coatsworth 1990, 181; López Pardo 1997, 25).

With the gradual nationalization of the majority of the rail system between 1902 and 1908, the Díaz administration proposed the creation of a single entity to coordinate and administer the railroads. Approved in 1908, FNM incorporated the two most important rail lines in the country, Ferrocarril Nacional and Ferrocarril Central, both founded by U.S. capital. Yet while the state held 51 percent of the shares of FNM, it left preferred stock to private shareholders, granting them rights to dividends and to control the company. The function of FNM was primarily to turn the debt of private firms over to the government (López Pardo 1997, 27, 29).

In 1910, civil war broke out, and the need to move troops and armaments placed the railroads at the center of the revolution. So essential were the railroads to whichever side could control them, that they quickly became the target of opposition attacks. By the end of the revolution, in 1917, President Venustiano Carranza reported that half of all locomotives and more than 40 percent of the rolling stock were out of service. An estimated 40 percent of tunnels and bridges were in need of repair and 80 percent of the rail was in poor condition (López Pardo 1997, 30). In addition to the physical destruction wrought by the war, the revolution left the railroads even more heavily indebted. Although FNM desperately needed investment to repair and rebuild from the war, by 1919 the company owed more than $93 million to its creditors (FNM 1987, 44).

Over the following half century, the government repeatedly reorganized the operating, financial, and ownership structures of the Mexican railroads, attempting to maintain rail service, limit the railroad's losses, meet foreign obligations, and pacify workers. During the early part of the century, 95 percent of all transportation in Mexico was carried by rail, making it virtually impossible for the government not to take an interest in the orderly operation and development of this sector (A. H. Chávez 1979, 131).

Foreign creditors continually pressured the Mexican government to reorganize the railroads. In 1922, Mexico signed the de la Huerta–Lamont pact with the United States, establishing a forty-year payment plan for Mexico's debt, one-quarter of which was owed by FNM (FNM 1987, 45, 46). Within two years Mexico suspended its payment on the debt, including that of the railroads. The Calles government worked out a new arrangement with foreign lenders in 1925, amending the 1922 agreement. Under the new agreement, called the Pani Amendment, the state retained a majority of the stock in FNM, but it placed control of the company back in the hands of foreign capital, guaranteed the company sufficient income to meet its financial obligations ($11 million annually), and created an Efficiency Commission to adjust employment and salary levels (López Pardo 1997, 35).

New agreements between Mexico and foreign creditors were reached—and breached—in 1930 and again in 1931. Following the collapse of commodity prices on world markets and the onset of the Great Depression, the Cárdenas administration attempted to reorganize the rail system, first creating a decentralized public organism in 1934 to manage various smaller railroads owned by the government and then establishing a Department of Railroads, Transit, and Tariffs in 1936 (López Pardo 1997, 51). A study by the U.S. Office of Federal Coordination of Transportation, conducted in 1934, stated simply that "Mexican railroad finance has been in a chaotic state for twenty years."

There is general agreement that the key to FNM's financial problems can be found in the company's rate structure. The fares set by the government were too low to cover costs. However, a proposal in 1938 to revise the fares for cargo generated immediate opposition from foreign mining companies, national business organizations, the mine workers' union, and even governors from states in which mining was an important part of the economy. Fares were ultimately left unchanged (Middlebrook 1995, 375 n. 97).

In June 1937, the Cárdenas administration announced the nationalization of FNM. Although the state already held a majority of the company's shares, the nationalization brought the operations of the railroad under the control of the state. In the year that it was nationalized, FNM controlled 66 percent of Mexico's 23,030 kilometers of rail and carried 77 percent of all cargo and 62 percent of all passengers. While nationalization ensured that the state would be able to use the railroad to supplement its larger development plans, it also made clear, once again, that the state, rather than private capital, would pay the costs of modernizing the rail system (López Pardo 1997, 52, 55).

With FNM now operating entirely under the control of the state, its continued failure to meet debt payments to foreign creditors strained relations between the Cárdenas administration and foreign creditors (Middlebrook 1995, 125). The onset of World War II delayed yet another renegotiation of FNM's debt, but now the Mexican government used FNM to contribute to the Allied war effort. Under an agreement with the U.S. Metals Reserve Company, the Mexican government guaranteed rates for the transportation of strategic ores and other minerals below the costs to FNM. The increased demand for subsidized freight during World War II meant that almost one-quarter of all the paid wartime freight that FNM carried was transported at a loss. Moreover, the war effort strained the capacity of the entire rail system. In 1941, FNM carried 16 million tons of cargo with a fleet of 911 steam locomotives, 17,563 tons per locomotive. Four years later, with only

877 steam locomotives and 17 diesel, FNM carried 21 million tons of cargo, or 23,490 tons per locomotive (FNM 1987, 50, 51; Middlebrook 1995, 130, 144).

The role of the state in the macroeconomy of the railroads evolved in a number of ways following World War II. As Table 6, in Chapter 4, shows, the percentage of public investment in railroads, though erratic, declines from approximately 50 percent of public spending on transportation during the 1950s to about 30 percent by the late 1960s.

Despite the uneven and declining support offered by the federal government for rail transportation, the amount of cargo carried by the national railroads grew steadily beginning in the early 1960s. Figure 12 shows that after dipping slightly in the early and late 1950s, the amount of cargo carried by the railroads increased from 30 million tons in 1959 to 48 million tons in 1969. After falling to 42 million tons in 1971, the growth in rail cargo transportation resumed, reaching 64 million tons in 1981 and 1985 before collapsing in the economic crisis of the mid-1980s. The transportation of passengers by rail, in contrast, rose and fell in spurts for almost thirty years between 1940 and 1969, when it reached a peak of 40 million passengers. After that, passenger transportation by rail entered into a sustained decline, dropping to fewer than a million people traveling by rail in 1999.

The one constant in the operation of the railroads was that it drained the resources of the state. Economic data on the railroads are generally even worse than data in other sectors of the economy. It is virtually impossible to find two sets of figures on the finances of the railroads that do not contradict each other. Even figures from the same source sometimes do not agree. This may be a consequence of the long history of state involvement in the sector, which tends to obscure revenues and losses behind the larger structure of the state, or it may reflect the ever changing organizational structure of FNM, a railroad that grew throughout its history by acquiring other rail lines.

Table 14 shows one of the longer series of data available for FNM's finances. According to these figures, FNM required a steadily growing amount of subsidization from the state to maintain its operations. Between 1965 and 1981, FNM progressively generated less and less of its own income and increasingly relied on transfers from the federal government. In 1965, FNM generated a little more than 88 percent of its own income. By 1981, it generated less than half of its own income, depending on the federal government for the remainder of its operating budget. The single, consistent theme that runs through every discussion of FNM's financial problems is

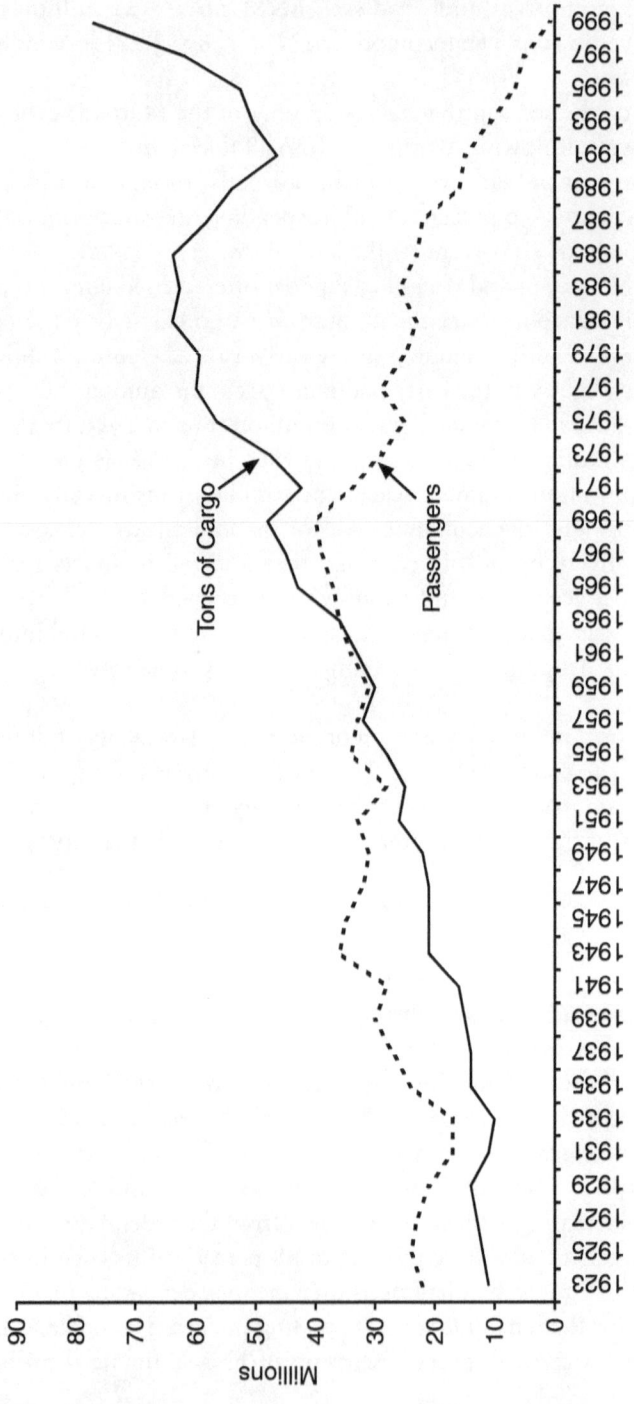

Fig. 12 Rail transportation, 1923–1999

TABLE 14 Income and transfers at FNM, 1965–1993 (millions of pesos)

Year	Total Income	Income Generated by FNM	Transfers from Federal Government	Percentage of Income Generated by FNM
1965	3.4	3	0.4	88.24
1966	3	2.6	0.4	86.67
1967	3.9	2.7	1.2	69.23
1968	4.4	2.7	1.7	61.36
1969	4.9	2.9	2	59.18
1970	5	3.1	1.9	62.00
1971	5.3	3	2.3	56.60
1972	5.6	3.1	2.5	55.36
1973	6.1	3.4	2.7	55.74
1974	7.5	4.1	3.4	54.67
1975	9.6	5.9	3.7	61.46
1976	11.1	5.7	5.4	51.35
1977	16.5	8.5	8	51.52
1978	18.6	9.8	8.8	52.69
1979	23	12	11	52.17
1980	31	15.7	15.3	50.65
1981	42.9	19	23.9	44.29
1982	65.6	31.3	34.3	47.71
1983	142.6	65.5	77.1	45.93
1984	211.2	143.7	67.5	68.04
1985	348.4	240.4	108	69.00
1986	638.2	439.1	199.1	68.80
1987	1,330.20	873.1	457.1	65.64
1988	2,259.80	1,717.80	542	76.02
1989	2,701.10	2,031.30	669.8	75.20
1990	3,301.00	2,423.80	877.2	73.43
1991	4,163.40	2,594.50	1,568.90	62.32
1992	4,685.60	2,653.00	2,032.60	56.62
1993	5,314.10	2,935.40	2,378.70	55.24

Source: Calculated from Salinas 1994, 57, 156.

that rates remained too low to cover operating expenses (see López Pardo 1997; Middlebrook 1995; Ruiz Dueñas 1988; Tamayo 1988a).[1]

Because of the history of foreign dominance in the sector, railroads were heavily dependent on imported capital goods. Backward linkages—locomotives, rail cars, and steel—reached directly into the international economy. As part of its larger policy to deepen Mexico's industrial base, the state pro-

---

1. Officials at FNM argued that passenger transportation had never been profitable and had always required subsidization. There was less agreement on the profitability of cargo transportation (Interview 62).

moted the development of domestic manufacture of essential inputs for the railroads. As discussed in Chapter 2, the Ávila Camacho administration created the national steel company, Altos Hornos de México, in 1942 and the Alemán administration founded National Rail Car Manufacturer in 1952 (Casar and Peres 1988, appendix 1; Novelo and Urteaga 1979, 51).

In the 1970s, as part of the effort to improve the process of planning, the state sought better coordination of the use of rail transportation services. Some of the main consumers of rail services were other parastate firms. In 1980, five such companies alone—Conasupo, Fertimex, Pemex, Sidermex, and Unpasa—accounted for more than a third of the cargo transported by railroad (Tamayo 1988a, 721). Between 1970 and 1976, the Echeverría administration attempted to rationalize the process of transporting the goods of parastate firms, further limiting the influence of the market on FNM's forward linkages (Dávila Capalleja 1999, 204).

Working on the Railroad and the Politics of Labor Resistance
As the Mexican state attempted to shape the development of the railroads from above, Mexican railroad workers struggled from within the institutions of the rail companies to influence the organization of this sector. The centrality of the railroads to Mexico's mineral exports and national development plans granted railroad workers considerable leverage in the debate over how to organize the sector. The strength of the Mexican Railroad Workers' Union (STFRM) and its active role in opposition politics also made it a target of government attempts to disorganize and deactivate the union, to limit the influence of labor over both the microeconomics of FNM and the larger political economy of Mexico.

Railroad workers had a long history of struggle in opposition to the foreign owners of the railroads. In the early 1900s, Mexican workers fought against a variety of workplace rules that favored U.S. workers. Mexican workers sought to improve their material conditions through basic guarantees—improved wages and working conditions, and through recognition by foreign companies of their labor organizations. They went out on strike to gain entry into skilled positions that were restricted to U.S. employees and to use Spanish in workplace communications. Railroad workers supported a policy of nationalization of the railroads from the beginning of the twentieth century and displayed a high level of radicalism in support of Mexico's revolutionary nationalism (Middlebrook 1995, 121).

The persistent economic problems in the railroads, the repeated attempts to reorganize FNM, and the delicate negotiations between foreign creditors and the Mexican government, made it impossible for workers to limit the focus of union politics to wages and benefits. Reorganization schemes invariably focused on the organization of work, not simply on the number of employees and their salaries. In the 1920s and 1930s, employment at Mexico's largest railroad oscillated by as many as ten thousand employees a year as FNM incorporated new rail lines into its structure or laid off employees (see Table 15; A. H. Chávez 1979, 133). Union leaders called for the government to take over the entire rail system and to increase rates for carrying cargo.

In addition to facing the challenges presented by the restructuring of the railroads, workers in this sector had to contend with the efforts of state-sponsored unions to co-opt their members. In the late 1920s, the leadership of the pro-government Regional Confederation of Mexican Workers attempted to infiltrate railroad workers' unions to prevent them from forming an independent power base among workers (A. H. Chávez 1979, 131). Railroad workers successfully resisted these efforts to undermine their unity and, in 1933, formed the STFRM, Mexico's first national industrial union (Middlebrook 1995, 120).

Labor solidarity among railroad workers, however, did not come easily. Many workers retained strong identification along craft lines. And although communists and members of other leftist groups occupied leadership positions within the STFRM until they were purged in 1948, STFRM support for independent, nationalist labor organizations and broader solidarity among Mexican workers came under attack from rank-and-file members worried about squandering STFRM resources on workers outside the railroads (Middlebrook 1995, 123, 140).

Still, by the mid-1930s, various attempts by management to reorganize the railroads and prevent the interference of the railroad workers had failed to produce a viable rail transportation system. Facing the same problems as his predecessors, President Cárdenas took an alternative approach. Consistent with his policy of national control of basic infrastructure and his alliance with organized labor, Cárdenas turned control of the management of FNM over to the STFRM in 1938.

Gaining control over the management of the company was, at best, a mixed blessing for workers. The company was still badly in debt and many of the more important elements of operational control were established by government decree. Although the STFRM named the entire seven-member executive board of FNM, the government placed a number of conditions on the railroad's operation: FNM expenses could not exceed 85 percent of its income, financial obligations of the company would be split evenly between the worker administration and the government, and FNM had to invest a fixed amount of total income to capital improvements and payments to the federal government. Nationalization redirected conflict between foreign

TABLE 15 Employment at FNM, 1910–1996

Year	Number of Workers	Year	Number of Workers	Year	Number of Workers
1910[a]	26,106	1940	43,377	1970[b]	73,268
1911	30,874	1941	43,414	1971	nd
1912	31,179	1942	46,938	1972	nd
1913	25,852	1943	47,807	1973	nd
1914	nd	1944	53,036	1974	nd
1915	nd	1945	59,192	1975	76,638
1916	nd	1946	46,941	1976	nd
1917	32,796	1947	65,536	1977	nd
1918	31,588	1948	57,487	1978	nd
1919	nd	1949	66,320	1979	nd
1920	31,500	1950	83,528	1980	78,838
1921	47,486	1951	76,595	1981	79,998
1922	49,426	1952	62,341	1982	79,713
1923	42,783	1953	65,127	1983	78,845
1924	42,783	1954	67,199	1984	79,669
1925	43,435	1955	66,567	1985	79,676
1926	42,576	1956	60,735	1986	81,132
1927	43,514	1957	nd	1987	81,670
1928	43,350	1958	nd	1988	81,248
1929	39,363	1959	nd	1989	82,928
1930	45,561	1960	nd	1990	83,290
1931	36,764	1961	nd	1991	78,114
1932	35,129	1962	nd	1992	58,626
1933	34,734	1963	nd	1993	55,664
1934	35,518	1964	nd	1994	49,323
1935	39,232	1965	nd	1995	46,283
1936	42,148	1966	nd	1996	44,169
1937	38,895	1967	nd	1997	nd
1938	44,773	1968	nd	1998	nd
1939	44,241	1969	nd	1999	nd

Sources: Years 1909–25 and 1926–37 are from López Pardo (1997, 31, 41, tables 4 and 9, respectively); years 1937–56 are from Middlebrook (1995, 126, table 4.1); years 1970–96 are from FNM (1997, 9).
Note: Changes in employment levels may reflect acquisition by FNM of other railroads.
nd = No data
[a] Years 1909–25 are for fiscal year; e.g., 1910 is fiscal year 1909–10.
[b] Years 1970–96 are reported as "Active Employees."

creditors and the state into conflict between labor and the state (López Pardo 1997, 56; Middlebrook 1995, 124).

During the brief experiment with worker administration of FNM, the union failed to meet most of its obligations. The STFRM reiterated its long-standing position that the railroads could not be solvent without raising the prices charged for carrying freight, but the government refused to raise rates (Middlebrook 1995, 130). A proposal in 1940 by STFRM managers to limit overtime payment, wages, and sick leave was rejected by the rank and file at a special congress (Middlebrook 1995, 125).

When Manuel Ávila Camacho became president of Mexico in 1940, he dissolved the worker administration of FNM and placed the company back under the direct control of the state (Middlebrook 1995, 125, 126). World War II, and Mexico's commitment to the Allied war effort, however, soon took precedence over attempts to make FNM solvent. As noted above, the Mexican government agreed to ship strategic minerals to the United States below cost, in quantities that constituted a significant portion of the total amount of cargo carried by FNM during the war years.

The growth in the amount of cargo carried by FNM between 1941 and 1945, and the absence of new equipment, however, required that the company hire still more workers. As Table 15 shows, during the war, the number of employees at FNM jumped by more than 25 percent, from 43,414 in 1941 to 59,192 in 1945. The STFRM showed its support for the war effort by agreeing to work with management and to suspend parts of its collective bargaining agreement during the war.

Although the experiment in worker administration ended in 1940, the collective bargaining agreement between railroad workers and FNM still contained important benefits for workers. These included basic material gains—overtime work, paid holidays, and health benefits—as well as restrictions on the authority that management had over labor, such as limitations on the right to transfer or fire workers and joint labor-management committees to preside over discipline (Middlebrook 1995, 128). According to León Guerrero, a railroad worker who chronicled the industry, there was no better collective bargaining agreement in Mexico than that of the railroad workers: "In this area, the Federal Labor Law has been left behind" (cited in Leyva Piña and Campos Rios 1990, 53).

Significantly, the proposals made by the Ávila Camacho administration for restructuring FNM focused less on reductions in employment or wage levels than they did on the need to increase managerial control over labor (Middlebrook 1995, 128). In contrast, the STFRM argued once again that FNM's problems could not be resolved without increasing the rates charged for freight. It also argued for the need to coordinate the construction and operation of railroads and highways to complement each other, and for the expropriation of privately operated rail companies (130).

In order to overcome worker opposition to the restructuring plans, the Ávila Camacho administration encouraged schisms within the STFRM. The government granted legal recognition to a faction of workers representing train crews and boilermakers, permitted automatic deductions of union dues for the splinter group, and allowed them to break the STFRM monopoly on the selection of candidates for employment with the railroads. The struggle within the STFRM enabled the government and FNM management to introduce major reforms at the company, such as eliminating the union's role in selecting administrative personnel, prohibiting ties between administrators and the union, and granting management increased flexibility in the assignment and disciplining of workers, measures it had sought from the earliest days of the Ávila Camacho administration (132, 133).

Yet even these reforms had to be softened with incentives designed to placate railroad workers. To make up for the changes in work rules, the state granted various improvements in wages and benefits in 1944. These benefits included subsidized recreational activities for workers, education for workers' children, credit and housing for workers, and special retail outlets with subsidized commodities for railroad workers (133).

Under President Alemán, the government continued to seek reforms in work rules at FNM. In 1948, a state-sponsored Tripartite Commission criticized the contract under which the STFRM worked, arguing that the provisions of the collective bargaining agreement "encourage waste of materials and fuel inefficiency in the use of equipment . . . satisfactory levels of productivity cannot be achieved without establishing effective administrative control in the workplace" (cited in Middlebrook 1995, 138). In addition to initiating this attempt to mandate greater efficiency, the Alemán administration provided support to a breakaway faction within the STFRM and succeeded in ousting radical leaders and reorganizing the union so that pro-government factions could centralize their control over the union (142).

This intervention in the internal organization of the STFRM and imposition of pro-government leaders, known as the *charrazo,* left the state with loyal allies in the leadership of the railroad workers union. Following the *charrazo,* STFRM leaders were rewarded with positions of power within the PRI, serving in the Mexican Congress as federal deputies and senators (142). The Alemán administration also implemented far-reaching contract changes under a federal labor law permitting an employer to change the terms of collective bargaining agreements "when altered economic conditions threaten a company's economic viability" (143).

Employment levels, however, actually jumped shortly after the *charrazo,* from 57,487 in 1948 to 83,528 by 1950, before falling back to 60,735 in 1956 (see Table 15). By the late 1950s, another major battle between railroad workers and the state took place. Leftist union leaders regained control of the STFRM in 1958 and their demands once again focused on the issue of rate reforms. In a report given to incoming president Adolfo López Mateos in early 1959, dissident STFRM leader, Demetrio Vallejo, emphasized the need to increase cargo rates so that none of the cargo would be carried at a loss. The report emphasized the subsidies that FNM continued to provide to foreign mining interests by transporting cargo at prices that failed to cover the railroad's costs. On average, FNM lost 2.3 centavos on every ton/kilometer of agricultural goods that it carried and 3.9 centavos on every ton/kilometer of minerals (Pellicer de Brody and Reyna 1978, 198).

The president rejected the price increases, and while negotiations between the union and management continued, STFRM leaders notified labor authorities that the union would go on strike in January, later postponing the strike date to February 25. As negotiations continued, labor demands began to focus on additional administrative issues. The STFRM presented demands that the administration of FNM be restructured to create a board of directors composed of "people with a broad understanding of the problems of transportation, patriotic, [and] without connections that influence the use of the firm for the goal of profit." The STFRM also proposed reducing the number of managers and administrative staff at the railroad (cited in Pellicer de Brody and Reyna 1978, 200, 205).

As the conflict with railroad workers began to spill over into other sectors of the economy (see Chapter 2), the government attempted to discredit the leadership of the STFRM, branding them communists and foreign agents. One day after the strike began, on February 25, the government and union reached a partial agreement, restricting the right of the company to leave positions open but focusing mostly on improvements in salaries and benefits. Continued negotiations over the Mexican, Pacific, and Western Railroads, however, began to undermine the union's solidarity. FNM and the state also adopted a harder line, firing eight thousand workers from the Pacific Railroad and five thousand from the Mexican (Pellicer de Brody and Reyna 1978, 209).

Federal police arrested the leadership of the STFRM along with fifteen hundred other workers in Guadalajara. Army troops took control of railroad installations and soldiers began escorting trains. By the end of the episode, FNM management acknowledged that ten thousand workers had been fired for "illegal agitation." Other estimates reached as high as twenty thousand layoffs. Many workers who lived in housing provided as part of their union benefits were thrown out of their homes (Pellicer de Brody and Reyna 1978, 210–13).

The purge of antigovernment dissidents from the STFRM in 1959 marked the beginning of a steady climb in the tonnage of cargo carried by the railroads, which grew from 30 million tons in 1959 to 64 million in 1981 (see Fig. 12). Pro-government union leaders never broached the topic of FNM administration and quickly announced their support for containing salaries. Still, the legitimation of the official leaders depended in part on providing real improvements to workers. STFRM leaders constantly struggled to strike a balance between the right level of repression and the incentives necessary to maintain the operation of rail transportation.

Labor unrest erupted again in the 1970s as President Echeverría opened the political system in response to the student movement of 1968. Dissident STFRM leaders, imprisoned in 1959, were released from jail in the late 1960s and almost immediately resumed their agitation in favor of an independent union (Leyva Piña 1995, 92). While Demetrio Vallejo and Valentín Campa organized among the rank and file, FNM's new general director, Victor Manuel Villaseñor, prepared his response to the company's perennial losses. Appointed to lead FNM in 1971, Villaseñor argued that workers were the source of FNM's problems because of "unjustified absenteeism in the workplace, as well as a lack of discipline, illicit practices among station managers, intentional reduction in the capacity of work in order to accumulate overtime, even pilfering of mechanical parts and apparatuses" (cited in Leyva Piña 1995, 22).

The attack on corruption within the labor force, however, hit close to home among pro-government leaders in the STFRM. Two days after Villaseñor took office, two trains crashed outside Guadalajara at a station named, not coincidentally, Villaseñor. Two weeks later, various locomotives took off without crews and eight locomotives crashed into the pit at the roundhouse in Mexico City. In San Luis Potosí, a train car was set on fire (Leyva Piña 1995, 23).

The recently freed dissident STFRM leaders were detained and questioned. They, in turn, blamed former STFRM general secretary Luis Gómez Zepeda, a loyal pro-government union leader who had himself hoped to be named director of FNM. In 1973, Gómez Zepeda was selected to replace Villaseñor as general director of FNM to contain both the dissident workers as well as his own followers, who were suspected of engineering the sabotage of FNM equipment. Although dissident workers once again insisted on raising rates to meet costs, the petroleum boom of the 1970s made financial restructuring less urgent (102). Transfers to FNM increased rapidly during the late 1970s and early 1980s as management sought to improve the operations of the company through investment in new equipment. Between 1970 and 1982, FNM acquired more than eight hundred new locomotives, increasing the stock of locomotives by almost 80 percent. During this period FNM also increased its investment in the preservation of existing track (FNM 1997, 6, 7).

MARKET FORCES IN A NONMARKET ENVIRONMENT

As the previous sections show, the Mexican railroads were shielded from market forces in a variety of ways. Railroads were originally built with subsidies from the state. Early in the century the state intervened to prevent the railroad companies from falling into bankruptcy, taking over a majority share of the largest railroads in the country. Prices were capped by the government, often below the actual cost of shipping, and profitability was never the dominant criterion upon which the railroads were operated. Instead, railroads were seen as a piece of the larger puzzle of economic development and were subsidized to promote economic growth, in particular the export of minerals.

The railroads were also protected from the market through the creation of nonmarket backward and forward linkages. Through publicly owned steel companies and the parastate firm National Rail Car Manufacturer (Concarril), the state attempted to create a buffer between the railroads and the international market for rail equipment. The state also sought to coordinate the provision of rail transportation services for parastate firms, which were among FNM's largest customers, taking the consumption of a large share of rail transportation services off of the market.

Yet even in this heavily shielded sector of the economy, the influences of price, profitability, and competition were not eliminated. Prices on international markets influenced the operation of the railroad in a number of different ways. To begin with, the origins of the railroads under the control of foreign capital made the price of money a crucial determinant of the operation of the railroads from the very earliest days. The foreign debt of the railroads was a source of constant concern, one that led to state intervention early in the twentieth century and to repeated reorganizations of the railroads throughout their history. Although the state managed to create soft budget constraints for the railroads, hard budget constraints remained a problem for the state, which had to negotiate lines of credit with the added burden of railroad losses on its balance sheet.

Attempts to limit the influence of international prices for capital goods through the creation of Concarril merely shifted the impact of price from one state firm to another. While the railroads may have avoided the effects of devaluations by buying locomotives and rail cars from a state-owned, domestic manufacturer, Concarril imported most of its capital goods.

Moreover, the parastate builder of rail cars and rail equipment never managed to provide all the inputs that FNM needed. In the 1970s, FNM management acknowledged that its imports of parts and equipment had grown more than tenfold in just four years, from approximately $22.6 million in 1973 to more than $263.9 million in 1977 (*Expansión* 1978, 62).

The prices of goods that the railroads carried also mattered. The collapse of international commodity prices at the end of the nineteenth century and again during the Great Depression reduced demand for rail service and created financial problems for the railroads. Nor was it possible to delink national prices entirely from international prices. With rates for exports set below those of domestic transportation, the railroads sought to make up the cost of shipping exports through elevated rates for carrying goods within the country.

Perhaps the most important form of market influence on the railroads, however, was the force of competition. Just as we saw with passenger transportation in the airlines, alternative means of transportation were available for cargo. Trucking companies were the most important source of competition for carrying cargo. Although the amount of cargo carried by the railroads grew steadily from the late 1940s until the late 1970s, that cargo represented a declining share of the total cargo shipped in Mexico. Between 56 and 70 percent of all cargo shipped in the country from 1952 to 1982 was carried by highway. During the same period, the percentage of cargo shipped by rail declined from a high of 40 percent in 1952 to 13.4 percent in 1982 (Islas Rivera 1990, 63; Salinas de Gortari 1994, 275).

The competition faced by the railroads might be interpreted as a sign of market forces escaping the fortress of the command economy. Yet the existence of competition from highway and maritime transportation depended on federal investment in highways and ports. Although the numbers shift quite drastically from year to year, as shown in Table 6 there is a long-term trend of federal investment among different forms of transportation until the economic crisis in 1982. Throughout the 1950s, the federal government devoted approximately half of all federal investment in transportation to the railroads and a little more than 40 percent to highways. In the 1960s, the percentage of federal investment devoted to highways began increasing, surpassing 60 percent of all investment in transportation in the early 1970s. Federal investment in railroads, in contrast, declined to around 30 percent of all federal investment in transportation in the 1960s, falling as low as 15 percent in the early 1970s.

Federal investment in highway transportation can also be seen in the growth in the number of kilometers of paved highway capable of carrying cargo transportation compared with the kilometers of rail. Because rail transportation is not practical over short distances or on steep grades, the lengths of the highway network and the rail network are not directly comparable. Figure 13, however, does show a steady growth in the kilometers of paved highways at the same time as the kilometers of rail stagnated. Between 1952 and 2000, the number of kilometers of rail remained almost constant at about twenty-five thousand. During the same period, the kilometers of paved highway grew more than fivefold, from a little more than sixteen thousand to more than ninety thousand kilometers in 1994.

Highway transportation, while also heavily regulated in Mexico, has very different characteristics from those of railroads. In addition to the technical considerations that distinguish rail and highway transportation, the two forms of transportation were regulated very differently. Trucking routes were granted as concessions by the SCT and rates were established through a process of review with input from a committee of concessionaires operating on particular routes. Rates were set under the premise that prices "cover average fixed and variable costs of service and provide a 'reasonable' margin of profit" (Dávila Capelleja 1991, 123–27). Significantly, these costs did not take into account the very infrastructure upon which trucking and bus services depend. Even after eight years of price hikes on tolls to cover a growing percentage of the costs of highway construction, by 1988, income from tolls covered barely 40 percent of investments in highway infrastructure (Sales, Sclar, and Videgaray 1999, 269).

The percentage of federal investment in ports and maritime shipping also grew to represent a significant percentage of total investment in transportation, especially in the 1970s and early 1980s. The growth in marine transportation coincides with the petroleum boom and reflects the increasing use of tankers to carry crude oil from offshore sites. Approximately two-thirds of all maritime cargo is international traffic, among which exports predominate over imports: exports make up more than 80 percent of Mexico's total international traffic. Petroleum products represent almost 80 percent of all maritime exports and more than half of all maritime transportation in Mexico, with petroleum products making up slightly less than two-thirds of domestic cargo (SCT 1997a, tables 5.5–5.10).[2]

The competition that the railroads faced, then, was among differently regulated industries. The reasons for the differences in regulation touch on

---

2. These numbers are intended only to illustrate broad trends. The extent to which alternative forms of transportation compete with or complement rail transportation cannot be inferred directly from these figures because in some cases shippers will use highway or maritime transportation *instead* of rail while in other cases, rail will make up one leg of a journey that includes highway or maritime transportation or both.

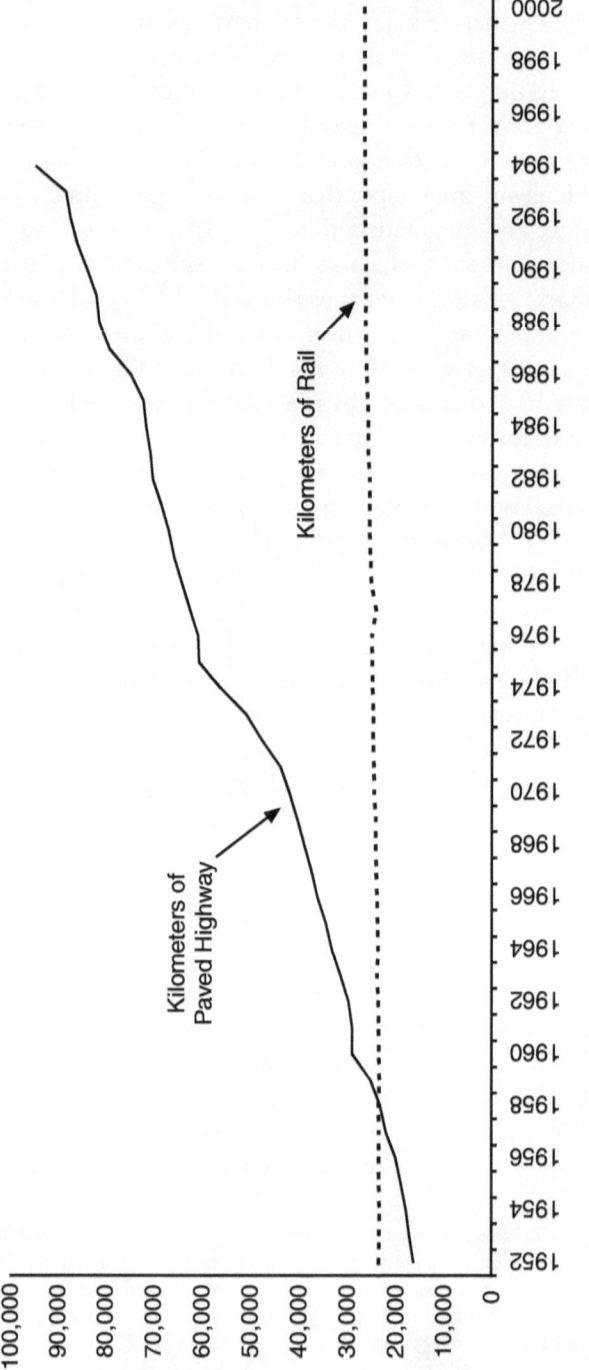

FIG. 13 Growth of rail and highway infrastructure, 1952–2000

Sources: Islas Rivera (1992, table 2.1), Zedillo (1996), Fox (2002). Data series on kilometers of paved highway ends in 1994 due to a change in methodology adopted by Fox (2002, 353).

virtually every element of the technical and social characteristics of the sectors and the interaction between these material and social forces. As discussed at the outset, railroad transportation requires heavier investment in fixed capital and generally favors a single carrier. Highway transportation, in contrast, while also dependent on long-term fixed investment, can support the operation of many small users of the highways. The concentration of ownership in railroads supported the organization of labor into a single, powerful industry union. Truckers, by contrast, are organized primarily as small owners.

The Consolidation and Reorganization of the Mexican Railroads

Unlike the other privatizations examined here, it is not clear that the initial steps taken to restructure FNM were intended actually to privatize the railroads. Throughout its long history, the rail system suffered from a wide range of dysfunctions. Over the course of the century, the state slowly acquired more and more of the track and rolling stock. Still, early in the de la Madrid administration, as the new president took the final steps toward consolidating the rail system and began yet another reorganization, it was far from certain that the president planned to go as far as to sell the railroads.

OPERATIONAL RESTRUCTURING AT FNM: CONSOLIDATION AND
PARTIAL PRIVATIZATION

Barely two months after president de la Madrid took office, the Mexican Congress amended Article 28 of the Constitution, adding railroads to the list of strategic sectors of the economy. As discussed in Chapter 2, a strategic sector of the economy is defined as an area of the economy over which the federal government is charged with maintaining both property and control of the sector (*Diario Oficial* 1983). More important, in contrast to "priority" sectors of the economy, in which both public and private ownership is permitted, private capital is excluded from participation in strategic sectors of the economy.[3]

3. Strangely, the incorporation of railroads into Article 28 of the Constitution is almost never discussed in the literature that looks at the restructuring of FNM. The official history of FNM, *Breve reseña histórica de los ferrocarriles mexicanos*, completely overlooks the constitutional reform and in the summary of the period from 1982 to 1987 simply declares that President de la Madrid continued the work of rehabilitation and modernization of the diverse rail lines (FNM 1987, 58). Leyva and Campos Rios (1990) and Leyva (1998) also omit any discussion of the constitutional amendment.

President de la Madrid's National Development Plan, published in May 1983, elaborated on the specific goals of the new administration but still fell short of declaring the railroads a target of privatization efforts. Pointing to the use value of transportation in general, the plan argues that the transportation system "enables the economic, political, social, and cultural integration of the country and permits the exercise of sovereignty over the national territory" (SPP 1983, 351). The plan goes on to argue for the need to incorporate greater participation of rail into national transportation. The list of specific goals for the rail system includes the need to "complete the merger of the four rail companies in existence, modernize their administration and promote actions that help to increase productivity" (358).

With a presidential decree in November 1986, the de la Madrid administration liquidated the last remaining railroads outside FNM—Ferrocarril del Pacífico, Ferrocarril Chihuahua al Pacífico, and Ferrocarril Sonora-Baja California—and folded their assets, along with a sleeping car service, into the administrative structure of FNM (FNM 1987, 60). In its report on the restructuring, the SCT points to the merger of the remaining rail companies as a means of "eliminating the obstacles to the flow of cargo traffic and the simplification of paperwork in the transportation of people and merchandise" (SCT 1994c, 4).

Part of the reorganization focused on improving productivity through targeted investment in infrastructure and the retirement of outdated equipment. The number of locomotives in service began a steady decline in 1985, falling from a high of 1,878 in 1984 to 1,318 by 1996. The average horsepower per locomotive, however, increased by a little more than 12 percent throughout the same period, indicating that the FNM administration was retiring older, less powerful locomotives and investing in the maintenance of newer ones. Thus, while the number of locomotives in service had fallen by almost 30 percent between 1984 and 1996, the total horsepower available fell by only 20 percent (FNM 1997, 7).

In the mid-1980s, FNM actually began to increase the amount of rail it replaced and maintained. From 1980 to 1984, it added an average of 236 kilometers of new rail every year. In the second half of the decade, FNM more than doubled the amount of new rail, laying an average of 534 kilometers of track annually from 1985 to 1989. In the 1990s, the amount of new rail laid and old rail recovered fell off again and hit all-time lows at the end of the Salinas *sexenio* and the beginning of president Zedillo's term in office (FNM 1997).

In 1986, the de la Madrid administration pushed through the Mexican Congress a new law governing the organization of FNM. The Organic Law of the Mexican National Railroad, like the Federal Law of Parastate Entities discussed in Chapter 3 and passed in the same year, brought increased financial oversight to FNM. The railroad law named the SHCP, SPP, and minister of commerce and industrial promotion to the company's board of directors and handed authority to monitor the company to a superintendent (*comisario*) named by the controller general. The law also directs the controller general to designate external auditors to FNM (*Diario Oficial* 1985, artículos 6, 16).

Another similarity between the new law on railroads and the Federal Law of Parastate Entities was that the railroad law failed to wrest operational control of FNM from employees and public officials with an interest in maintaining the firm under state control. Thus, although the law enabled the financial branch of the state to monitor the operations of the company more closely, it also checked their power by placing on the board of directors representatives of FNM's labor union and other ministries who were dependent on the railroad. The law also provides for the SCT to name two representatives to the board of directors from "organizations of users of services" (*Diario Oficial* 1985, artículo 6).

It was not until the publication of the National Development Plan in the first year of the Salinas presidency that the outlines of a strategy for privatization of the railroads began to emerge. President Salinas's National Development Plan, published in May 1989, makes the same assertions as other official declarations about the essential role of communications and transportation in national development and the importance of state rectorship over these sectors. However, he also goes on to argue that "[t]his function of rectorship is complemented and strengthened by the participation and resources of private actors in the construction and operation of infrastructure" (SPP 1989, 5.3.4.1). In addition, the plan emphasizes the need for greater competition in highway transportation, improved coordination among different modes of transportation, and the need for rates to cover the costs of operation.

Organizational restructuring during the Salinas administration consisted of the gradual introduction of private capital into peripheral areas of FNM's operations. While the introduction of new services to target the changing composition of rail traffic followed naturally from the restructuring begun under President de la Madrid, FNM also began working with the private cargo transportation company Mexican Maritime Transportation (TMM) to improve its service. Together, FNM and TMM increased the railroad's use of containerized shipping and began providing double-loading rail cars for containers and for shipping automobiles that were produced for Mexico's growing export market (SCT 1994c, 6, 7).

Under the Salinas administration, private companies were encouraged to organize the loading and unloading of private rail cars, with FNM providing only the locomotives. In 1991, the SCT signed an Agreement on Actions to Modernize the Mexican Rail System with representatives of three of Mexico's major business chambers, who were devoted to promoting foreign trade with Mexico (SCT 1994c, 8; American Chamber of Commerce of Mexico 1995, 36–38). The National Association of Exporters, the National Confederation of Chambers of Industry, and the National Council for Foreign Trade each entered into the agreement with FNM, which allowed private companies to begin using their own equipment and installations in conjunction with the still-public railroad. By 1995, private investment in this area is estimated to have surpassed $130 million (*Expansión* 1996, 25; SCT 1995a, 5).

FNM also increased its cooperation with private, foreign companies during the Salinas administration. The state railroad entered into service agreements with Burlington Northern and the South Orient Railroad to operate a short line running from the city of Chihuahua to the Texas border at Ojinaga. And FNM signed agreements with North American railroads to rent and exchange equipment (SCT 1994c, 12).

Throughout the 1980s, however, FNM continued to lose ground relative to other forms of cargo transportation in Mexico. Moreover, the amount of cargo carried by FNM declined in absolute terms. As Figure 12 shows, after the amount of cargo hauled by rail peaked in the early 1980s, tonnage began to fall and by 1991 declined to levels not seen since 1973. Although the tons of cargo carried by FNM began to rise again in 1992, FNM's percentage of all cargo hauled in Mexico continued to decline, falling from 13.4 percent in 1982 to 8.2 percent in 1994. The percentage of total cargo weight carried by maritime shipping jumped in the 1970s, growing from 14.9 to 27.9 percent of all cargo carried in Mexico between 1970 and 1980, and then stabilized at between 27 and 31 percent of cargo transportation in Mexico during the 1980s. At the same time, the cargo weight shipped by highway grew slowly but steadily through the decade, from 57 percent of national cargo transportation in 1980 to 62 percent in 1994 (Salinas de Gortari 1994, 275).

There are two exceptions to the overall decline in the amount of cargo carried by FNM: agricultural products and industrial goods. During the 1980s and early 1990s, the absolute and relative amount of petroleum products, minerals, and inorganic products carried by FNM all fell. In contrast, the amount of agricultural cargo hauled by FNM grew slightly from 12.7 million tons in 1982 to 14.8 million in 1996. The most significant growth, however, was registered in industrial products, composed primarily

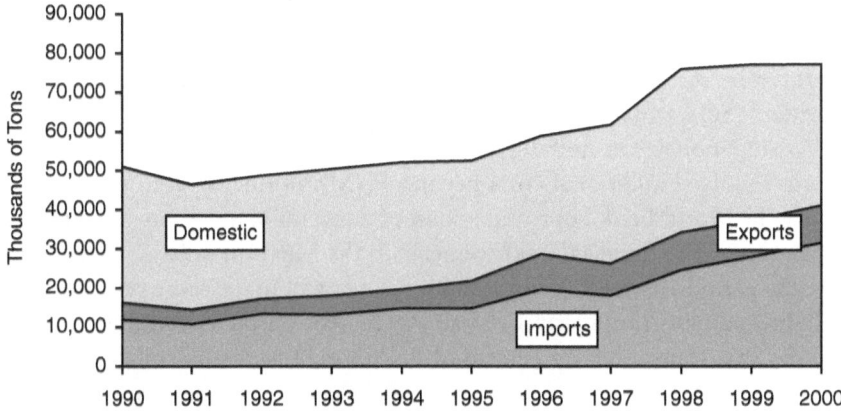

FIG. 14 Foreign and domestic cargo carried by rail, 1990–2000
Source: Fox 2002, 354.

of automobiles and automobile parts. The amount of industrial products carried by FNM grew from 17.9 million tons in 1982 to more than 26 million in 1996. Consequently, this part of FNM's traffic made up a growing portion of the total cargo carried by the company, rising from a little over 31 percent of FNM's freight to more than 45 percent between 1982 and 1996 (FNM 1997, 14–19).

The changing composition of FNM's traffic is part of a larger reorientation of Mexican production toward foreign markets and can be seen in Figure 14. Between 1990 and 1996, the amount of domestic cargo carried by FNM declined slowly from 34.7 million tons to 30.2 million. Although domestic cargo transportation by rail recovered in 1997, jumping to 41.7 million tons by 1998, domestic hauling slipped again in 1999, and fell back to 36.2 million tons in 2000. At the same time, the amount of international cargo carried grew by more than 75 percent between 1990 and 1996, from 16.2 million tons to 28.6 million. By the end of the decade, the total tonnage of foreign cargo had surpassed domestic, reaching almost 41 million tons. The steady growth in international hauling is driven primarily by rapidly growing imports, which made up more than three times the amount of cargo exported by the year 2000.

FINANCIAL RESTRUCTURING

In 1983 the de la Madrid administration published a National Rail System Modernization Program. FNM continued to drain resources from the state and the Modernization Program articulated the need "to achieve self-sufficiency through the rationalization and selective application of transfers granted by the state; [and] to increase productivity and modernize the rate system" (SCT 1994c, 4).

Losing money was nothing new at FNM. However, restructuring the railroads involved additional costs beyond FNM's normal operating losses. In 1985 the World Bank approved a loan of $300 million for "investment and operational and financial improvements to the Mexican railway system during the period 1985-88" (World Bank 1985, 142). In the same year, the de la Madrid administration approved an Agreement on the Assumption of Liabilities that transferred 47 percent of FNM's debt to the federal government at a cost of $627 million. Two years later, the state absorbed an additional $380 million in debt, most of which came from the consolidation of the remaining private rail companies and their losses into FNM (SCT 1994c, 4).

As in the past, a key problem with the finances of the railroads was the rate structure. The failure of FNM to adjust its prices for inflation meant that in 1982 the rate it charged for cargo had fallen in real terms to less than half the rate in 1960 (Tamayo 1988a, 735). The de la Madrid administration authorized FNM to establish fares that would cover operating costs, leading to a series of price hikes through the Salinas administration. Between 1982 and 1993, FNM increased its prices per ton of cargo enough to offset in part the losses in income from the reduction in cargo it carried. The price per ton of cargo charged by FNM actually surpassed the rate of inflation during this period (SCT 1994a, 12, 13). Thus, despite the fact that FNM was carrying almost 25 percent less cargo in 1993 than it hauled in 1982, its cargo revenue had fallen by only a little more than 10 percent.

Still, FNM continued to lose money and market share. In 1996, the Mexican business magazine *Expansión* reported that the price of railroad cargo transportation was only 32 percent the price of highway transportation (*Expansión* 1996, 27). And although FNM attempted to improve its financial condition by limiting the number of partially full cars that it hauled and carrying higher-value cargo, the company still needed an infusion of $765.8 million from the state in 1993.

RESTRUCTURING INDUSTRIAL RELATIONS

Despite the fact that the leadership of the STFRM supported the government and its efforts to reform FNM, restructuring in the railroads was a much more prolonged process than in the other cases examined here. Bankruptcy at Aeroméxico provided the justification for firing workers and restructuring the labor agreement. At Telmex, a protected market and the promise of strong growth in the telecommunications sector facilitated compromises between labor and capital. Restructuring of the railroads, in contrast, involved a slow bleeding of workers from FNM until, finally, in 1997, all but a handful of STFRM workers were severed completely from the firm.

Like the operational restructuring of the railroads, the transformation of industrial relations at FNM had been a source of constant struggle since the late 1920s. Already one of the largest employers in the country, FNM grew still larger in the 1980s through the consolidation of the railroads. Between 1982 and 1990 the number of workers at the company increased from 78,838 to 83,290. Even so, FNM limited the growth of employees by encouraging "voluntary" retirement. While the number of active employees grew by almost four thousand between 1982 and 1990, the number of retired employees grew by more than thirteen thousand during the same period.

After the total number of employees peaked in 1990, employment began to fall rapidly at FNM. As Table 15 shows, the number of active employees at FNM fell by more than 5,000 workers in 1991, then collapsed from 78,114 workers to 58,626 workers in 1992. By 1996, the number of active workers at FNM had fallen to almost half the number employed in 1990 and below the number of employees at FNM in any year since 1941.

The acquisition of other railroads by FNM in the 1980s resulted in a reduction of benefits to workers joining FNM as collective bargaining agreements were adjusted to conform to the collective bargaining agreement at FNM. In 1982, workers who joined FNM from Ferrocarriles Unidos de Sureste lost three holidays, lodging on trips that end away from home, death and maternity benefits, overtime payment, and dental and corrective vision benefits. In the 1987 mergers, workers who joined FNM lost two holidays, travel and survivor benefits, and company sponsorship of sporting events. In the case of Ferrocarril del Pacífico, workers lost the rights to consult on matters of corporate reorganization (Leyva and Campos 1990, 63, 64).

Dissident workers at FNM were slow to develop a response to the restructuring efforts at FNM. It was not until July 1984—the same month that the government entered into an agreement with official union leaders and the management of FNM to modernize the railroads—that two different opposition groups at FNM joined forces to create the Democratic Railroad Worker Movement (MFD) (Leyva 1995, 115). The opposition movement attempted to generate support among the rank and file with meetings and rallies protesting the restructuring. The MFD argued that they were not opposed to the modernization of FNM but, instead, were against the transformation of workplace rules and job categories (116).

In 1985, the MFD leadership toured the entire rail system, seeking support for their demand that FNM continue providing passenger service, speaking in defense of the collective bargaining agreement, and advocating for better salaries and benefits. In particular, opposition leaders argued that the mergers of new firms into FNM should result in a leveling upward to bring the salaries of all workers into line with the highest wages. The MFD also denounced the subcontracting of repair work on engines to private companies (Leyva 1995, 117, 119).

As usual, however, the state overcame labor opposition using both incentives and coercion. Although workers who became FNM employees through the merger of their firms into the National Railroad lost important benefits, FNM employees were granted increases in salaries and benefits and productivity incentives during the same period (Leyva 1995, 116). During the Salinas administration, FNM provided additional incentives to FNM workers to maintain the peace. To encourage retirement, FNM increased the amount of pensions to match the salaries of active workers (Leyva and Campos 1990, 60).

It is a testament to the political skill of pro-government union leaders that there was so little open resistance to the restructuring of the railroads. It was also, very likely, a consequence of the depth of the economic crisis, and the sheer exhaustion of FNM workers. Middlebrook's characterization of organized labor's response to neoliberal reforms in Mexico as "the sounds of silence" is an apt description of the inability of railroad workers to offer any effective resistance to the transformation of the industry.

Getting FNM Ready for Market

At the end of 1994, despite a reduction of more than thirty thousand jobs, the cancellation of more than $1 billion of debt, and the introduction of new services and private capital into the operations of FNM, the railroad was still in no condition to be sold. The problem was not simply that FNM management had failed to staunch the financial bleeding at the company. Economic woes aside, the consolidation of the railroads into a single industrial behemoth created yet another problem for FNM management: devising a marketable structure for the railroads, some organizational model that would attract private investment.

As the sole provider of rail service in Mexico, FNM presented many of the same problems as Telmex—in particular, its size and market power—with none of the advantages. Moreover, the challenge of privatizing FNM became especially urgent following the economic crisis in the first month of the Zedillo presidency. President Zedillo hoped to show foreign investors that he remained committed to the economic model pursued over the previous two *sexenios*. However drastic the changes at FNM, they had nonetheless taken place deliberately over the course of more than twelve years. When the peso collapsed in December 1994, incrementalism was no longer an option.

PREPARING THE LEGAL FRAMEWORK

The first problem facing the Zedillo administration was the fact that the 1983 amendment to the Constitution designated the railroads a strategic sector, thus prohibiting private-sector involvement in their operation. Although the constitutional prohibition had not prevented President Salinas from introducing private capital into peripheral areas of rail operations, ridding the state of FNM would require that the Constitution be amended once more.

Barely two months into his term, in February 1995, the new president had moved the constitutional amendment through the Congress (*Expansión* 1996, 26). Two months later, the implementing legislation was passed. While the preface to the new law reiterates the state's "rectorship and authority for ensuring that the offering of this important means of transportation be realized in conditions that benefit society" it also points to the inability of the state to make the investments needed to modernize the railroads (*Diario Oficial* 1995b).

Instead of authorizing an outright sale of the railroad, however, the new law created a framework for establishing concessions for private firms to operate rail companies. Thus, Article 8 of the law indicates that "the general means of rail communication will be maintained in all instances within the public domain of the Federation" while Article 9 establishes the bases on which concessions can be granted. The law grants the SCT the authority to solicit bids for rail concessions and specifies the terms on which the bidding will take place.

With the scandals of the Salinas administration and the collapse of recently privatized banks appearing in the daily news, government officials emphasized nonmarket criteria in the new legislation. Rather than make price the principal consideration in accepting bids, Article 9 indicates that technical, administrative, and financial capacity will all be taken into account as well as the proposed investment, the quality of service, the projected sales volumes, and the formula for establishing rates. Article 11 establishes fifty years as the maximum duration of a concession but permits an extension of an additional fifty years if the concessionaire has met the conditions of the original concession, accepts any new conditions determined by the SCT, and has improved the operating quality and efficiency of rail service.

The law also places limits on foreign investment in railroads. As in the other cases examined here, foreign investment is restricted to 49 percent of the ownership of firms that buy concessions. However, Article 17 allows for exceptions if the National Commission on Foreign Investment determines that majority foreign ownership will promote regional and technological development and that the sovereignty of the nation will be safeguarded. Concessionaires are required to notify the SCT if they modify the statutes of their company in terms of their business plans or ownership structure.

While the law places these limits on the property rights of concessionaires, it also permits the sale of rolling stock. Article 13 grants new owners the freedom to sell property apart from track and rights of way. At the end of the concession, all rail, rights of way, and centers for control of traffic will revert to the federal government (Article 14).

The fifth transitory to the law states that concessions and permits granted under the law "will not affect the rights of workers, active, retired, and pensioners of the public, decentralized organism, Ferrocarriles Nacionales de México, which will be respected in accordance with that which is established in Article 123 of the Constitution and Federal Labor Law" (Transitorio Quinto). However, Article 56 reasserts the right of the state to use the *requisa* and intervene in the railroads to maintain their operation in case of "natural disaster, war, serious disturbance of public order, or when some imminent danger to national security, domestic peace in the country, or to the national economy is foreseen."

PREPARING TO SELL THE RAILROADS

Creating a new legal framework for privatization of the railroads ultimately proved easier than creating a new organizational structure for FNM. After President Zedillo reshuffled his cabinet in early 1995 (see Chapter 3), Carlos Ruiz Sacristán was named minister of communications and transportation. As the Unit for the Support of Structural Change (UACE) set up shop in the basement of the SCT's main offices, it formed a Restructuring Committee, specifically for railroads. The committee included veterans from the privatizations of the Salinas administration, who wasted no time in reviving the model used under President Salinas: agent banks and high-profile international consulting firms were brought in to handle the details while the Restructuring Committee oversaw major strategic decisions. The principal difference was that these officials no longer worked out of the Ministry of the Treasury: now they operated from within the ministry in which the privatization was taking place.

The Restructuring Committee contracted with Mercer Management Consulting, C. S. First Boston, and Price Waterhouse, along with Banco Serfin as the Mexican agent bank. These firms began the process of assessing FNM's assets, including 26,445 kilometers of rail, 1,370 locomotives, and more than 36,000 rail cars, attempting to categorize and value them in some way that could be reflected in a price (Chaffin 1997, 39). A leading official from the government's team remarked that at one point in the process an auditor from FNM sheepishly reported to him that he had only been able to account for 94 percent of the rolling stock in the system: "That much?!" was the official's response (Interview 65).

In order to assess FNM's assets and devise a plan for privatization, the consultants, accountants, and agent banks put together a "data room" that would provide an overview of the system and eventually serve as the key source of information for prospective buyers. The consultants soon realized that the physical condition of the railroad was not as bad as had been assumed. Still, what exactly a marketable railroad would look like was unclear (Chaffin 1997, 39).

Because a relatively small number of firms consumed the majority of rail service—FNM officials claimed that approximately three hundred users represented 80 percent of FNM's cargo traffic and income—government officials and their advisors were able to confer with these firms in the process of assessing the structure of FNM. Representatives of the automobile industry were especially concerned that the lines running through the middle of the country not be left in the hands of a single rail operator. These lines run through Saltillo, a growing industrial center near the site of a Chrysler factory (Interview 51; see Fig. 15).

In November 1995, the Restructuring Committee and its consultants settled on a plan to divide FNM into three main lines and numerous short lines (see Table 16 and Fig. 15). The Northeastern Railroad (Ferrocarril del Noreste) connects Mexico City with the U.S. border, at Laredo and Brownsville, Texas, traveling through the rapidly industrializing state of Nuevo Leon. The Northeastern Railroad also runs east-west, connecting to the Pacific port of Lázaro Cárdenas and the gulf ports of Tampico and Veracruz. The

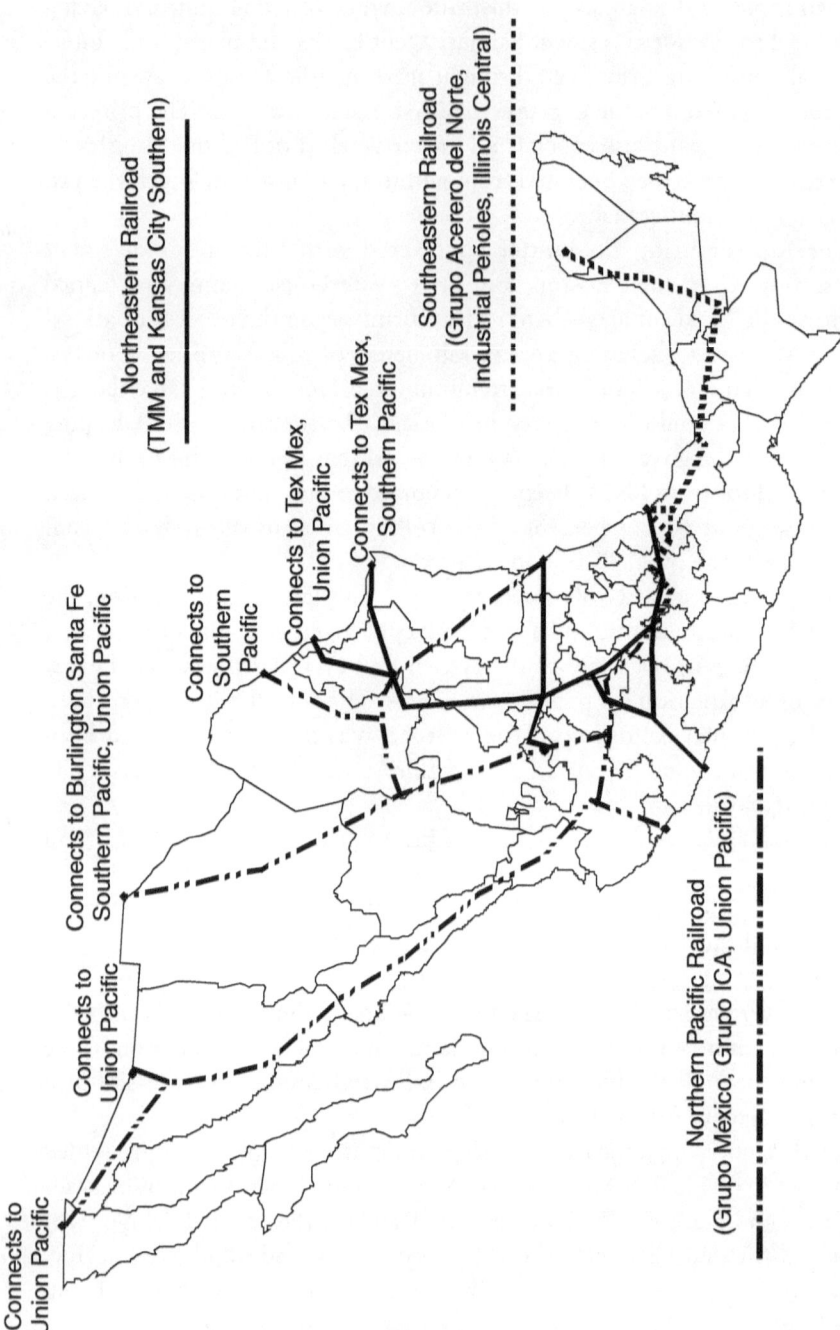

FIG. 15 Restructured rail system
Source: SCT 1995a.

Northern Pacific Railroad (Ferrocarril del Pacífico Norte) consists of longer routes to the U.S. border, with one line following the Pacific coast and touching the ports of Manzanillo and Mazatlán before connecting with Mexicali, California, and Nogales, Arizona. The Northern Pacific Railroad also stretches north, up the center of Mexico through mining regions and into El Paso and Eagle Pass, Texas. Another line of the Northern Pacific Railroad splits off from the north-south route near Torreón and passes through Monterrey, Nuevo Leon, on its way to the gulf port of Tampico (SCT 1995a).

The third main line, the Southeastern Railroad (Ferrocarril del Sureste), was originally designed to connect Mexico City with its principal gulf port, Veracruz; extend south into the Yucatán Peninsula; and connect the Gulf of Mexico with the Pacific port of Salinas Cruz, crossing the Isthmus of Tehuantepec. Controversy over the proposed Southeastern Railroad eventually forced officials to change the plan. Domestic freight makes up a much larger percentage of this line's total operations than it does of either of the other lines. And the gulf-Pacific route is considered strategic because of its proximity to Mexico's oil deposits and the fact that the Isthmus of Tehuantepec is a potential competitor with the Panama Canal. Finally, the government's conflict with guerrillas in this region led officials to reserve the gulf-Pacific crossing in the hands of the state while selling a concession for the remainder of this route (Interviews 51, 62).

According to a UACE official involved in the process, "[D]ividing FNM was a nightmare." Accountants first had to combine the financial and operating records of the multiple different railroads that FNM had absorbed to create some sense of what the company's total assets actually were. Then this data had to be redivided in order to create the sales prospectuses that would give potential buyers a sense of the property and operations that made up the separate routes (Interview 48). A technical evaluation assessed which lines used which types of cars to determine how to assign rolling stock in accordance with the profile of the traffic. With the exception of a maintenance shop in Jalapa, Veracruz, which went to the Southern line, shops retained their association with the lines to which they were connected (Interview 51).

In the sales prospectus for the privatization of FNM, the SCT points to the growing importance of industrial cargo, noting that "in the first eight months of 1995, the number of automobiles transported by railway increased by 19 percent over the same period in 1994. Also for the same period, 49,000 containers of automotive parts were handled, 33 percent more than last year" (SCT 1995a, 4). Although the prospectus does not say

TABLE 16 Principal characteristics of newly created railroads

	Northern Pacific	Northeastern	Southeastern	Short Lines	Totals
Length (Km)	6,200	3,960	2,200	7,950	20,310
Percentage of FNM	30.3	19.3	10.7	38.7	100
Total Traffic in 1994 (Millions Ton/Km)	17,200	14,000	3,200	2,900	37,300
Percentage of FNM	46.2	37.6	8.6	7.8	100
Domestic Traffic	10,600	6,000	2,300	nd	18,900
Percentage of FNM	49.7	28.1	10.8	nd	88.6
International Traffic	6,600	8,000	900	nd	15,500
Percentage of FNM	41.5	50.3	5.6	nd	97.4
Revenue in 1994 (Millions of New Pesos)	1,230	1,020	270	230	2,750
Percentage of FNM	44.7	37.1	9.8	8.4	100
Revenue/Km. in 1994 (New Pesos)	198,387	257,575	122,727	25,136	603,825

Source: SCT 1995a.
nd = No data

so, the Restructuring Committee arranged to sell the lines free of debt and without a single worker. Significantly, in its profile of the three main lines, the prospectus omits any mention of the workforce. The collective bargaining agreement at FNM, which once included 3,045 clauses, was renegotiated by the firm that won the bid for the Northeastern Railroad to contain only 38 clauses and allow for geographic mobility and subcontracting of work if needed (Correa 1997, 17; Interview 75).

THE PRIVATE SECTOR AND THE RAILROADS

Government officials and their international consultants were not the only ones attempting to assess the value of the railroads. The Mexican cargo transportation firm TMM, had already participated in restructuring FNM's operations during the de la Madrid and Salinas administrations. TMM had a special interest in the operation of the railroads because the company had recently begun to diversify from its primary business, carrying maritime freight, into ground transportation and logistics for Mexico's growing automobile industry. Railroads would permit TMM to operate multimodal transportation networks, connecting rail, highway, and maritime transportation with containerized shipping. During the Zedillo administration, TMM purchased concessions to operate three ports and provide stevedoring services at a fourth, further strengthening its position in the transportation sector (TMM 1996b, 16).

According to one official working in TMM's land operations area, the company determined in 1991 that it needed to take advantage of the growing trade between the United States and Mexico. The negotiation of NAFTA made it even more clear that the future of transportation in Mexico would increasingly run north and south rather than east and west. Initially, TMM entered into trucking with the goal of making itself into an integral logistics company, offering door-to-door service, warehousing, and other logistics operations (Interview 33).

The official went on to note that the crisis of 1994 presented a major opportunity for TMM. With the exception of petroleum, Mexico normally imports much more than it exports. This causes real problems for transportation firms, especially with the growing use of containerized shipping, because "you end up transporting air." Historically, TMM dealt with the problem by sending containers to the United States from Asian and European routes to make up for the imbalance. With the collapse of the peso at the end of 1994, TMM enjoyed one of its best years in 1995 as containers of exports and imports balanced out to almost a one-to-one ratio. The crisis was also a boon to TMM because the majority of its income came in dollars, while the majority of its costs were in pesos (Interview 37).

Still, the historical record of the railroads—their labor problems, financial insolvency, and operational difficulties—could hardly inspire confidence in private investors. Thus, while the Restructuring Committee worked with its consultants to create a new organizational structure for railroads, TMM brought in its own experts. In 1994, TMM hired Brad Skinner, a former vice president of Intermodal Operations at Southern Pacific, to help TMM break into trucking. Skinner subsequently led the research effort to determine whether, how much, and on which railroad TMM should bid for a concession (Interview 75).

Under Skinner's direction, TMM hired the consulting firm Kingsely of Mexico to assist in the evaluation of the railroads. According to one official close to the process, the analysis was exhaustive and can be broken into two main categories: market analysis and operational evaluation. The market analysis consisted of 189 interviews with customers of transportation services in which the customers were asked about their "modal preferences" and what it would take for them to use the railroad. The operational analysis involved close inspection of the track—driving the length of the railroad in a Chevrolet Suburban fitted with rail wheels and checking the quality of the track bed, the rails, and even the nails—as well as assessment of other physical assets, labor, and potential environmental problems. Another sixteen experts were hired to analyze the cost side and looked at technological innovations that might be introduced (Interview 75).

By December 1996, TMM signed a letter of intent to join a relatively small U.S. rail company, Kansas City Southern (KCS), to create Mexican Rail Transportation (TFM). Even before the SCT announced the opening of the bidding process, TMM was ready.

SELLING THE RAILROADS

Unlike the alliance that created TFM, other joint ventures that emerged to participate in the privatization of the railroads were slow to materialize. Because the legal framework governing the privatization program allowed the National Commission on Foreign Investment to approve majority foreign ownership in the railroad, many foreign companies registered to take part in the bidding without a Mexican partner. Of twenty-four different companies, Mexican and foreign, registered to review the information contained in the data room by May 1996, only one other joint venture had been negotiated apart from the TMM-KCS alliance. The remaining companies were registered individually and included seven U.S. rail companies, a Chilean finance company, a Belgian equipment repair firm, and Mexican petroleum, construction, mining, heavy industry and financial corporations (*Expansión* 1996, 36–38).

By May 1996, the SCT solicited bids for the first concession—the shortline Chihuahua-Pacific. The real interest from private companies and the real test for the privatization program, however, would be in the three main lines carved out of the system. Shortly after soliciting bids for the Chihuahua-Pacific line, the SCT solicited bids for the first of the main lines, the Northeastern Railroad. Three groups bid on the Northeastern Railroad: TMM-KCS; Grupo México, a Mexican mining conglomerate; and a partnership that included Grupo ICA, a major Mexican construction firm, Union Pacific, and Southwestern Bell (Estrada 1997a).

As part of its effort to improve public confidence in the privatization program, the Restructuring Committee invited the media to cover the opening of the bids. In October 1996, when none of the bids for the Chihuahua Pacific short line met the minimum price, the director of the UACE, Jorge Silberstein, was forced, before the television cameras, to declare the bidding for the line null and void. The *Wall Street Journal* declared the privatization a "public relations fiasco" (Chaffin 1997, 33). One month later, at the opening of the Northeastern bids, Silberstein was once again in front of the cameras.

Grupo México and the partnership of Grupo ICA, Union Pacific, and Southwestern Bell each made bids above the minimum asking price, around $550 million. TMM-KCS made a bid that was almost three times the amount of the other offers, $1.4 billion. According to Chaffin (1997, 34), the sale "marked a turning point for privatization in Mexico in general, and the Zedillo administration in particular." The sale of the Northeastern Railroad promised to bring in almost six times as much revenue as the Zedillo administration had taken in from all of its previous privatizations combined.

The disparity between the bids, however, created immediate problems for TMM-KCS. In addition to the $1.4 billion that TMM-KCS offered for the Northeastern Railroad, the consortium promised to invest an additional $800 million over the following five years. When investors in TMM-KCS realized that they had bid almost triple the amount of other companies, they began to question the wisdom of the offer. Kansas City Southern's stock fell 5 percent on the announcement and Moody's Investors Service placed KCS under review, to consider downgrading its long-term debt (Machalaba and Mathews 1996).

A high-ranking UACE official said that he knew immediately that the size of the winning bid would cause problems. The official indicated that following the bid, TMM-KCS officials did everything they could to try to pressure the government to lower the price (Interview 65). TMM-KCS faced pressure of its own as the financial agents, Chase Manhattan and Merrill Lynch, backed out of the deal. With the first payment of 40 percent due on January 31, 1997, the new owners of the Northeastern Railroad scrambled to renegotiate with the government and to find a new financial agent.

Fearing that TMM-KCS might walk away from its bid, government officials renegotiated the details of the purchase. The original agreement had TMM-KCS taking 80 percent of the stock of the Northeastern Railroad, of which TMM would hold 51 percent and KCS the other 49 percent. Negotiating for the government, the Interministerial Committee for Divestment (CID) stepped in to restructure the deal. The Mexican government agreed to assist TMM with its financing, holding on to 24 percent of TMM's shares of the Northeastern Railroad in exchange for $200 million in financing. In addition, FNM agreed to help TMM-KCS meet its obligations to invest in the railroad with the help of a $70 million loan. TMM retained the first rights to purchase the 24 percent stake that the state controlled (Interview 65; Estrada 1997b).

Government officials and an executive from TMM-KCS suggested that Union Pacific was attempting to sabotage the deal. One TMM-KCS official even claimed that Union Pacific sought to bribe public officials to nullify the sale of the Northeastern Railroad. Trying to put the best spin possible on the bid that TMM-KCS had made on the railroad, the executive joked that the auction is all about winning the bid; once the bid has been won, you "try to screw the government" (Interview 75).

Finally, the STFRM did win some important concessions from the sale of FNM. Revenues from the sale were used to recapitalize the union's bankrupt pension fund. The new owners of the restructured railroads also had to sign agreements with STFRM workers. However, at the moment of the sale, the government retired or liquidated all of the company's workers, paying the entire cost of severance (Interview 62). TMM-KCS hired forty-five hundred STFRM members but expected to reduce that number to three thousand by the end of their first year in operation. According to an official of the newly created railroad, the collective bargaining agreement between TMM-KCS and the workers contains approximately thirty-six clauses and was "designed for sustainability for the employees," creating universal workers and allowing for geographic mobility and subcontracting of work (Interview 75).

Discussion

The particular combination of government regulation and private investment that characterized the early development of the Mexican railroads failed to produce either social benefits for the country or profits for foreign capitalists. Although the railroads linked some of the larger cities in Mexico with one another, they also left large swaths of the country untouched while connecting Mexico primarily to foreign commodity markets. The Yucatán Peninsula and Baja California, in particular, remained largely disconnected from the remainder of the country. Moreover, the array of different firms operating concessions failed to coordinate their efforts and created a chaotic system of duplication, incompatible equipment, and vast gaps in national rail transportation (López Pardo 1997, 22, 23). As Coatsworth (1979, 939) argues, "Railroads promoted both growth and underdevelopment."

State policy toward the railroads largely undermined the influence of price and profitability in the allocation of rail services by depressing cargo rates below the cost of shipping. Yet we can still find important areas in which exchange value and market allocations exercised a strong influence on the operation of the railroads. Competition with alternative means of transportation—trucking, in particular—prevented the monopolization of cargo and passenger transportation by the railroads. Fluctuations of prices on international markets also exercised a strong influence over the railroads, especially in prices for capital goods as well as the prices of raw materials that were carried by rail. And despite the government's willingness to continue subsidizing the railroads, mounting losses in the industry were a source of constant concern for international investors and the Mexican government.

The subsidization of Mexico's railroads provided a general subsidy to national commerce and foreign exports, especially to large commercial shippers and other parastate firms. Efforts to raise shipping rates to improve the profitability of the railroads consistently met with resistance from groups that benefited from subsidized rail service. And because Pemex, Conasupo, Fertimex, and other large public firms were among the most important users of the rail system, any improvement in FNM's balance sheet from rate increases would have translated into higher costs for other parastate firms.

By the mid-1940s, government policy on railroads as well as other public firms began to emphasize the importance of operating along commercial rather than political lines (Torres 1984, 31). Yet FNM was part of a larger economic and political calculation. The ultimate goals of promoting sustained economic growth for the Mexican economy as a whole could certainly withstand some losses at the level of individual industries. Such losses may even have contributed to that growth. And while the costs of maintaining political allies within the STFRM may have introduced "political" criteria into the operations of FNM, Mexico's leaders appear to have calculated that the benefits of pacifying labor outweighed the costs of allowing a radical, antigovernment union to lead workers in an industry as important as the railroads.

SEVEN

Economic Transformation and the Limits of Neoliberal Reform

What the market does is to reduce greatly the range of issues that must be decided through political means, and thereby to minimize the extent to which government need participate directly in the game.

—*Milton Friedman,* Capitalism and Freedom

[T]he exercise of the very right to organize production itself—to choose both its means and its ends—is an exercise of political power, an instance of domination. The deployment of the legal authority of the capitalist within the confines of his business enterprise thus constitutes an unrecognized transfer of political power from the state into private hands.

—*Robert Heilbroner,* The Nature and Logic of Capitalism

Research emphasizing the essential role of the state in the economy appeared at first to pose a serious challenge to the dominance of neoliberal theory. In recent years, however, this insight has been incorporated into mainstream policy recommendations and now risks slipping into triviality. Markets and states are interdependent: in a capitalist economy, markets cannot allocate resources without the institutional underpinning of the state any more than states can survive without tapping into resources that are generated on the market. That they are interdependent does not, however, warrant the conclusion reached by the World Bank that "markets and governments are complementary" (1997, 4). Rather, markets and states are organized along deeply contradictory lines.

One of the principal expressions of this contradiction is the partition of the economy into public and private spheres. The distinction between public and private under capitalism has nothing to do with whether or not production, exchange, or consumption are organized as social activities. Indeed, if anything, the expansion of the market tends to make economic life more social. Growth in a capitalist economy draws increasing numbers of people into the orbit of exchange relations, extending the division of labor and creating broader and deeper networks of material connections among people. In the process, this expansion creates new social units defined by the widening social division of labor.

Yet within a capitalist framework the expansion of society through market exchange is a unique form of social integration in that it simultaneously erects barriers among actors through rights of exclusion. Although actors find themselves increasingly connected to one another, their economic interdependence is matched by legal, political, and social autonomy. Property rights and hard budget constraints are two pillars of this autonomy. Property rights allow actors to deny resources to other actors. Hard budget constraints force actors to operate profitably, to ignore the social ramifications of engaging in particular activities and, instead, produce goods and services by which exchange value can be increased. Under these conditions, failing to exclude other actors from one's property, or allocating resources without regard for individual profit maximization, can only be considered irrational.

To capitalism's enthusiasts, it is one of the virtues of the market that exchange relations are structured to prevent the intrusion of social considerations into the allocation of resources. Within the liberal, and now neoliberal, worldview, the capitalist system produces the material wealth needed for society not by fostering bonds of solidarity, loyalty, kinship, duty, or any other common interest among actors. Rather, it produces material abundance specifically because it permits individuals to pursue their self-interest. As Adam Smith argued, "It is not from the benevolence of the butcher, the brewer, or the baker, that we expect our dinner, but from their regard to their own interest" (1981, 26, 27).

Moreover, according to liberal theory, the market supports individual freedom by limiting the range of decisions over which collective—in other words, political—authority may be exercised. Although liberal theorists stop short of arguing that the organization of the market is synonymous with democracy, from their perspective the economic freedom that actors enjoy in the market is an essential component of other freedoms (Friedman 1982, 24). As Hayek (1994, 77) argues, "It is the price of democracy that the possibilities of conscious control are restricted to the fields where true agreement exists and that in some fields things must be left to chance." Hayek goes on to argue that the exclusion of collective decision-making and "conscious control" from the economy is essential to individual freedom because "[e]conomic control is not merely control of a sector of human life which can be separated from the rest; it is the control of the means for all our ends. And whoever has sole control of the means must also determine which ends are to be served, which values are to be rated higher and which lower—in short, what men should believe and strive for" (101).

Recent research by economic sociologists has turned the liberal formulation on its head. As Block (1990, 23) argues, "The tradition of economic sociology challenges the idea that the economy is an analytically separate realm of society that has its own internal dynamics." Given the unity of economy and society, "economic transactions are seen to be continuously shaped and influenced by social, political, and cultural factors" (23; see also Portes and Sensenbrenner 1993; Swedberg, Himmelstrand, and Brulin 1990, 64–66). Rather than viewing the inevitable social influence on economic action as a danger to growth or individual freedom, sociologists tend to emphasize the complementarity of states and markets that was recently embraced by the World Bank. As Evans (1995, 26) argues, "[M]arkets operate well only when they are supported by other kinds of social networks."

In the preceding chapters I have attempted to show how economic control—despite its interdependence with political and social institutions—can be established as a separate realm of society: one that is shielded from certain political and social obligations. Although markets are embedded in political and social ties that may limit the extent to which actors can operate on principles of price, profitability, and competition, they are also structured to ensure that these same principles apply to a wide range of allocations. This does not mean—as Hayek and Friedman would have it—that markets remove politics from economic allocations. In every economy there are unavoidable losses—goods and services that are necessary but that cannot for either technical or social reasons be produced for a profit. Furthermore, as Heilbroner points out in the epigraph to this chapter, the organization of production within the confines of the firm represents "an unrecognized transfer of political power from the state into private hands." To say that the market removes politics from economic allocations is merely to ignore the social structural underpinnings of market exchange: the use of collective resources to correct market failure and overcome transaction costs, and the authority that capitalists wield within the firm.

At the same time, to argue that the institutions that have been created to correct market failure or overcome transaction costs complement market allocations is to ignore the uneasy coexistence of markets with the nonmarket institutions on which they depend. By enforcing property rights and hard budget constraints, the state creates a form of dual sovereignty in which it effectively cedes partial control, to, in Hayek's words, "determine which ends are to be served, which values are to be rated higher and which lower." It is not only the exercise of "sole control" but also the exercise of unequal control over resources that grants actors the power to determine "what men should believe and strive for." The capitalist market creates a social space in which actors with unequal endowments of resources can bargain over the exchange and distribution of their property without incurring obligations on the uses of their property. Actors in the market retain the right to withdraw their property from circulation if they do not like the terms of exchange. While the importance of the freedom to exchange under these conditions should not be underestimated, it is a freedom that depends, critically, on two conditions: first, that actors possess some threshold level of resources that they can exchange, and second, that they possess sufficient resources in reserve not to exchange.

Absent these conditions, the opportunities and freedoms of the market become constraints and imperatives compelling actors to participate in production and exchange from which they may not, themselves, benefit. A capitalist market achieves this, in part, by establishing price, profitability, and competition as the rules of the game, thus locking into place a specific type of rationality that grants social power to private actors through the apparently impersonal mechanism of accumulation. The political and legal framework that enforces property rights and hard budget constraints transforms the ideological construction of the market into a social fact: creating the appearance of scarcity by shielding private property from social obligations, and obscuring the mechanisms through which decisions are made about the relative importance of different needs.

Consistent with the perspective of economic sociology, many of the social and political institutions within and around which markets operate stabilize exchange relations by counteracting the market mechanism. Families, firms, even entire nations produce and allocate some goods on the basis of their collective use value rather than for the expansion of exchange value. Nonmarket institutions require that actors defer their utility-maximizing behaviors, spreading cost-benefit calculations over larger spatial and temporal structures: families pass resources across generations, firms spread expenses and earnings among divisions, and states redistribute collective resources to particular actors. The suspension of market exchanges in one area of the economy may even contribute to the development of market exchanges in other areas of the economy or on a different scale. By absorbing necessary losses in some sector of the economy or cross-subsidizing the production of some good or service that would not be produced if left to market forces, nonmarket institutions can promote the development of one market by eliminating market allocations in another.

Still, in order to channel resources into areas not dictated by considerations of profitability, nonmarket institutions must either siphon resources from profitable actors in the economy or exclude private participation in some sector of the economy. The same structures that support a particular market may also constitute barriers to its further expansion. As Schumpeter (1942, 82) argues, "The essential point to grasp is that in dealing with capitalism we are dealing with an evolutionary process.... Capitalism, then, is by its very nature a form or method of economic change and not only never is but never can be stationary." Schumpeter elaborates on the implications of this conception of capitalism for understanding the relationship between markets and states. "In other words, the problem that is usually being visualized is how capitalism administers existing structures, whereas the relevant problem is how it creates and destroys them. As long as this is not recognized, the investigator does a meaningless job. As soon as it is recognized, his outlook on capitalist practice and its social results changes considerably" (84).

In the pages that remain, I will highlight some of the specific elements of Mexico's economic transformation and show how they illustrate the theoretical concepts discussed here. These concluding comments are organized around three themes that emerge from studying the process of privatization in Mexico. First, I discuss the structural reconfiguration of economy and society in Mexico. In this section, I focus on three key dimensions along which social relations have been transformed to accommodate neoliberal reforms: interstate, intrafirm, and interfirm. Second, I turn to the issue of change within the economy. I explore how and why institutional change occurred in Mexico, the extent to which failure was a real possibility in the market, and the means by which actors sought to soften their budget constraints. Finally, I discuss the relationship between political and economic change, looking at the influence of Mexico's neoliberal reform program on the important political changes that occurred in Mexico during the final decades of the twentieth century and exploring possibilities for further change in the future.

Separating Economy and Society: Social Boundaries and Economic Action

In the introductory chapter, I claim that markets in a capitalist economy require the structural separation of actors from one another. Even before actors can operate on the principles ascribed to them in neoclassical theory—utility maximization based on prices, profitability, and competition—they must be constituted as distinct entities. Removing public firms from the control of the state is the most obvious sense in which privatization began to create differentiated social entities. By itself, however, the separation of firms from the state did not necessarily entail the creation of a market. Privatization was only one part of a larger set of reforms needed to impose a new economic model and involved the transformation of social relations along three other key dimensions: the relationship between Mexico and the global economy, relations between workers and management within firms, and relations among firms in the market.

REINTEGRATING MEXICO INTO THE GLOBAL ECONOMY

The Revolution of 1910 marked the starting point of a major reorganization of Mexico's relationship with the international economy. Prior to 1910, the government of Porfirio Díaz had been content to allow foreign capital wide latitude in its economic dealings within Mexico's borders. The Constitution of 1917 and the creation of the National Revolutionary Party in 1929 provided the legal and political foundation for increased state intervention in and direction of economic activity. With foreign firms dominating key sectors of the economy, the efforts of the state to direct the process of production toward specific political and economic goals inevitably came into conflict with foreign capital.

Breaking the ties that connected it to the international economy, however, required that the Mexican state bind actors more closely to one another within the domestic economy. The Mexican state justified its efforts to restrict foreign participation in the economy by appealing to the collective interests of Mexicans. In its Declaration of Principles and Program of Action of 1938, for example, the ruling party emphasized the need to "liberate the country from outside economic influence" and "organize the economy of the country on the principle that production and distribution are oriented toward the true satisfaction of popular needs" (see Chapter 2). Workers, capitalists, peasants, and landowners, were brought together by the PRI and encouraged to set aside their particularistic interests for the collective good of national development.

Despite growing intervention by the Mexican state between the early 1930s and 1982, however, Mexico never came close to achieving autarky or escaping the influence of the international economy. Public firms allowed the Mexican state to appease various domestic constituencies and promote the production of specific goods and services for the goal of national development rather than for their exchange value. Still, the Mexican state was buffeted by the international economy in a variety of ways. It remained dependent on the sale of commodities in the world market, required access to foreign capital, and suffered from repeated balance-of-payments crises when it attempted to open its markets.

Perhaps the most important influence of the international environment on the Mexican state, however, was that it allowed domestic capital to escape the embedded relations of the national development project. Although theories of hard budget constraints and property rights emphasize the role of the state in enforcing these rules, domestic capitalists imposed hard budget constraints on the Mexican state. They were able to exclude the state from their property through capital flight because they could take advantage of the legal protection that was provided to property in other countries.

Thus, while the Mexican economy was embedded within a specific set of domestic political and social relations, these relations were, themselves, conditioned by Mexico's position in the global economy. Examining the interrelationship between Mexico's position in the international economy and the ability of the state to reconcile competing interests within the domestic economy does not contradict the notion of embeddedness. It does, however, force analysts to consider the various levels at which economic action is embedded. Although markets always exist within a specific institutional framework, the global economy provides something of a master framework that, to varying degrees, influences regional, national, and local economies. Within each of these lower-level institutions, the ability of certain actors to exit can enhance their voice in negotiations with other actors.

The economic crisis of 1982 forced Mexico to renegotiate the terms on which it relates to the global economy. The new model of integration into the North American economy has involved the disintegration of many of the social ties that formerly linked actors within Mexico—not the least of which has been the erosion of loyalty to the PRI by workers and middle-class Mexicans. Although NAFTA creates a regional governance structure linking the economies of Canada, the United States, and Mexico, this agreement falls considerably short of creating a single, North American market. In the previous chapters, we have found a wide range of restrictions on the influence of price, profitability, and competition *across* borders, many of which are necessary to the creation of market exchange *within* Mexico.

Privatization and liberalization in Mexico's telecom market provides the clearest example of the continued importance of national borders in the North American economy. Although Mexico has opened its long-distance market to investment from foreign capital, these companies may not own a majority stake in a long-distance carrier but must, instead, operate as minority partners to Mexican firms. Even in partnership with Mexican capital, foreign firms are prohibited from operating as subsidiaries of their parent companies and, instead, must function *as if* they were independent entities. All long-distance calls entering Mexico are redistributed to companies in the domestic market in the same proportion that these companies send outgoing calls. Without these barriers to integration, the Mexican market would largely disappear into the corporate structure of AT&T, which would connect calls from the United States through its network, which now extends into Mexico.

The market for air transportation in Mexico also depends upon specific barriers to the international market. As with telecommunications, foreign owners are prohibited from holding a majority share of a Mexican airline and foreign carriers are still prohibited from operating on domestic routes. Yet the size of the Mexican market makes it difficult for Mexico's airlines to achieve the economies of scale needed to remain competitive with carriers from advanced industrial countries on international routes, and competition on these routes makes it difficult to cross-subsidize domestic service, for which there may be no clear profit motive to serve. Although Mexican owners have attempted to create a business model that would incorporate the airlines into a larger corporate structure that includes reservations systems, jet maintenance, and hotels in the tourism sector, so far this has not proved profitable.

REALIGNING INDUSTRIAL RELATIONS: SEPARATING WORKERS
FROM MANAGEMENT

Between the early 1930s and 1982, workers in large parastate firms acquired important benefits and a certain degree of leverage over the process of production. These workers were generally organized into large, national unions prior to the nationalization of their industries, although in the case of Telmex the consolidation of telecommunications under state guidance created a single, national union. Conflict between workers and foreign owners sometimes precipitated state intervention, to prevent disruption of the production of essential goods and services or to buy loyalty to the ruling party.

The incorporation of firms into the public sector transformed relations between workers and management. Workers often made significant gains on paper, including rights to participate in basic decisions over the organization of production. In the railroads these rights came to include a brief period during which the labor union was placed in charge of the administration of the firm. In telecommunications, workers gained rights of consultation in the introduction of new technologies. Many workers also gained important material benefits as the state used improvements in wages and benefits to buy labor peace in strategic industries. At the same time, the state repeatedly intervened to suppress worker demands for democratic representation, diminishing the value of some of the contractual benefits that workers had won.

Thus, state ownership of the means of production was far from an unqualified blessing for workers in public firms. Although state ownership restricted the rights of capitalists to profit from these firms, the same justification that limited the influence of the market over these industries—that production should be geared toward the collective use of Mexico—was also used to restrict the rights of labor. In a public firm, improved wages and benefits for workers appear to come more directly at the expense of the consuming public, allowing the state to portray public-sector workers' efforts to improve wages and working conditions as unpatriotic. Workers in state-owned firms were frequently required to limit their profit maximization for the greater good of Mexico. When moral suasion and economic incentives failed, the state intervened directly in union politics, imposing progovernment leaders, and using the *requisa* to order workers back to the job. In extreme cases, the state called on the army and police to limit workers' demands.

The process of privatization has involved a profound restructuring of the relationship between workers and owners. Workers have generally lost rights that they enjoyed under previous collective bargaining agreements to participate in decisions over the organization of production. To be clear, there is not a necessary relationship between privatization and the separation of workers from the management of a firm. It might have been possible to create a market of firms in which labor and capital collaborated in the management of newly privatized firms. Prospective buyers, however, insisted that collective bargaining agreements be renegotiated to restrict the rights of workers within the firm. In addition, state actors had an ideological preference for placing control over firms in the hands of specific groups of capitalists. One high-ranking official from the Salinas administration argued that the "democratization of capital" is inefficient: "You have to identify a controlling group in order for a company to operate efficiently: there has to be responsibility" (Interview 77).

Large privatizations, therefore, were almost invariably accompanied by the restructuring of labor contracts on terms more favorable to capital. Among the case studies presented here, the restructuring of the collective bargaining agreement was most dramatic in the airlines and the railroads. Privatization had the most catastrophic effect on Mexico's railroad workers. Between 1990 and 1996, the year before the first privatization in the railroads, almost one-half of FNM's eighty-three thousand active employees had been fired or retired. When the concessions for the three principal companies were sold, they did not include a single worker on their payrolls. Private firms were able to recontract union members at will under a collective bargaining agreement that removed all vestiges of worker control over the process of production in the railroads. Although precise numbers are not available, perhaps fewer than ten thousand employees remain employed in this sector.

The declaration of bankruptcy at Aeroméxico allowed the state to fire workers, suspend the collective bargaining agreement, and cut benefits to workers. The reorganization of Aeroméxico in receivership and its resurrection under a new name allowed the state to maintain the new work rules and organizational structure, making the firm more attractive to potential buyers. Even though the pilots' union purchased 25 percent of Aeroméxico as part of its reorganization, pilots agreed to various concessions in their contracts. Pilots accepted a 10 percent pay cut and agreed to work longer hours. Flight attendants and ground crews—whose unions could not afford to buy stock in the company—fared worse, losing more jobs and making broader concessions than the pilots.

It is interesting to note that the separation of workers from the management of newly privatized firms has not necessarily led to increased conflict between workers and management. Instead, concerns about job security in a more competitive environment have led airline pilots to focus on the importance of productivity at the firm. Pilots have also becomes staunch defenders of the right of Grupo Cintra to organize and coordinate the activities of both Aeroméxico and Mexicana, speaking out against efforts by the Federal Competition Commission to split the two firms apart.

In telecommunications, workers fared much better in their negotiations with the government and the new owners. Even at Telmex, however, worker rights over the production process were cut back. Agreements on the introduction of new technologies and mixed committees of labor and management to oversee production were all scrapped prior to the privatization of the firm. Yet like the airline pilots' union, the telecom workers' union has demonstrated loyalty to its employer, and its members have defended Telmex's dominance of the telecom market, out of concern that heightened competition will undermine job security.

CREATING A MARKET: SEPARATING AND RECOMBINING FIRMS

Even under state ownership, both the railroads and the airlines faced competition from domestic transportation alternatives, especially buses and trucking. In addition, the airlines competed against foreign firms on international routes. Of the case studies examined here, only Telmex operated without any competition to its service. Still, even in the railroads and airlines, structurally separating firms from one another in the market was a central concern for state officials involved in the privatization process. The separation of firms from one another made it possible to encourage price-based competition in these industries. As the scale and scope of the market changed, so did the boundaries of cooperation and competition among firms.

In the railroads, it was necessary first to consolidate into FNM all the track in the country, before devising a new organizational structure for the firm. Breaking the railroads into three main lines served at least two objectives that were advanced by state officials. First, it created firms small enough for private investors to buy. Second, it provided transportation alternatives for major freight customers in the industrial heartland around Mexico City. Automobile manufacturers were especially concerned that the privatization of the railroads not leave them dependent on a single railroad company to export cars. In order to achieve competition between the two main lines running north to the U.S. border, the two railroads were actually required to share track north of Mexico City.

Although it is probably still too soon to say, the railroads may succeed in operating as discrete entities because their separate operations have each become part of a larger, international economic calculus. All three of the main lines operate with the participation of foreign partners, who provide some measure of insulation from the forces of competition. In the case of the Northeastern Railroad, the multimodal transportation services of Mexico's principal investor create yet another buffer to protect the firm from excessive competition. Thus, even if the Mexican railroads are not profitable by themselves, their contribution to the operations of larger corporate structures within which they are currently embedded may allow them to continue to operate outside state control.

Maintaining the separation of firms in air transportation has been more difficult. Increased competition from new domestic carriers as well as international competition made it difficult for the newly privatized airlines to operate profitably. Competition led Mexico's major carriers to coordinate their operations and eventually brought about the consolidation of the two firms and a number of smaller firms under the control of a holding company. Although both Aeroméxico and Mexicana have entered into international agreements with major foreign carriers, these agreements are less comprehensive than the partnerships that the railroads have with foreign capital. Nor is it clear that Mexico is a viable market for air transportation. Given the persistence of poverty, the limited territorial reach of domestic air carriers, and the current turmoil in the industry worldwide, market failure will probably continue to plague this sector.

The process of separating firms from one another was most complicated in the area of telecommunications. Here, the government's decision to sell Telmex as a single firm—rather than break the company into separate local, long-distance, and cellular companies—represented more an effort to create a Mexican firm capable of competing against international telecommunications' firms than an attempt to produce a Mexican market for telecommunications. The consolidation of the telecom giant's holdings prior to privatization strengthened Telmex and allowed for a partial opening of the long-distance market without major upheaval or economic crisis for Telmex or other firms in this sector. The structure of this sector, however, is one in which the influence of price and competition are both highly restricted.

Moreover, the dominance of Telmex made it impossible to carve out a long-distance market in Mexico without increased coordination among firms. Telmex's monopoly control of local phone service means that all domestic calls begin and end on Telmex's network, granting the former state monopoly an enormous advantage over its competitors and a clear conflict of interest in its relationships with other long-distance providers. Much like the regulations that prohibit foreign telecom firms from coordinating their operations with their Mexican counterparts, regulations governing long-distance require Telmex local to treat its long-distance counterpart as if it were a separate company.

By enforcing legal divisions between firms and forcing them to operate as functionally separate units, these regulations seek to create the anonymous, atomized market structure found in neoliberal theory. They attempt to prohibit discrimination by Telmex's local division in favor of Telmex's long-distance division by imposing price as the determinant of allocations for access to the network. Yet while prices are supposed to reflect costs in a market, regulators are uncertain what exactly the costs are. Telmex and its competitors make very different claims about the actual cost of connecting to the network. As with the price of completing a call across national borders, the prices charged by companies for interconnection have been established by regulatory fiat rather than through competition in the market, and they reflect different assumptions about how to identify the appropriate economic unit to which profitability should apply.

Thus, while the state has prohibited cross-subsidization and discriminatory practices in the market, these regulations merely highlight the fact that firms continue to internalize a wide range of transactions within their own structure and among related firms. Since there is no logical terminus to the commodification of exchanges, no point at which economic transactions could not be further subdivided to create more competition, regulators need to select some viable scale and scope of internalized transactions and permit these, lest the market self-destruct. Dominant firms in this sector have also recognized this and have limited the implementation of access-code dialing so that the market will be one of customers, not calls (see Chapter 5).

Sources and Consequences of Economic Change

The search for agency in the study of structural change has led many researchers to emphasize the importance of state autonomy in the implementation of neoliberal reforms. This literature tends to focus on the collective action problem, specifically, the difficulty of implementing reforms when benefits are diffuse and costs concentrated (see Haggard and Kaufman 1992, 18; Schneider 1999, 55). I argue that the collective action problem provides little insight into the process of economic reform in Mexico. Indeed, the premises underlying the collective action problem are the very opposite of those that prevailed in Mexico: neoliberal reforms in Mexico produced highly concentrated benefits and very diffuse costs. Mexico's largest and most powerful private groups reaped enormous benefits from the neoliberal reforms that were implemented under the three PRI presidencies studied here while living standards fell for the vast majority of Mexico's citizens. Conservative estimates indicate that 45 percent of Mexico's 100 million citizens now live in poverty. Other estimates range as high as 75 percent (Otero 2000, 4).

Another important explanation for economic reform in Mexico is crisis. Although not as prominent in the literature as state autonomy, economic crisis is cited by most observers as the cause that precipitated the wave of neoliberal reforms that were begun in the developing world in the early 1980s. While I agree that crisis was an important catalyst for the implementation of neoliberal reforms in Mexico, one caveat is worth mentioning.

It is important to note that Mexico faced *multiple* crises, not just a single shock to the system. The crisis of 1982 led to the first wave of reform. Subsequent crises spurred the government to intensify its restructuring efforts. The collapse of oil prices in 1986, followed by runaway inflation, provided the impetus for the Salinas administration to deepen structural adjustment, beginning in 1988. Immediately following the Salinas *sexenio,* the collapse of the peso in December 1994 sparked further reforms under President Zedillo. Note, however, that while the crisis of 1982 led the Mexican government to reverse the course of state economic policy, subsequent crises had the very opposite effect. Instead of changing the direction of policy in response to the crises of 1986 and 1994, state managers took the crises as a signal that they had not gone far enough in implementing neoliberal reforms. As the international business magazine *Privatisation International* reported in February 1995, "Mexico's currency crisis has given new urgency to the government's plans to privatise several key infrastructure areas" (Malkin 1997).

Thus, while crisis may indicate something about the timing of economic reform, the fact of crisis alone appears to provide very little guidance for the direction reform may take. The fact of crisis says nothing about why policy makers decided to abandon fifty years of state-led development rather than adopt some other type of economic reforms, just as subsequent crises do not indicate why state managers chose to intensify neoliberal reforms. The limitations of both state autonomy and crisis as explanations for why change occurs and the direction it takes suggest that we need to look elsewhere to understand the causes and consequences of economic restructuring.

Polanyi's *The Great Transformation* chronicles a period of economic change with striking parallels to those examined here. As in the reforms of the past two decades, reformers of this earlier period emphasized free trade, stabilization of national currencies, and repayment of foreign loans (Polanyi 1957, 142). Nineteenth-century justifications for poor laws echo in the formulation of hard budget constraints, in which permitting individual failure is necessary to encourage industry and promote aggregate welfare. For Polanyi, the motive force behind the great transformation was social struggle between proponents of creating a self-regulating market and defenders of social protection (132).

Do these ideal types help identify real actors in Mexico who sought to create a self-regulating market, or who fought for social protection? Clearly Mexico's largest financial groups weighed in heavily on the side of economic liberalization. These groups actually helped precipitate the debt crisis, borrowing heavily on international markets and engaging in speculation against the peso and capital flight. Following the crash of 1982, Mexico's largest capitalists continued to send capital abroad while they used the major business associations to organize in support of the new economic model. Business leaders continued to push for additional privatizations through the CCE and the Employer's Confederation of Mexico (Rodríguez 1988). They also pressured the PRI by sponsoring candidates for public office through the PAN.

These same groups were the primary beneficiaries of economic reform. While the Mexican government imposed austerity upon the vast majority of the Mexican people, it used surpluses and foreign exchange generated by public firms to recapitalize the largest private firms. Major beneficiaries of these policies included Grupos Alfa, Cemex, Visa and Vitro (see Chapter 3 and Jáquez 1998; Maxfield 1990, 160). Bailing these firms out of the mess that they had created in PMT, however, was not enough. The privatization program served to strengthen these groups still further. Privatizations helped Mexico's largest capitalists to consolidate their position in markets in which they already operated and diversify into new areas of the economy.

The SHCP and SPP, where supporters of a more business-friendly economic model already operated, strengthened their ties to the private sector during the early period of retrenchment. Private firms gained access to government officials though the Subministry of Banking and the Council of Advisors within the SHCP. Private actors gained additional access to the de la Madrid and Salinas administrations through agent banks that were contracted to handle the financial evaluation and engineering involved in the privatization program. Although the banking system had been nationalized, only the top management changed. Thus, through the nationalization of the banks, the state incorporated into its very heart a cadre of private actors.

The United States government, private international banks, the IMF, the World Bank, and private consulting firms also pressured Mexico to implement neoliberal liberal reforms. While private international banks refused to provide fresh credit, the United States and multilateral lending institutions imposed conditions on renewed lending to Mexico. International consulting firms provided expertise and guidance to assist in the implementation of reforms.

While the proponents and beneficiaries of economic reform are easily identified, the specific forms that the "countermovement" took are less clear. Part of the difficulty of using Polanyi's formulation to predict the alliances or social groupings that emerge in opposition to neoliberal restructuring is that the introduction of market mechanisms into resource allocation has such varied and often unpredictable consequences. As Polanyi (1957, 145) argues, "The great variety of forms in which the 'collectivist' countermovement appeared was not due to any preference for socialism or nationalism on the part of concerted interests, but exclusively to the broader range of the vital social interests affected by the expanding market mechanism." Given the breadth and depth of the reforms undertaken in Mexico, virtually no group was left untouched by neoliberal restructuring.

It may be useful, then, to focus on the areas that Polanyi refers to as the "false commodities": money, land, and labor.

The attempt to create a self-regulating market for money was perhaps the largest single failure of the privatization program. After consolidating the Mexican banking system and absorbing at least $21 billion in private-sector debt, the state sold eighteen banks for approximately $12 billion between 1991 and 1993. A little more than two years later, the peso crisis wiped out the banking system, forcing the government to step in to rescue the private bankers who had so recently clamored for the state to get out of the market. Although the final tally is still not known, by 2001 the cost of the bailout was being estimated at $115 billion (see Chapter 3 and Sánchez 2001b).

Without actually opposing the creation of a self-regulating market for money, both the state and the international financial community were forced to step in to prevent the market from destroying itself. While maintaining the importance of private control over money, proponents of neoliberal reforms called upon the state to soften their budget constraints. As Polanyi argued, "[T]he organization of capitalistic production itself had to be sheltered from the devastating effects of a self-regulating market" (1957, 132). Rather than permit Mexico's largest and most powerful financial groups to fail, the Mexican state, with the assistance of a loan from the United States, promptly intervened to rescue its private sector. The state also absorbed private losses by partially privatizing the social security system, funneling individual retirement savings into private financial companies. And the Mexican state used its own petroleum reserves as collateral for the loan from the United States (see Chapter 3).

Opposition to the commodification of land under President Salinas came to the attention of the world dramatically on January 1, 1994. On that day, a peasant rebellion in the southern Mexican state of Chiapas shook a complacent international consensus that Mexico was a showcase for the successful implementation of neoliberal reforms. The rebellion in Chiapas was directed specifically at land reforms that allowed the buying and selling of communal holdings known as *ejidos*. Although their military capacity was quite limited, the Zapatista Army of National Liberation (EZLN) benefited from the increased international attention that the negotiation of NAFTA had focused on the nation. In addition to opposing the constitutional land reform passed under President Salinas, the EZLN highlighted its concerns that foreign imports of grain under NAFTA would undermine farm communities.

Opposition by workers to reforms that seek to extend market control over the allocation of labor should have been the easiest to predict. Yet here the web of political relations within which workers are embedded at the national level has blunted a reaction of equal and opposite force against neoliberal reforms. National labor law still provides real benefits to workers, even if these benefits have diminished in recent years. The potential for gain from economic integration with a more wealthy neighbor probably also impeded solid labor opposition to NAFTA within Mexico. Continued manipulation of workers' organizations by the PRI and the creation of new political deals, such as those with the telecommunications' workers, made the major labor organizations sympathetic to certain elements of the neoliberal agenda.

Still, despite the enduring corporatist ties that bound labor and the PRI, the official unions mobilized and managed to defeat comprehensive labor reform that President Salinas sought. Moreover, efforts among workers to organize across borders have grown in recent years. The political battle to pass NAFTA in the United States raised awareness among U.S. workers about working conditions in Mexico. The liberalization of rules governing investment and the increased mobility of capital across Canada, the United States, and Mexico have underlined the need for international solidarity to the AFL-CIO. Even if viewed in the most cynical light, as simply an effort to protect U.S. jobs, the campaign to incorporate labor and environmental standards into NAFTA represents a qualitatively different formulation of social protection by U.S. workers.

From Mexico, the threat of foreign investment by firms that are hostile to organized labor and the corporate practices of foreign shareholders have created unexpected alliances. Following the announcement of an agreement between Telmex and the U.S. long-distance company Sprint, Telmex workers promptly filed the first complaint under the labor side agreement of NAFTA charging Sprint with illegally firing workers in a San Francisco–based subsidiary. Although largely a symbolic gesture, Telmex workers and the Communications Workers of America have signed agreements pledging mutual support. The unions representing Mexico's pilots and flight attendants have also engaged in efforts to build solidarity with workers in the United States. Again, while economic reform has broken apart some social relations at the national level, it appears also to be facilitating the development of new relations across national boundaries.

The Unintended Consequences of Economic Restructuring:
Toward a More Democratic Mexico?

One of the few bright spots in an otherwise bleak picture of economic reform in Mexico has been the gradual democratization of the country's political institutions. Under President Salinas, in 1989, the PAN won the governorship of North Baja California, the first to go to an opposition candidate since the creation of the forerunner to the PRI in 1929. The PAN and, to a lesser degree, the left-wing PRD continued to make gains in state and national elections throughout the Salinas and Zedillo presidencies. Along with a number of minor parties, the combined opposition eventually gained a slight majority over the PRI in the lower house of the legislature in midterm elections in 1997. This process of democratization appears to have culminated with the election of Vicente Fox Quesada of the PAN to the presidency in July 2000. What effect, if any, have the economic reforms enacted under the PRI had on these political developments?

As noted above, liberal theorists see economic freedom as supportive of other freedoms. The gradual democratization of Mexico during the past three *sexenios* would seem to support the contention that political and economic liberalization go hand in hand. As formulated by Hayek, however, there is something of a tautology built into the liberal equation. If, as Hayek (1994, 77) argues, "[i]t is the price of democracy that the possibilities of conscious control are restricted to the fields where true agreement exists," and if neoliberal policies restrict the role of the state in the economy to only those things on which "true agreement exists," then, by definition, restricting the role of the state in the economy is compatible with democracy. To the extent that economic reforms have cut back on the resources available to political leaders and restricted the sphere of activity of the state, they have made the state safe for democratic control. Reforms implemented between 1983 and 2000 gradually transferred the balance of social power to private actors. With this power more firmly lodged in private hands, the danger to private property of collective action by the state is considerably diminished.

Although Amartya Sen argues that economic and political freedoms are compatible, it is interesting to note that his formulation does not refer to property rights in terms of rights of exclusions. Rather, Sen argues for the compatibility of economic and political freedom by referring to the use value of goods, not their exchange value. Rights to "economic facilities," according to Sen, "refer to the opportunities that individuals respectively enjoy to utilize economic resources for the purpose of consumption, or production, or exchange" (1999, 39).

I argue that the extension of political democracy in Mexico has been one of the unintended consequences of economic reform. More complete democratization in Mexico will require the extension of democratic principles to the private sphere—the extension of economic facilities—not their exclusive application within the realm of political institutions. Given that tens of millions of Mexicans still lack adequate nutrition, basic sanitation, access to medical care, and gainful employment, rights to exclude others from property and rights to economic facilities will necessarily come into conflict.

For the present, let me sketch out what I believe to be the chain of events that led the PRI to open the political system and as a result of which it eventually lost control of the presidency. To begin with, evaluating the relationship between economic reform and political reform must take into account the timing and sequencing of these policies. Had political reform taken place before 1988, PRI defector Cuauhtémoc Cárdenas, and not Carlos Salinas de Gortari, might very well have won the presidency. Although many populist campaigners have converted to the gospel of neoliberal reform when faced with the challenges of governance, it is unlikely that a President Cárdenas would have implemented neoliberal reforms with the same intensity as President Salinas.

And while the gains made by the PAN under President Salinas appeared to represent an important political opening, it is useful to note how selective President Salinas's political reforms were. Most significantly, the PRI ceded control primarily to the party whose economic policies it had largely adopted. At the same time, it fiercely contested any advance of opponents of neoliberal reforms. The alliance between the PAN and PRI actually buttressed Salinas's economic agenda by providing him support for constitutional reform in the Congress, improving his image as a reformer abroad, and simultaneously, further isolating the principal opponents to neoliberal reforms in the PRD. Another key limitation to the political opening under Salinas was that it remained highly personalized. Rather than implement reforms of the institutions of elections and governance, Salinas personally intervened to set aside fraudulent PRI victories and turn control of the states of Chihuahua and Guanajuato over to PAN candidates (MacLeod 1997).

It was the Zapatista uprising, and the economic crisis of 1994–95, that led to the institutionalization of democratic reforms. The Zapatista uprising could not have come at a worse time for the Salinas administration, which, in the final year of the *sexenio*, was preparing to pass presidential power to Salinas's handpicked successor, Luis Donaldo Colosio. The election season generally involves additional state spending to guarantee a prosperous economy and minimize the amount of fraud necessary to grant victory to the PRI's candidate. The transition in 1994 would be especially challenging, given the increased scrutiny of Mexico's political system that the signing of

NAFTA had generated. Salinas faced opposition to the economic reform program from within the PRI, in addition to the opposition of the Zapatista rebels. More cracks in the foundation of Salinas's political-economic program were revealed with Colosio's assassination in the spring of 1994. Although the PRI managed to win the election, the price appears to have been the economic downturn of 1995 (Oppenheimer 1996).

Entering office under the cloud of corruption that was coming to light following the Salinas administration, and forced to preside over yet another meltdown of the Mexican economy, President Zedillo could offer the Mexican people almost nothing. Zedillo moved to privatize additional firms to lure back private capital. He also focused on repayment of the loan from the United States. With limited material benefits left to reward loyalty to the PRI, Zedillo gambled. In exchange for the previous twelve years of declining economic fortunes, he would offer the Mexican people political reform. At the time, it appeared to be the least expensive reform that he could implement.

Thus, the importance of the election of Vicente Fox should not be under- or overestimated. Fox's election is especially significant because so much power is concentrated in the presidency in Mexico. Without the control of the executive, it is not at all clear how the PRI will maintain the loyalty of its various constituencies. At the same time, the PAN will very certainly attempt to continue the economic policies that have defined the past three presidencies. Thus, the election of Vicente Fox may serve to relieve some of the political pressures that would have accompanied continued neoliberal reforms under the PRI, drawing attention away from the economy and placing it on the state.

Once again, Karl Polanyi's work is instructive here. Pointing to the Chartist Movement in the nineteenth century, Polanyi notes, "The more viciously the labor market contorted the lives of the workers, the more insistently they clamored for the vote." Yet, as with political reform in Mexico, concessions were granted to the Chartists only *after* the reforms against which they had organized, had been implemented: "The Chartists had fought for the right to stop the mill of the market which ground the lives of the people. But the people were granted rights only when the awful adjustment had been made" (1957, 225, 226).

Concluding Comments

Despite the critique that I have made of economic sociology in the preceding chapters, the evidence from Mexico's privatization program provides support for an important point that has been made repeatedly by economic sociologists: simple assessments of "how much" the state intervenes in the economy tell us very little about the relationship between state and economy. The critical questions for understanding how the state influences economic activity are qualitative, not quantitative. These questions refer to *how* the state intervenes in the economy: What mechanisms does it use to intervene? Where does it enforce property rights and hard budget constraints? Where does it redistribute resources to subsidize actors and activities that are not individually profitable? Who benefits from state intervention? And what social groups support or oppose different forms of state intervention?

In answering these questions, I have attempted to respect the complexity of the process of neoliberal restructuring in Mexico. I have emphasized both the withdrawal of the state from certain forms of intervention as well as its extension and reentry in other forms. With these concluding comments, I review the means by which the Mexican state withdrew from the economy and discuss the limits of neoliberal reforms.

THREE AREAS WHERE THE STATE HAS WITHDRAWN FROM THE ECONOMY

There are at least three ways in which the Mexican state actually removed itself from the economy over the past two decades without necessarily changing how much it intervenes. The first of these refers to the apparent inability of state managers to resist rescuing large, bankrupt firms. By propping up the banking system and rescuing the airlines in 1995, the state demonstrated its continued unwillingness to allow large private firms to fail. It should be noted that throughout the de la Madrid, Salinas, and Zedillo administrations, there did appear to be a greater willingness to permit the failure of small firms. Moreover, the massive rescue package that channeled public funds to the banking system under the Zedillo administration sparked widespread protests from small debtors, who did not receive such generous treatment in the handling of their debt.

But while the state clearly indicated its continued willingness to absorb the losses of Mexico's major corporations following the peso crisis of 1994, these interventions were qualitatively different from previous bailouts or nationalizations of private firms. While the state continues to underwrite the losses of these firms, it now appears to be less inclined to participate in management decisions regarding these firms. According to one official from the Zedillo administration, the state does not involve itself in the management of firms even when it retains ownership of stock and has representation on boards of directors, because it does not want to give the appearance of its approval or disapproval of a company's actions (Interview 49). Thus, even while the state has acquired shares of previously privatized firms, it now makes fewer claims on these firms' property. The Zedillo administration also accepted foreign investment into sectors of the economy that had previously been reserved for domestic capital.

A second important way in which the state has taken itself out of the economy in recent years is by establishing or yielding to alternative governance structures within markets. This form of state withdrawal from the market was especially important in the telecommunications sector. A private, third party was used to implement the program of presubscription that allocated long-distance subscribers to new companies in the market. Third-party mediation was also established through committees of experts who are assembled to mediate specific disputes. In order for private firms to adjudicate disputes between companies, they must demonstrate their independence from parties to the disputes much as a state's judicial branch is assumed in liberal theory to provide for impartial adjudication of conflicts (see Chapter 5).

Paradoxically, in order to establish independent, third-party mediation, the state had to foster greater cooperation among firms in the market. In the telecommunications' market it sought to provide this by creating the Committee of Operators. The formation of this committee was a conscious attempt to prevent the telecom market from degenerating in ruinous competition, the intention being to establish in advance the body that would resolve collective action problems. And it is this collaborative group that recognizes sufficient common interests to choose an independent agent to mediate conflicts. Although the state remains the arbiter of last resort, it has accepted and encouraged the use of private firms to provide some of the regulatory infrastructure for the market.

The third way in which the state has removed itself from the economy is by contracting with private vendors to perform specific functions and channeling money to private companies. The growth of contracting to nonstate entities actually gained momentum in the privatization program as the Council of Advisors under the de la Madrid administration and later the Unit for the Divestiture of Parastate Entities farmed out the financial and technical evaluation of firms to agent banks. Although still publicly owned when this process began, domestic agent banks provided a crucial link from the state to private actors. In the larger, more complex privatizations, international banks and private consulting firms also took increased responsibility for implementing state policies.

This privatization of the privatization process contradicts assertions of researchers who focus on the importance of state autonomy to economic reform. For example, Haggard and Kaufman (1992, 25) argue that privatization requires that states develop "technical expertise in financial restructuring, rehabilitating companies, and preparing them for divestiture, as well as procedures that guarantee equal access by potential buyers." While all these are necessary components of privatization, the state need not actually perform these functions. As the Mexican case shows, the state can purchase the expertise that it needs or devolve the authority over some aspect of privatization to private firms (see Chapter 3).

It might be argued that subcontracting work to an agent does not diminish the autonomy of the state. After all, private firms subcontract work without any apparent sacrifice of their capacity. The important difference here is that the Mexican state did not independently develop a reform agenda and then seek out assistance for its implementation. To the contrary. Private firms played an important role in the evolution of the crisis that forced some form of economic restructuring. Many business leaders then pushed for the adoption of neoliberal policies in general and the privatization of parastate firms in particular. Under these conditions, the use of agent banks, domestic and international, as well as the assistance provided by private consulting firms and multilateral organizations all call into question the actual capacity of the Mexican state.

Public contracting with private firms has grown in a number of other areas of state activity as well. The state contracted with private firms to help restructure the operations of the railroads and also in an ambitious private highway construction project that subsequently failed and was rescued at a cost of approximately $3 billion. Public-private partnerships are also a key component of the privatization of the ports and their operation under private management. Although the amount of state investment in these areas may even increase, its level of control over and ability to direct this economic activity can, at the same time, decrease. Social security payroll deductions that once went into public coffers are now transferred by the state directly into the accounts of private fund managers to recapitalize investment and insurance firms.

THE LIMITS TO NEOLIBERAL REFORMS

Robert Heilbroner asserts that "the division between state and economy is not one of extrinsic function—the political realm concerned with 'public' needs, the economic realm with 'private' ones. The essential difference is rather one of the possibility of the recapture of expenditure in the marketplace" (1985, 104, 105). Consistent with this formulation, since 1983, the Mexican state has generally sought to relinquish control over those areas of the economy where profits can be made. In some cases it has badly misjudged the abilities of private firms to recapture their expenditure in the marketplace, leading it right back in to rescue failing firms. Nonetheless, the general principle appears to hold, with the notable exception of the state petroleum company, Pemex, which for political reasons will remain difficult to privatize.

Neoliberal guidelines for state intervention in the economy are not especially helpful. The inadequacy of the neoliberal framework can be traced back to Adam Smith. In *The Wealth of Nations,* Smith admonishes the state to do as little as possible (1981, 687) while still doing what is necessary (723), begging the questions, What is possible, and what is necessary? Writing more than two centuries later, the World Bank is similarly unhelpful on this question. In its jeremiad against public ownership, the World Bank (1995a, 3) argues that "requiring bureaucrats to oversee businesses better handled by private entrepreneurs places a heavy toll on bureaucracies in developing economies." It sheds little light on what those businesses are that are "better handled by private entrepreneurs." Instead, the bank assumes away the problem by defining the state-owned enterprise sector—the sector that should be turned over to private entrepreneurs—as "economic entities that generate the bulk of the revenues from selling goods and services." It then arbitrarily sets aside "education, health services, and road construction and maintenance" for the state simply because they are financed from general revenues (26).

The role of the state in the neoliberal model—and, therefore, in practice in much of the world today—is that of a loss leader in the economy. Private firms operate as profit centers, absorbing losses only where these can be justified by the potential for some long-run gain. The state, however, acts as a loss leader to the entire national economy, absorbing losses according to political pressures and the conceptions of state managers of what is necessary. If these are the limits to state involvement in the economy, what are the limits of the market in the economy, that is, what are the limits to neoliberal reforms?

There are two specific areas where neoliberal reforms may be especially inappropriate. The first of these might be thought of as sectors where society is unwilling to allow for arbitrary uses of resources. I use the word *arbitrary* here in precisely the *opposite* way in which it is generally used in neoclassical and institutional economics. As I show in the introductory chapter, the arbitrary use of property in economic literature has nothing to do with the use value of property. Rather, *arbitrary* in the neoclassical world refers to the exchange value of property. Property is used arbitrarily in these formulations if *(a)* it is not used to generate increased exchange value or *(b)* nonowners gain rights over its use. I have attempted to show how this type of rationality can be socially constructed and I have also attempted to show how the allocation of resources based on this type of rationality can produce disastrous results.

Thus, in claiming that certain resources should not be used arbitrarily, I refer to substantive rationality and the use value of property. Within the neoliberal framework, self-interest can motivate the butcher, brewer, or baker to provide our dinner only if we have sufficient resources to induce such exchanges. When large numbers of people in society lack those resources, markets can create exceptionally irrational outcomes, creating a system of disproportionate representation of those with means and malignant neglect of those without.

The second limit to neoliberal reforms refers to those areas where society is concerned about the democratic accountability of power. Private property is not merely about the control of resources. The market transforms private property into social power, allowing actors to enter into exchange relations without imposing social obligations on property. The market does not eliminate coercion. Instead, coercion takes on the appearance of economic necessity for those without resources, while the organization of production within the firm allows capitalists to direct the economic activities of large numbers of people without securing their agreement.

The appearance of impersonal forces governing market allocations may actually serve to legitimate social inequalities in a capitalist economy by removing the authority over allocations from political leaders. Deprivation and abundance can coexist much more easily—even among those with clear material connections such as workers and owners—if the inequality can be attributed to the natural laws of the market. As Hayek (1994, 117) argues, "Inequality is undoubtedly more readily borne, and affects the dignity of the person much less, if it is determined by impersonal forces than when it is due to design."

These limits to neoliberal reform, however, depend upon social action. The restrictions that are ultimately placed on the market are determined through social struggle. Polanyi (1957, 3) argues that out of the destruction wrought by the creation of a self-regulating market, "[i]nevitably, society took measures to protect itself." Yet the societies that sought self-protection were not predetermined social groups. Rather, attempts to extend the reach of the market defined both the groups that would emerge in opposition to its influence as well as the terms on which those groups would resist.

Polanyi argues that in the 1920s, "[t]he fascist attack on popular democracy merely revived the issue of political interventionism that haunted the history of market economy, since that issue was hardly more than another name for the separation of the economic from the political sphere" (1957, 223). Writing at the end of World War II, however, Polanyi concluded that both Marxist and fascist assumptions regarding the centrality of the economy proved wrong because "the social unit of the nation proved, in the long run, even more relevant than the economic unit of class" (1957, 246).

As the postwar accommodation progressively embedded the global economy in national welfare states within a restricted international trading system, Polanyi's work appeared prophetic. Yet the wave of economic restructuring documented in the preceding pages also suggests the importance of heeding Schumpeter's (1942, 82) words that capitalism "not only never is but never can be stationary." Economic transformation at the dawn of the twenty-first century has begun to dissolve some of the social bonds in which national economies had been embedded, establishing new international social relations. Whether or not the social unit of the nation continues to prove more relevant than new regional economic groupings such as the European Union or NAFTA remains to be seen.

For the present, our perspective is limited by the proximity of these changes. The disembedding of markets from domestic social relations is further advanced than the embedding of international markets in new regulatory structures. Probably the most important international governance structure is the transnational firm, whose activities are regulated and monitored through the private control of capitalists. Regional and international trade regimes lag somewhat behind these private fortresses of regulation within the international economy. The creation of international social groups, seeking to impose democratic accountability on international transactions, lag still further behind. The task of documenting the creation of an international society that matches the reach of the international economy will be the challenge for future researchers.

APPENDIX

Methodology

KEY-INFORMANT INTERVIEWS

Although the theoretical framework advanced in this book is intended as a critique and supplement to recent work in economic sociology, the empirical focus and methodology are entirely consistent with this literature. If the key to understanding the economy is, as Block argues, "to focus on some of the specific ways in which states and economies intersect and begin examining the variations in these intersections across space and time" (1994, 698), then the privatization of public firms should provide a wealth of insight into economic organization.

Many aspects of the privatization process are highly public. Press reports, trade journals, legislation, government statistics, and corporate reports are all readily available. My research draws upon all of these sources and, especially for historical background, incorporates secondary literature on Mexican history, the state, and specific industries. But another part of the process of economic change is less visible. The actual decision-making, the structure and development of state and private institutions, social relations among actors, and their motivations cannot always be gleaned from published reports.

For this reason, in this book I rely on key-informant interviews to provide insights into the less public side of privatization. For these interviews, I used primarily open-ended questions in an unstandardized format, to elicit information that is unique to individuals occupying strategic locations within a government or corporate structure. This type of interview is especially appropriate for studying periods of change and helps to uncover underlying rationales and complex relationships among actors (Schoenberger 1991). As

TABLE 17 Interview summary

	Labor Unions[a]	Business[b]	Government[c]	Other[d]	Total
Unique Informants	6	23	26	3	58
Interviews/Events	13	29	32	4	78

[a]Includes tour of labor union headquarters and attendance at annual convention of union.
[b]Includes business associations and joint meeting with representatives of regulatory agency and business official.
[c]Includes former government officials now working in private sector, retired, or working in multilateral agencies.
[d]Includes attendance at presentation of bids for satellites and interviews of journalist.

Evans (1995, 19) argues, much of the social structure of relations among actors in the economy "can only be appreciated by talking to individual state managers and private executives."

The interviews cited in the text of the book are drawn from seventy-eight key-informant interviews conducted between 1995 and 1999. Because follow-up interviews were conducted with some of the sources, the interviews represent only fifty-seven unique informants. Most of the informants were government officials; the remainder were business officials and representatives of labor unions (see Table 17). The distinction between government and business officials was not always clear cut, because some of the business officials whom I interviewed had worked in government and were familiar with aspects of the privatization process from both the public and private sectors. Less frequently, public-sector officials had worked in the private sector before entering government.

Because access to high-level officials is often difficult to attain, the method of selecting informants for interviews was a "convenience" or "snowball" sample: contact with one official would often open doors to another official and so on. While this method of selecting data sources presents the potential for sample bias in more traditional types of data collection, the purpose of key-informant interviews is quite different from that of quantitative research, in which random assignment is critical to data integrity. Key informants are valuable for the perspective that they bring to a particular event based on their position within a specific institution—whether a firm or government agency. The information that these interviews provide is useful not because it is representative of some larger group but because the informant was present during a particular event or process.

Each interview was unique in that it drew on information gleaned from previous interviews and involved active probing of the answers provided by informants. In this sense, *snowball* is an apt description not only of the procedure for sampling but also of the substance of the interviews. As my familiarity with the process of privatization grew, the questions that I asked changed. In later interviews I was able to draw on information gathered in previous interviews, and questions could be formulated to cross-check an interpretation, invite comment, or challenge the position taken by an informant. When I was able to interview the same informant more than once, I was also able to cross-check statements from the previous interview and ask follow-up questions.

All but one of the interviews were conducted in Mexico City, most of these during a year of fieldwork in 1997, which was funded by a Fulbright scholarship. Two of the interviews were carried out by phone, while all the rest were conducted in person. Most interviews were one-on-one, although a small number of interviews included more than one representative of a firm or government agency. Most of the interviews were conducted in Spanish except when an informant was a native English speaker or in a few cases when the informant preferred to speak in English. I chose not to use a tape recorder and, instead, took notes by hand during the interviews. These notes were transcribed following the interview. The transcribed interviews make up 230 pages of single-spaced notes based on approximately 106 hours of interviews.

CONFIDENTIALITY OF INFORMANTS' IDENTITIES

Although some of the informants were open to my using their names, others wished to remain anonymous. In Table 18, I identify all the interviewees in generic terms as "government," "business," or "labor union" officials. Although I provide information on the division within the government or sector of the economy in which informants worked, companies are referred to by a generic name. For informants from the public sector, I have omitted information on the presidential administration during which the official served. Reference within the text to the presidency during which informants worked is used only when informants agreed to be identified.

Although the rank of the informants varied, the interviews with government officials included meetings with ministerial-level state officials, subministerial-level officials, as well as division directors and subdirectors. In the text, I use the term *high ranking* to refer only to ministerial level, subministerial, and division directors. Informants in the private sector included directors of investor relations, directors of governmental relations, and general directors and secretaries-general of business and trade associations. Interviews with labor union officials included advisors to the secretary-general of unions, press secretaries, and directors of research and statistics within labor unions.

TABLE 18 List of interviews

No.	Date	Informant/Event	Agency/Company/Sector
1	23 May 1995	Labor Union Official 1	Union representing workers in large firm privatized in communications sector
2	29 May 1995	Labor Union Official 2	Union representing workers in large firm privatized in communications sector
3	6 March 1997	Labor Union Official 2	Union representing workers in large firm privatized in communications sector
4	14 March 1997	Business Official 1	Privatized company in communications sector
5	17 March 1997	Business Official 2	Supplier A of equipment in communications sector
6	21 March 1997	Former Government Official 1, retired	Ministry of Communications and Transportation
7	24 March 1997	Business Officials 3 and 4	Privatized company in communications sector
8	26 March 1997	Business Official 1, Government Officials 2 and 3	Privatized company in communications sector and regulatory agency overseeing telecommunications sector
9	25 April 1997	Business Official 5	Company A operating in market recently opened to competition against onetime parastate monopoly
10	30 April 1997	Business Association Official 1	Business group representing firms operating in newly privatized and deregulated market A
11	7 May 1997	Government Official 4	Regulatory agency overseeing telecommunications sector
12	7 May 1997	Labor Union Official 2	Union representing workers in large firm privatized in communications sector
13	9 May 1997	Business Association Official 1	Business group representing firms operating in newly privatized and deregulated market A
14	9 May 1997	Business Official 6	Supplier B of equipment in communications sector
15	14 May 1997	Business Official 7	Company B operating in market recently opened to competition against onetime parastate monopoly
16	14 May 1997	Business Official 8	Supplier B of equipment in communications sector

TABLE 18 *(continued)* List of interviews

No.	Date	Informant/Event	Agency/Company/Sector
17	16 May 1997	Business Official 9	Company B operating in market recently opened to competition against onetime parastate monopoly
18	20 May 1997	Former Government Official 5, currently working in private firm	Ministry of Communications and Transportation; Supplier C of equipment in communications sector
19	26 May 1997	Former Government Official 6, currently working in private firm	Ministry of Communications and Transportation
20	27 May 1997	Government Official 4	Regulatory agency overseeing telecommunications sector
21	29 May 1997	Business Officials 10 and 11	Privatized company in communications sector
22	14 July 1997	Business Official 12	Supplier B of equipment in communications sector
23	14 July 1997	Business Official 13	Supplier B of equipment in communications sector
24	15 July 1997	Business Official 14	Supplier A of equipment in communications sector
25	22 July 1997	Business Official 15	Company B operating in market recently opened to competition against onetime parastate monopoly
26	22 July 1997	Business Official 16	Company C operating in market recently opened to competition against onetime parastate monopoly
27	23 July 1997	Business Official 17	Supplier C of equipment in communications sector
28	30 July 1997	Business Official 18	Company D operating in market recently opened to competition against onetime parastate monopoly
29	30 July 1997	Labor Union Official 1	Union representing workers in large firm privatized in communications sector
30	11 August 1997	Labor Union Official 1	Union representing workers in large firm privatized in communications sector
31	12 August 1997	Business Official 18	Company E operating in market recently opened to competition against onetime parastate monopoly
32	15 August 1997	Journalist 1	Daily newspaper in Mexico City
33	22 August 1997	Business Official 19	Company that purchased large firm in transportation sector

TABLE 18 (*continued*) List of interviews

No.	Date	Informant/Event	Agency/Company/Sector
34	24 August 1997	Journalist 1	Daily newspaper in Mexico City
35	26 August 1997	Labor Union Official 3	Union representing workers in large firm privatized in communications sector
36	26 August 1997	Labor Union Official 2	Union representing workers in large firm privatized in communications sector
37	27 August 1997	Business Official 20	Company that purchased large firm in transportation sector
38	29 August 1997	Business Official 14	Supplier A of equipment in communications sector
39	3 September 1997	Business Association Official 2	Business group representing firms operating in newly privatized and deregulated market B
40	17 September 1997	Labor Union Annual Convention	Union representing workers in large firm privatized in communications sector
41	22 September 1997	Business Official 21	Company B operating in market recently opened to competition against onetime parastate monopoly
42	22 September 1997	Tour of Labor Union Headquarters	Union representing workers in large firm A privatized in transportation sector
43	24 September 1997	Business Association Official 3	Business group representing firms operating in newly deregulated market C
44	2 October 1997	Government Official 7	Government regulatory agency
45	2 October 1997	Former Government Official 8, currently working as private consultant	Ministry of Communications and Transportation
46	7 October 1997	Government Official 9	Ministry of Communications and Transportation
47	7 October 1997	Government Official 10	Ministry of Communications and Transportation
48	7 October 1997	Government Official 11	Ministry of the Treasury
49	8 October 1997	Government Official 12	Ministry of Communications and Transportation
	9 October 1997	Government Official 13	Ministry of Communications and Transportation

TABLE 18 *(continued)* List of interviews

No.	Date	Informant/Event	Agency/Company/Sector
50	9 October 1997	Government Official 13	Ministry of Communications and Transportation
51	10 October 1997	Government Official 14	Large firm in transportation sector in process of being privatized
52	13 October 1997	Government Official 11	Ministry of the Treasury
53	14 October 1997	Business Association Official 4	Umbrella business organization A
54	15 October 1997	Labor Union Official 4	Union representing workers in large firm B privatized in transportation sector
55	16 October 1997	Government Officials 15 and 16	Ministry of Communications and Transportation
56	17 October 1997	Government Official 11	Ministry of the Treasury
57	22 October 1997	Government Official 17	Ministry of Communications and Transportation
58	23 October 1997	Government Official 18	Ministry of the Treasury
59	24 October 1997	Presentation of Bids for Satellite Privatization	Ministry of Communications and Transportation
60	24 October 1997	Government Official 11	Ministry of the Treasury
61	28 October 1997	Labor Union Official 5	Union representing workers in large firm B privatized in transportation sector
62	28 October 1997	Government Officials 14 and 19	Large firm in transportation sector in process of being privatized
63	30 October 1997	Business Association Official 5	Umbrella business organization B
64	30 October 1997	Government Official 20	Ministry of the Treasury
65	3 November 1997	Government Official 9	Ministry of Communications and Transportation
66	4 November 1997	Labor Union Official 6	Large public-sector labor union
67	4 November 1997	Government Official 21	Ministry of Communications and Transportation
68	5 November 1997	Former Government Official 8, currently working as private consultant	Ministry of Communications and Transportation
69	5 November 1997	Government Official 21	Ministry of Communications and Transportation
70	10 November 1997	Government Official 22	Ministry of Communications and Transportation

TABLE 18 *(continued)* List of interviews

No.	Date	Informant/Event	Agency/Company/Sector
71	10 November 1997	Government Official 21	Ministry of Communications and Transportation
72	11 November 1997	Government Official 18	Ministry of the Treasury
73	13 November 1997	Government Official 23	Ministry of Communications and Transportation
74	13 November 1997	Government Official 24	Ministry of the Treasury
75	14 November 1997	Business Official 22	Company that purchased large firm in transportation sector
76	17 November 1997	Business Official 23	Company operating concessions in newly privatized segments of transportation sector
77	3 March 1998	Former Government Official 25, currently working in multilateral development agency	Ministry of the Treasury
78	16 July 1999	Government Official 26	Ministry of the Treasury

REFERENCES

Abrams, Philip. 1982. *Historical Sociology.* Ithaca: Cornell University Press.
Acle Tomasini, Alfredo, and Juan Manuel Vega Hutchison. 1986. *La empresa pública: Desde fuera, desde dentro.* Mexico City: Instituto Nacional de Administración Pública, Editorial Limusa.
Aguileta, Fernando. 1988. "Inconformidad por el retiro de Aeroméxico en Chetumal, Q. Roo." *El Universal,* March 10, p. 1.
Aglietta, Michel. 1979. *A Theory of Capitalist Regulation: The US Experience.* London: Verso.
Aguilar, Gabriela. 1994. "AT&T continuará como proveedor de Telmex." *El Financiero,* November 14, p. 17.
Alcudia, José Luis. 1988. "Los energéticos y el desarrollo nacional." In *México: 75 años de revolución.* Vol. 2, *Desarrollo económico.* Mexico City: Fondo de Cultura Económica.
Alestra. n.d. "Architecture of Alestra's Network." Alestra, Mexico City. Photocopy.
Álvarez, Alejandro. 1987. *La crisis global del capitalismo en México.* Mexico City: Ediciones Era.
American Chamber of Commerce of Mexico. 1995. *The Guide to Mexico for Business.* Mexico City: American Chamber of Commerce of Mexico.
———. 1997. *Investment Promotion Task Force: Report on the Realities and Perceptions of the Investment Climate in Mexico, Final Report,* Mexico City.
Amozurrutia, José Antonio. 2002. "Impacto del proceso de desincorporación y políticas de cambio estructural en la economía mexicana 1982–2000." Unpublished manuscript, Mexico City.
Amsden, Alice. 1989. *Asia's Next Giant: South Korea and Late Industrialization.* New York: Oxford University Press.
Anderson, Charles W. 1963. "Bankers as Revolutionaries: Politics and Development Banking in Mexico." In *The Political Economy of Mexico,* by William P. Glade Jr. and Charles W. Anderson. Madison: University of Wisconsin Press.
Andrews, Edmund L. 1995. "In Global Sales, Breakup Plan Will Give Units Room to Roam." *New York Times,* September 22, pp. D1, D4.

Aranda, Enrique, and Manuel Alonso. 1989. "19 empresarios, acusados de fraude y evasión fiscal." *El Universal,* February 2, p. 1.
Ardavín Migoni, Bernardo. 1988. "Diferencias y semejanzas entre el neoliberalismo y la economía social de mercado." Paper presented at LIII Asamblea Nacional, Coparmex, "México nuevo compromiso personal," Acapulco, Guerrero, October 13–15.
Aronson, Jonathan David, and Peter F. Cowhey. 1988. *When Countries Talk: International Trade in Telecommunications Services.* Cambridge, Mass.: American Enterprise Institute/Ballinger.
Arrighi, Giovanni. 1994. *The Long Twentieth Century: Money, Power, and the Origins of Our Times.* London: Verso.
Arrighi, Giovanni, and Beverly Silver. 1999. *Chaos and Governance in the Modern World System.* Minneapolis: University of Minnesota Press.
Arrow, Kenneth J. 1969. "The Organization of Economic Activity: Issues Pertinent to the Choice of Market Versus Nonmarket Allocation." In *The Analysis and Evaluation of Public Expenditure: The PPB System.* Vol. 1, 59–73. U.S. Joint Economic Committee, 91st Congress, 1st Session. Washington, D.C.: U.S. Government Printing Office.
Aspe, Pedro. 1993. *El camino mexicano de la transformación económica.* Mexico City: Fondo de Cultura Económica.
Aspe, Pedro, and Javier Beristain. 1984. "The Evolution of Income Distribution Policies During the Post-Revolutionary Period in Mexico." In *The Political Economy of Income Distribution in Mexico,* ed. Pedro Aspe and Paul E. Sigmund. New York: Holmes and Meier.
Avantel. 1996. "Complete Fiber Optic Network." Avantel, Mexico City. Photocopy.
Avantel. n.d. "Cronología." Mexico City: Avantel.
Babai, Don. 1988. "The World Bank and the IMF: Rolling Back the State or Backing Its Role?" In *The Promise of Privatization: A Challenge for U.S. Policy,* ed. Raymond Vernon. New York: Council on Foreign Relations.
Babb, Sarah. 2001. *Managing Mexico: Economists from Nationalism to Neoliberalism.* Princeton: Princeton University Press.
Bagli, Charles V. 1999. "A $6 Billion Merger Would Create a Transcontinental Railroad." *New York Times,* December 21, pp. C1, C4.
Bailey, John J. 1988. *Governing Mexico: The Statecraft of Crisis Management.* New York: St. Martin's Press.
Balassa, Bela, Gerardo M. Bueno, Pedro-Pablo Kuczynski, and Mario Henriqui Simonsen. 1986. *Toward Renewed Economic Growth in Latin America.* Washington, D.C.: Institute for International Development.
Banamex. 1990. "Constitutional Reform Bill Reestablishing a Regime of Mixed Ownership for Banking and Credit Services." *Review of the Economic Situation of Mexico* (Banco Nacional de México) 66, no. 775:317–24.
———. 1991. "Privatization: Goals and Achievements." *Review of the Economic Situation of Mexico* (Banco Nacional de México) 67, no. 787:231–37.
Baran, Paul A., and Paul Marlor Sweezy. 1966. *Monopoly Capital: An Essay on the American Economic and Social Order.* New York: Monthly Review Press.

Barbosa, Isidro. 1997. "Plagadas de anomalías y quejas, la presuscripción telefónica." *El Financiero*, March 11, p. 14.
Baudrillard, Jean. 1981. *For a Critique of the Political Economy of the Sign*. New York: Telos Press.
Bazdresch, Carlos, and Santiago Levy. 1991. "Populism and Economic Policy in Mexico, 1970–1982." In *The Macroeconomics of Populism in Latin America*, ed. Rudiger Dornbusch and Sebastian Edwards. Chicago: University of Chicago Press,.
Bennett, Douglas C., and Kenneth E. Sharpe. 1985. *Transnational Corporations Versus the State: The Political Economy of the Mexican Auto Industry*. Princeton: Princeton University Press.
Bensusán, Graciela A. 1990. "Políticas de modernización y relaciones laborales en el sector paraestatal." In *Relaciones laborales en las empresas paraestatales*, ed. Graciela Bensusán and Carlos García. Mexico City: Fundación Friedrich Ebert.
Besen, Stanley M., and Garth Saloner. 1989. "The Economics of Telecommunications Standards." In *Changing the Rules: Technological Change, International Competition, and Regulation in Communications*, ed. Robert W. Crandall and Kenneth Flamm. Washington, D.C.: Brookings Institution.
Blair, Calvin P. 1964. "Nacional Financiera: Entrepreneurship in a Mixed Economy." In *Public Policy and Private Enterprise in Mexico*, ed. Raymond Vernon. Cambridge: Harvard University Press.
Block, Fred. 1990. *Postindustrial Possibilities: A Critique of Economic Discourse*. Berkeley and Los Angeles: University of California Press.
———. 1994. "The Roles of the State in the Economy." In *The Handbook of Economic Sociology*, ed. Neil J. Smelser and Richard Swedberg. Princeton: Princeton University Press.
Boeker, Paul H. 1993. "Latin America's Economic Opening and the Rediscovery of Foreign Investment." In *Latin America's Turnaround: Privatization, Foreign Investment, and Growth*. San Francisco: Institute for Contemporary Studies.
Booth, Robert C., and Donald Garvett. 1998. "Latin American Strategies for the Global Marketplace." In *Handbook of Airline Marketing*, ed. Gail F. Butler and Martin R. Keller. New York: Aviation Week Group, McGraw-Hill.
Bourdieu, Pierre. 1980. *The Logic of Practice*. Stanford: Stanford University Press.
Bowles, Samuel, and Herbert Gintis. 1986. *Democracy and Capitalism: Property, Community, and the Contradictions of Modern Social Thought*. New York: Basic Books.
Boyer, Robert. 1990. *The Regulation School: A Critical Introduction*. New York: Columbia University Press.
Brachet-Márquez, Viviane. 1996. *El pacto de dominación: Estado, clase y reforma social en México (1910–1995)*. Mexico City: El Colegio de México.
Brandenburg, Frank. 1964. *The Making of Modern Mexico*. Englewood Cliffs, N.J.: Prentice-Hall.
Bueno, Aurelio. 1997a. "Balanza comercial: Distorcionan empresas maquiladoras el comercio de telecomunicaciones." *Dígito Cero* (biweekly telecommunications supplement of *El Financiero*), March 18, pp. 8, 9.

———. 1997b. "Inversión a *cuentagotas*." *Dígito Cero* (biweekly telecommunications supplement of *El Financiero*), May 27, p. 8.
———. 1997c. "Ronda el fantasma de la 'Chilenización' en larga distancia en México." *Dígito Cero* (biweekly telecommunications supplement of *El Financiero*), January 7, pp. 8, 9.
Cairncross, Frances. 1996. "The Death of Distance: A Survey of Telecommunications." *The Economist,* September 30.
———. 1997. "A Connected World: A Survey of Telecommunications." *The Economist,* September 13.
Callaghy, Thomas M. 1990. "Lost Between State and Market: The Politics of Economic Adjustment in Ghana, Zambia, and Nigeria." In *Economic Crisis and Policy Choice: The Politics of Adjustment in the Third World,* ed. Joan M. Nelson. Princeton: Princeton University Press.
Calleja, Eugenia. 1988. "Dispuesta ASPA, a comprar la empresa o sólo trece aviones." *El Universal,* March 9, p. 25.
Camp, Roderic Ai. 1996. *Politics in Mexico.* 2d ed. New York: Oxford University Press.
Campa, Homero, and Salvador Corro. 1988. "Con un contrato colectivo leonino, aerovías aplasta a sus trabajadores." *Proceso,* no. 629 (November 21): 10.
Cardoso, Víctor. 1988. "Anunció Aeroméxico un programa de reestructuración financiera." *La Jornada,* March 8, p. 23.
———. 1997. "Alestra Intenta Corromper a Empleados de Correos: Sepomex." *La Jornada,* January 18, p. 42.
Carrillo Castro, Alejandro, and Sergio García Ramírez. 1983. *Las empresas públicas en México.* Mexico City: Miguel Ángel Porrúa.
Casar, María Amparo, and Wilson Peres. 1988. *El estado empresario en México: ¿Agotamiento o renovación?* Mexico City: Siglo Veintiuno Editores.
Casas Guzmán, Francisco Javier. 1994. *La modernización de la empresa pública en México: Logros y resultados 1988–1994.* Mexico City: Fondo de Cultura Económico.
Centeno, Miguel Ángel. 1994. *Democracy Within Reason: Technocratic Revolution in Mexico.* University Park: Pennsylvania State University Press.
Chaffin, Joshua. 1997. "The Great Railway Sale." *US/Mexico Business,* November, 33–43.
Chandler, Alfred D., Jr. 1977. *The Visible Hand: The Managerial Revolution in American Business.* Cambridge: Harvard University Press, Belknap Press.
———. 1990. *Scale and Scope: The Dynamics of Industrial Capitalism.* Cambridge: Harvard University Press, Belknap Press.
Chase-Dunn, Christopher. 1981. "Interstate System and Capitalist World Economy: One Logic or Two?" In *World System Structure: Continuity and Change.* Beverly Hills, Ca.: Sage.
———. 1989. *Global Formation: Structures of the World Economy.* Cambridge, Mass.: Basil Blackwell.
Chávez, Alicia Hernández. 1979. *Historia de la revolución mexicana 1934–1940: La mecánica cardenista.* Mexico City: El Colegio de México.
Chávez, Carlos. 1988. "Desechar 'ideas como las socialistas' propone el nuevo presidente de Coparmex." *Uno Más Uno,* April 20, p. 3.

Cintra. 1996. "Offering Circular: Public Offer to Purchase Exhibit B." Casa de Bolsa InverMexico, Casa de Bolsa Bancomer, Mexico City, May 30.
———. 1997. *Annual Report*. Mexico City: Cintra, S.A. de C.V.
———. 1998. *Annual Report*. Mexico City: Cintra, S.A. de C.V.
———. 1999. *Annual Report*. Mexico City: Cintra, S.A. de C.V.
———. 2000. *Annual Report*. Mexico City: Cintra, S.A. de C.V.
Clarke, Thomas, ed. 1994. *International Privatisation: Strategies and Practices*. Berlin: Walter de Gruyter.
Coase, Ronald. 1937. "The Nature of the Firm." *Economica* 4:386–405.
Coatsworth, John H. 1979. "Indespensible Railroads in a Backward Economy: The Case of Mexico." *Journal of Economic History* 39, no. 4:939–60.
———. 1990. *Los orígenes del atraso: Nueve ensayos de historia económica de México en los siglos XVIII y XIX*. Mexico City: Alianza Editorial Mexicana.
Cobián, Felipe, and Salvador Corro. 1999. "Desde 1998, las autoridades conocían la inseguridad en TAESA." *Proceso*, no. 1203 (November 21).
Cole, Robert E. 1979. *Work, Mobility, and Participation: A Comparative Study of American and Japanese Industry*. Berkeley and Los Angeles: University of California Press.
Coll, Steve. 1986. *The Deal of the Century: The Breakup of AT&T*. New York: Simon and Schuster.
Cook, Maria Lorena. 1997. *Organizing Dissent: Unions, the State, and the Democratic Teachers' Movement in Mexico*. University Park: Pennsylvania State University Press.
Córdova, Arnaldo. 1972. *La formación del poder político en México*. Mexico City: Ediciones Era.
———. 1974. *La política de masas de cardenismo*. Mexico City: Ediciones Era.
———. 1997. "El legado de Salinas." *Nexos*, no. 234 (June).
Córdova, Carlos Acosta. 1998. "La mitad del Fobaproa, en el 0.13% de los créditos de banqueros a parientes, amigos, socios y ellos mismos." *Proceso*, no. 1125 (May 24).
Cornelius, Wayne A., Judith Gentleman, and Peter H. Smith. 1989. "Overview: The Dynamics of Political Change in Mexico." In *Mexico's Alternative Political Futures*, ed. Wayne A. Cornelius, Judith Gentleman and Peter H. Smith. Monograph Series 30. San Diego: Center for U.S.-Mexican Studies, University of California.
Correa, Guillermo. 1997. "Ferrocarriles: En aras de la privatización, despidos, 'retiros voluntarios' y mutilación bárbara del contrato colectivo." *Proceso*, no. 1053 (January 5): 16–18.
Correa, Guillermo, and Salvador Corro. 1989. "El gobierno configura sindicatos a su conveniencia e indigna a los trabajadores." *Proceso*, no. 652 (May 1): 6–13.
Corro, Salvador. 1988a. "Bolaños, el socio de la Quina, ofreció por Mexicana 200 millones de dolares; el gobierno rechazó su oferta." *Proceso*, no. 629 (November 21): 12.
———. 1988b. "El gobierno cede el transporte aéreo a particulares y empresas extranjeras." *Proceso*, no. 598 (April 18): 30.
———. 1991. "Los empresarios se quejaron ante josé cordoba y vino la requisa." *Proceso*, no. 762 (June 10): 22–24.

Cowhey, Peter F., and Jonathan D. Aronson. 1989. "Trade in Services and Changes in the World Telecommunications System." In *Changing Networks: Mexico's Telecommunications Options*, ed. Peter F. Cowhey, Jonathan D. Aronson, and Gabriel Székely. Monograph Series 32. San Diego: Center for U.S.-Mexican Studies.

Crandall, Robert W., and Kenneth Flamm, eds. 1989. *Changing the Rules: Technological Change, International Competition, and Regulation in Communications*. Washington, D.C.: Brookings Institution,

Crook, Clive. 1997. "The Future of the State: A Survey of the State." *The Economist*, September 20.

Dávila Capelleja, Enrique Rafael. 1991. "La reglamentación del autotransporte de carga en México." In *El efecto de la regulación en algunos sectores de la economía mexicana*, ed. Francisco Gil Díaz and Arturo M. Fernández. Mexico City: Instituto Tecnológico Autónomo de México.

———. 1999. "Privatización del transporte ferroviario en México." In *Regulación de los sectores de infraestructura y energéticos en México*, ed. Pablo T. Spiller and Carlos Sales. Mexico City: Instituto Tecnológico Autónomo de México.

Daza, Germán Sánchez, Jorge Sandoval, and Enrique de la Garza. 1988. "La digitalización en Telmex, una transformación global." *El Cotidiano*, no. 21 (January–February): 63–70.

de Buen, Néstor. 1989. "El convenio de modernización en Teléfonos de México." *El Cotidiano*, no. 30 (July–August): 59–61.

de la Garza, Enrique. 1989. "¿Quién ganó en Telmex?" *El Cotidiano*, no. 32 (November–December): 49–56.

———, ed. 1998. *La privatización en México: Consecuencias sociales y laborales*. Instituto de Estudios de la Revolución Democrática: Mexico City.

de la Garza, Enrique, and Fernando Herrera. 1997. "Mexico." In *Telecommunications: Restructuring Work and Employment Relations Worldwide*, ed. Harry C. Katz. Ithaca: Cornell University Press.

de la Madrid, Miguel. 1981. *La empresa pública en México factor de desarrollo económico y social del país*. Secretaría de Trabajo y Previsión Social, Centro Nacional de Productividad, Serie Memorias, no. 1. Mexico City: Secretaría de Trabajo y Previsión Social.

de León, María Eugenia. 1987. "El reto de la modernización ferroviaria." Paper from the conference "Comunicaciones y transportes: Modernización y decentralización," Instituto de Estudios Políticos Económicos y Sociales del PRI. *Diálogo Nacional: Revista de la Consulta Popular*, no. 12 (December 30).

DePalma, Anthony. 1996a. "2 Decisions in Mexico May Aid U.S. Companies." *New York Times*, April 27, p. D2.

———. 1996b. "2 Mexican Phone Start-Ups Agree to Merge." *New York Times*, April 23, p. D6.

Deyo, Frederic, C., ed. 1987. *The Political Economy of the New Asian Industrialism*. Ithaca: Cornell University Press.

El Día. 1988a. "Censuras a la pretensión de privatizar Telmex." January 22, p. 2.

———. 1988b. "Genera confianza la decisión de liquidar a Aeroméxico: Chapa." April 20, p. 9.

Diario Oficial. 1947. "Ley para el control por parte del gobierno federal de los organismos descentralizados y empresas de participación estatal." December 31.
———. 1948. "Comisión nacional de inversiones." January 31.
———. 1983. "Decreto que reforma y adiciona los artículos 16, 25, 26, 27, fracciones XIX y XX; 28, 73, fracciones XXIX-D; XXIX-E y XXIX-F de la Constitución Política de los Estados Unidos Mexicanos." February 3.
———. 1985. "Ley orgánica de Ferrocarriles Nacionales de México." January 14.
———. 1990a. "Decreto que reforma, adiciona y deroga diversas disposiciones del reglamento interior de la Secretaría de Hacienda y Crédito Público." October 30.
———. 1990b. "Modificación al título de concesión de Teléfonos de México, S.A. de C.V." December 10.
———. 1990c. "Reglamento a la ley federal de las entidades paraestatales." January 26.
———. 1993. "Ley de puertos." July 19.
———. 1995a. "Ley Federal de telecomunicaciones." June 7.
———. 1995b. "Ley reglamentaria del servicio ferroviario." April 28.
———. 1996. "Resolución administrativa por la que la Secretaría de Comunicaciones y Transportes establece la regulación tarifaria aplicable a los servicios de interconexión de redes públicas de telecomunicaciones, autorizadas para prestar servicios de larga distancia." April 26.
Díaz, Francisco Gil, and Arturo M. Fernández, eds. 1991. *El efecto de la regulación en algunos sectores de la economía mexicana.* Mexico City: Instituto Tecnológico Autónomo de México.
Dígito Cero. 1997. "Cuatro grupos dominantes: Industria de las telecomunicaciones." In *El Financiero* (biweekly telecommunications supplement), September 17, p. 9.
Domhoff, G. William. 1967. *Who Rules America?* Englewood Cliffs, N.J.: Prentice-Hall.
Dubb, Steve, 1992. "Trozos de Cristal: Privatization and Union Politics at Teléfonos de México." Paper presented at the Seventeenth International Congress of the Latin American Studies Association, Los Angeles, September 24–27.
———. 1998. "Mexican Corporatism Reconsidered: Labor, Dissent, the STRM, and the New Reform Movement." Paper presented at the Twenty-first International Congress of the Latin American Studies Association, Chicago, September 24–26.
The Economist. 1993a. "The Greatest Assets Ever Sold." August 21, pp. 13, 14.
———. 1993b. "The Return of the Railroads?" November 27, pp. 65, 66.
———. 1993c. "Shooting for the Moon." October 30, pp. 69, 70.
———. 1994. "Reach for the Open Sky." November 5, p. 15.
———. 1996a. "Dial M for Merger." November 9, pp. 20, 21.
———. 1996b. "A Marriage of Convenience." November 9, pp. 103, 104.
———. 1997a. "The Future of the State: A Survey of the State." September 20.
———. 1997b. "A Survey of Business in Latin America." December 6.
Estrada, Darío Celis. 1997a. "Corporativo: El Chase y Merrill Lynch dejaron colgado a Serrano." *El Financiero,* February 4, p. 18.

———. 1997b. "Corporativo: Rescata el gobierno a TMM en la ruta noreste de FNM." *El Financiero,* January 31, p. 14.
Evans, Peter. 1995. *Embedded Autonomy: States and Industrial Transformation.* Princeton: Princeton University Press.
Evans, Peter, Dietrich Rueschemeyer, and Theda Skocpol. 1985. *Bringing the State Back In.* New York: Cambridge University Press.
Excelsior. 1989a. "A la opinion pública: STRM." December 27, p. 20.
———. 1989b. "'Es inevitable la privatización de Teléfonos de México.'" September 14, p. 4.
Expansión. 1978. "Los ferrocarriles no terminan de encarrilarse." June 7, pp. 60–63.
———. 1979. "Las 500 empresas más importantes de México." August 22, pp. 92–158.
———. 1985. "Las 500 empresas individuales más importantes de México." August 21, pp. 96–191.
———. 1988. "El enredo de Aeroméxico." August 3, pp. 36–41.
———. 1989. "Las apuestas de Xabre." April 12, pp. 28–32.
———. 1991. "Las 500 empresas más importantes de México." August 21, pp. 312–391.
———. 1993a. "Mexicana de Aviación: Vuela, vuela." December 22, pp. 90–92.
———. 1993b. "La 'privatización' de la política." July 7, pp. 54–79.
———. 1994a. "Las empresas más importantes de México." August 17, pp. 299–384.
———. 1994b. "¿Quedó algo por vender?" November 23, pp. 71, 72.
———. 1995. "Transporte: Todavía sin rumbo." June 7, pp. 24–51.
———. 1996. "Ferrocarriles: Vía libre al capital privado." May 22, pp. 24–43.
———. 1997. "Las empresas más importantes de México." August 13, pp. 337–450.
———. 1999a. "Aeroméxico: Depués de las turbulencias." July 7–21, pp. 36–41.
———. 1999b. "Mexicana: A levantar el vuelo." July 7–21, pp. 28–35.
———. 1999c. "Por qué conviene Cintra." July 7–21, pp. 43–47.
———. 2000. "Las 500, al umbral del nuevo siglo." July 19–August 2, pp. 416–22.
———. 2001. "Las 500 en 2000." July 25–August 8, pp. 306–404.
FCC (Federal Communications Commission). 1996. "In the Matter of IB Docket No. 96-261 International Settlement Rates Notice of Proposed Rulemaking." Federal Communications Commission, FCC 96-484, December 19, Washington, D.C.
———. 2002. IMTS Accounting Rates of the United States, 1985–2002.
Feigenbaum, Harvey B., and Jeffrey R. Henig. 1997. "Privatization and Political Theory." *Journal of International Affairs* 50, no. 2:338–418.
Fernández Varela, Félix Vélez. 1991. "Condicionamientos estructurales del derecho económico en México." In *El efecto de la regulación en algunos sectores de la economía mexicana,* ed. Francisco Gil Díaz and Arturo M. Fernández. Mexico City: Instituto Tecnológico Autónomo de México.
El Financiero. 1989a. "La modernización y eficiencia, son los compromisos de la empresa y el sindicato." April 17, p. 52.

———. 1989b. "Telefonistas, el próximo objetivo de la reconversión." March 8, p. 54.
———. 1994. "Alianza de alfa 'sorprende' al medio financiero." November 11, p. 10.
———. 1997a. "Conserva Telmex 74% del mercado de LD: Cofetel." July 4, p. 12.
———. 1997b. "Empresarios: Informe especial." June 8, pp. 45–49.
Fligstein, Neil. 1990. *The Transformation of Corporate Control*. Cambridge: Harvard University Press.
———. 1996a. "The Economic Sociology of the Transitions from Socialism." *American Journal of Sociology* 101, no. 4:1074–81.
———. 1996b. "Markets as Politics: A Political-Cultural Approach to Market Institutions." *American Sociological Review* 61, no. 4:656–73.
———. 2001. *The Architecture of Markets: An Economic Sociology of Twenty-first-Century Capitalist Societies*. Princeton: Princeton University Press.
Fligstein, Neil, and Iona Mara-Drita. 1996. "How to Make a Market: Reflections on the Attempt to Create a Single Market in the European Union." *American Journal of Sociology* 102, no. 1:1–33.
FNM (Ferrocarriles Nacionales de México). 1987. *Breve reseña histórica de los ferrocarriles mexicanos*. Mexico City: FNM.
———. 1997. *Series estadísticas 1996*. Mexico City: FNM.
Fox, Vicente Quesada. 2002. *Segundo informe de gobierno, anexo estadístico*. Mexico City: Poder Ejecutivo Federal.
Frank, André Gunder. 1967. *Capitalism and Underdevelopment in Latin America*. New York: Monthly Review Press.
Fraser, Robert C., Alan D. Donheiser, and Thomas G. Miller, Jr. 1972. *Civil Aviation Development: A Policy and Operations Analysis*. New York: Praeger Publishers.
Friedman, Milton. 1982. *Capitalism and Freedom*. Chicago: University of Chicago Press.
Galal, Ahmed, and Mary Shirley. "Overview." In *Does Privatization Deliver? Highlights from a World Bank Conference*, ed. Ahmed Galal and Mary Shirley. Washington, D.C.: World Bank.
García, María Angélica. 1988. "Riesgosa, la propuesta de salinas sobre el comercio exterior." *El Financiero*, February 15, p. 60.
García, Martínez. 1988. "Muchas reprivatizaciones el próximo sexenio: CCE." *Excelsior*, February 24, p. 1.
Garrido, Luis Javier. 1989. "The Crisis of *Presidencialismo*." In *Mexico's Alternative Political Futures*, ed. Wayne A. Cornelius, Judith Gentleman and Peter H. Smith. Monograph Series 30. San Diego, Ca.: Center for U.S.-Mexican Studies.
Genel García, Julio Alfredo. 1989. "Aeroméxico: Metamorfosis creativa." In *Privatización: Alcance e implicaciones*, ed. Carlos Bazdresch P. and Víctor L. Urquidi. Morelos: Documento de Trabajo, Centro de Investigación y Docencia Económicas, a.c., Centro Tepoztlán.
Gereffi, Gary. 1994. "The International Economy and Economic Development." In *The Handbook of Economic Sociology*, ed. Neil J. Smelser and Richard Swedberg. Princeton: Princeton University Press.

Gidwitz, Betsy. 1980. *The Politics of International Air Transport.* Lexington, Ky.: Lexington Books.
Glade, William P., Jr. 1963. "Revolution and Economic Development: A Mexican Reprise." In *The Political Economy of Mexico,* by William P. Glade Jr. and Charles W. Anderson. Madison: University of Wisconsin Press.
———. 1996. "Privatization: Pictures of a Process." In *Bigger Economies, Smaller Governments: Privatization in Latin America,* ed. William Glade. Boulder, Colo.: Westview Press.
Goldthorpe, John H., ed. 1984. *Order and Conflict in Contemporary Capitalism: Studies in the Political Economy of Western European Nations.* Oxford, Eng.: Clarendon Press.
Gómez Maza, Francisco, and Gustavo Lomelín. 1988. "Garantiza el nuevo gabinete económico lª estrategia delineada en el sexenio de MM." *El Financiero,* December 1, p. 12.
González, A., and A. Bueno. 1997. "Telmex mantuvo el liderazgo en la primera etapa." *Dígito Cero* (biweekly telecommunications supplement of *El Financiero*), July 8, p. 4.
González, Héctor A. 1990. "Discriminación sobre los delitos fiscales; Otorga SHCP el perdón tributario a ICA y al Grupo Visa." *El Financiero,* November 21, p. 8.
González, Maribel. 2000. "Acusan a TAESA en EU." *Reforma,* March 11.
Granovetter, Mark. 1985. "Economic Action and Social Structure: The Problem of Embeddedness." *American Journal of Sociology* 91, no. 3:481–510.
———. 1990. "The Old and the New Economic Sociology: A History and an Agenda." In *Beyond the Marketplace: Rethinking Economy and Society,* ed. Roger Friedland and A. F. Robertson. New York: Aldine de Gruyter.
———. 1995a. "Coase Revisited: Business Groups in the Modern Economy." *Industrial and Corporate Change* 4, no. 1:93–130.
———. 1995b. "The Economic Sociology of Firms and Entrepreneurs." In *The Economic Sociology of Immigration: Essays on Networks, Ethnicity and Entrepreneurship.* New York: Russell Sage Foundation.
Grindle, Merilee S. 1996. *Challenging the State: Crisis and Innovation in Latin America and Africa.* New York: Cambridge University Press.
Guerrero, Rodolfo. 1988. "Programa inmediato para reponer los vuelos que hacía Aeroméxico." *El Universal,* April 23, p. 1.
Guerrero, Rodolfo, and Emilio Viale. 1988. "Declara oficial la quiebra de Aeroméxico, el juez." *El Universal,* April 19, p. 1.
Guthrie, Doug. 1999. *Dragon in a Three-Piece Suit: The Emergence of Capitalism in China.* Princeton: Princeton University Press.
Haggard, Stephan, and Robert R. Kaufman. 1992. "Institutions and Economic Adjustment." In *The Politics of Economic Adjustment: International Constraints, Distributive Conflicts, and the State,* ed. Stephan Haggard and Robert R. Kaufman. Princeton: Princeton University Press.
Hamilton, Nora. 1982. *The Limits of State Autonomy: Post-Revolutionary Mexico.* Princeton: Princeton University Press.
Hanlon, J. P. 1996. *Global Airlines: Competition in a Transnational Industry.* Oxford, Eng.: Butterworth-Heinemann.

Hansell, Saul. 2002. "Fare Idea Returns to Haunt Airlines." *New York Times,* October 27.
Hansen, Roger D. 1971. *The Politics of Mexican Development.* Baltimore: Johns Hopkins University Press.
Harvey, David. 1982. *The Limits to Capital.* Oxford, Eng.: Basil Blackwell.
———. 1985. "The Geopolitics of Capitalism." In *Social Relations and Spatial Structures,* ed. D. Gregory and J. Urry. New York: St. Martin's Press.
Hayek, F. A. 1994. *The Road to Serfdom.* 50th anniversary ed. Chicago: University of Chicago Press.
Heilbroner, Robert L. 1985. *The Nature and Logic of Capitalism.* New York: W. W. Norton.
Hernández Juárez, Francisco, and María Xelhuantzi López. 1993. *El sindicalismo en la reforma del estado.* Mexico City: Fondo de Cultura Económico.
Hernández López, Julio. 1998. "Astillero." *La Jornada,* February 19.
Hernández Rodríguez, Rogelio. 1986. "La política y los empresarios después de la nacionalización bancaria." *Foro Internacional* 27, no. 2:247–65.
———. 1989. "Las relaciones entre el empresariado y el estado: La genesis de un conflicto." In *La economía mexicana y sus empresarios,* ed. Javier Elguea Solís. Mexico City: Universidad Anáhuac del Sur.
Hernández Vélez, Avelino. 1988a. "'Desventajoso' para México, el convenio aéreo con EU." *El Financiero,* April 25, p. 46.
———. 1988b. "Irreversible, el convenio bilateral aéreo México-EU: Alba Z." *El Financiero,* February 23, p. 35.
Hill, Raymond. 1984. "State Enterprise and Income Distribution in Mexico." In *The Political Economy of Income Distribution in Mexico,* ed. Pedro Aspe and Paul E. Sigmund. New York: Holmes and Meier.
Hirsch, Paul M. 1993. "Undoing the Managerial Revolution? Needed Research on the Decline of Middle Management and Internal Labor Markets." In *Explorations in Economic Sociology,* ed. Richard Swedberg. New York: Russell Sage Foundation.
INEGI (Instituto Nacional de Estadística Geografía e Informática). 2002. *Sistema de cuentas nacionales, indicadores macroeconómicos del sector público 1988–2000.* Mexico City: INEGI.
Islas Rivera, Víctor. 1990. *Estructura y desarrollo del sector transporte en México.* Mexico City: El Colegio de México.
Jáquez, Antonio. 1998. "Con los mismos rescatados y los mismos rescatistas, Ficorca y Fobaproa formaron 'el hoyo negro de la economía' en favor de los oligopolios triunfantes." *Proceso,* no. 1127 (June 7).
———. 1999. "Sin política aeronáutica, el gobierno llevó a las aerolíneas de la protección a ultranza a la desregulación caótica." *Proceso,* no. 1164 (February 21).
Johnson, Bryan T., Kim R. Holmes, and Melanie Kirkpatrick, eds. 1998. *1998 Index of Economic Freedom.* New York: The Heritage Foundation and the Wall Street Journal.
Jones, Leroy P., and Edward S. Mason. 1982. "Role of Economic Factors in Determining the Size and Structure of the Public-Enterprise Sector in Less-Developed Countries with Mixed Economies." In *Public Enterprise*

in Less-Developed Countries, ed. Leroy Jones. New York: Cambridge University Press.
Jones, Leroy P., Pankaj Tandon, and Ingo Vogelsang. 1990. *Selling Public Enterprises: A Cost-Benefit Methodology.* Cambridge: MIT Press.
La Jornada. 1989a. "Modernización de Telmex con rectoría estatal, anuncia CSG." September 19, p. 1.
———. 1989b. "Se indexarán a la inflación las tarifas de comunicaciones." November 1, p. 17.
Josephson, Matthew. 1934. *The Robber Barons.* Orlando, Fla.: Harcourt, Brace.
Kahler, Miles. 1990. "Orthodoxy and Its Alternatives: Explaining Approaches to Stabilization and Adjustment." In *Economic Crisis and Policy Choice: The Politics of Adjustment in the Third World,* ed. Joan M. Nelson. Princeton: Princeton University Press.
Kaufman, Robert R. 1990. "Stabilization and Adjustment in Argentina, Brazil, and Mexico." In *Economic Crisis and Policy Choice: The Politics of Adjustment in the Third World,* ed. Joan M. Nelson. Princeton: Princeton University Press.
Keister, Lisa. 2001. "Exchange Structures in Transition: Lending and Trade Relations in Chinese Business Groups." *American Sociological Review* 66 (June): 336–60.
Kollock, Peter. 1994. "The Emergence of Exchange Structures: An Experimental Study of Uncertainty, Commitment, and Trust." *American Journal of Sociology* 100, no. 2:313–45.
Kornai, János. 1990. "The Soft Budget Constraint." In *Vision and Reality, Market and State: Contradictions and Dilemmas Revisited.* New York: Routledge.
———. 1992. *The Socialist System: The Political Economy of Communism.* Princeton: Princeton University Press.
Labaton, Stephen. 2001. "Airlines and Antitrust: A New World. Or Not." *New York Times,* November 18.
La Botz, Dan. 1992. *Mask of Democracy: Labor Suppression in Mexico Today.* Boston: South End Press.
Lasswell, H. R. 1950. *Politics: Who Gets What, When, and How.* New York: P. Smith.
La Porta, Rafael, and Florencio López-de-Silanes. 1999. "Benefits of Privatization: Evidence from Mexico." *Quarterly Journal of Economics* 114, no. 4:1193–242.
Ledesma, Matilda Luna. 1992. *Los empresarios y el cambio político: México, 1970–1987.* Mexico City: Ediciones Era.
Lewis, W. David, ed. 2000. *Airline Executives and Federal Regulation: Case Studies in American Enterprise from the Airmail Era to the Dawn of the Jet Age.* Columbus: Ohio State University Press.
Ley, Concepción, and Carlos Velasco. 1988. "Anuncia mexicana que todos sus vuelos están saturados hasta junio." *Heraldo,* April 19, p. 2.
Leyva, Marco Antonio. 1995. *Poder y dominación en Ferrocarriles Nacionales de México: 1970–1988.* Mexico City: Universidad Autónoma Metropolitana, Unidad Iztapalapa, División de Ciencias Sociales y Humanidades.
———. 1998. "Ferrocarriles Nacionales de México." In *La privatización en México: Consecuencias sociales y laborales,* ed. Enrique de la Garza Toledo. Mexico City: Instituto de Estudios de la Revolución Democrática.

Leyva, Marco Antonio, and Guillermo Campos Rios. 1990. "Modernización y relaciones laborales en Ferrocarriles Nacionales de México." In *Relaciones laborales en las empresas paraestatales,* ed. Graciela Bensusán and Carlos García. Mexico City: Fundación Friedrich Ebert.

Leyva, Marco Antonio, Moisés Mecalco López, and Rogelio Mendoza Molina. 1998. "Mexicana y Aeroméxico." In *La privatización en México: Consecuencias sociales y laborales,* ed. Enrique de la Garza Toledo. Mexico City: Instituto de Estudios de la Revolución Democrática.

Lipietz, Alain. 1982. "Towards Global Fordism." *New Left Review,* no. 132 (March–April): 33–58.

Lizárraga R., Rebeca. 1988. "Divide a los empresarios la apertura a la inversión externa." *Uno Más Uno,* February 5.

López Pardo, Gustavo. 1997. *La administración obrera de los Ferrocarriles Nacionales de México.* Universidad Nacional Autónoma de México, Instituto de Investigaciones Económicas. Mexico City: Ediciones el Caballito.

Lukács, Georg. 1971. *History and Class Consciousness: Studies in Marxist Dialectics.* Cambridge: MIT Press.

Lustig, Nora. 1994. *Hacia la reconstrucción de una economía.* Mexico City: Fondo de Cultura Económica and Colegio de México.

Machalaba, Daniel, and Anna Wilde Mathews. 1996. "Kansas City Southern Takes Many by Surprise with Mexican Rail Bid." *Wall Street Journal,* December 9, p. B6.

McGuire, Patrick, Mark Granovetter, and Michael Schwartz. 1993. "Thomas Edison and the Social Construction of the Early Electricity Industry in America." In *Explorations in Economic Sociology,* ed. Richard Swedberg. New York: Russell Sage Foundation.

MacLeod, Dag. 1997. "Political Realignment in Mexico: The PRD Moves from 'Show' to 'Place.'" Working paper no. 20, Program in Comparative and International Development, Department of Sociology, Johns Hopkins University. Online. Available: http://www.firstdraft.org/mexico/1997elections.htm.

———. 2001. "Taking the State Back Out? Privatization and the Limits of Neoliberal Reform in Mexico." Ph.D. diss., Johns Hopkins University.

Malkin, Elisabeth. 1995. "Mexico's Problems Will Expedite Privatisations." *Privatisation International: The Monthly Intelligence Report on Privatisation and Private Infrastructure Worldwide,* no. 77 (February): 1, 26.

Márquez, Alfredo. 1988. "Franca alianza gobierno IP en este sexenio: Garza Treviño." *El Financiero,* November 9, p. 16.

Martínez Varga, Lorenzo. 1988. "La línea aérea que fue para abajo: Aeroméxico." *Novedades,* May 5.

Marx, Karl. 1967a. *Capital.* Vol. 1. New York: International Publishers.

———. 1967b. *Capital.* Vol. 3. New York: International Publishers.

Mathews, Anna Wilde. 1998. "North American Trade Blasts Old Limits: Railways Shift Gears to Form 3-Nation Network." *Wall Street Journal,* February 11, p. A2.

Mattelart, Armand. 1994. *Mapping World Communication: War, Progress, Culture,* trans. Susan Emanuel and James A. Cohen. Minneapolis: University of Minnesota Press.

Matus, María Fernanda. 1997a. "Amenaza una 'guerra' entre Telmex y Avantel." *Reforma,* May 12, p. 27A.
——. 1997b. "'Es México el más irregular.'" *Reforma,* April 16, p. 35A.
——. 1997c. "Invalida jueza decisión de Cofetel contra Avantel." *Reforma,* July 15, p. 1A.
——. 1997d. "Trascienden críticas a Telmex: Llegan a oídos de la FCC de Estados Unidos." *Reforma,* May 12, p. 39A.
Maxfield, Sylvia. 1989. "International Economic Opening and Government-Business Relations." In *Mexico's Alternative Political Futures,* ed. Wayne A. Cornelius, Judith Gentleman, and Peter H. Smith. Monograph Series 30. San Diego, Ca.: Center for U.S.-Mexican Studies.
——. 1990. *Governing Capital: International Finance and Mexican Politics.* Ithaca: Cornell University Press.
——. 1992. "The International Political Economy of Bank Nationalization: Mexico in Comparative Perspective." *Latin American Research Review* 27, no. 1:75–103.
Medina Peña, Luis. 1994. *Hacia el nuevo estado: México, 1920–1994.* Mexico City: Fondo de Cultura Económico.
Megginson, William L., and Jeffry M. Netter. 2001. "From State to Market: A Survey of Empirical Studies on Privatization." *Journal of Economic Literature* 39 (June): 321–89.
Mendoza Sanchez, Juan Manuel. 1994. *La liberalización del transporte aéreo en Canadá y Estados Unidos de América y su impacto en la economía mexicana (1988–1993).* Tesis para obtener el título de licenciado en relaciones internacionales, Universidad Nacional Autonoma de México, Escuela Nacional de Estudios Profesionales "Aragon."
El Mercado de Valores. 1995. "Desincorporación de entidades paraestatales, 1994–2000." No. 8 (August): 17–22.
Mexicana. 2000. "Seguimos haciendo historia." Online. Available: http://www.mexicana.com.mx/mx2/spanish/mexicana/historia/.
Meyer, John, John Boli, and George M. Thomas. 1987. "Ontology and Rationalization in the Western Cultural Account." In *Institutional Structure: Constituting State, Society, and the Individual.* Newbury Park, Ca.: Sage.
Meyer, Lorenzo. 1977. *Mexico and the United States in the Oil Controversy, 1917–1942.* Austin: University of Texas Press.
——. 1992. *La segunda muerte de la revolución mexicana.* Mexico City: Cal y Arena.
——. 1995. *Liberalismo autoritario: Las contradicciones del sistema político mexicano.* Mexico City: Editorial Oceano de México.
——. 1997. "Danza con lobos o el PAN y la presidencia." *Reforma,* February 13, p. 15A.
Middlebrook, Kevin. 1989. "The Sounds of Silence: Organised Labour's Response to Economic Crisis in Mexico." *Journal of Latin American Studies* 21, no. 2:195–220.
Middlebrook, Kevin J. 1995. *The Paradox of Revolution: Labor, the State, and Authoritarianism in Mexico.* Baltimore: Johns Hopkins University Press.
Miliband, Ralph. 1969. *The State in Capitalist Society.* New York: Basic Books.

Mondragón, Luz Ma. 1988. "No vamos a crecer aprisa así está planeado, Señaló Eugenio Garza Lagüera." *El Sol*, October 22, p. 1.
Murphy, Craig N. 1994. *International Organization and Industrial Change: Global Governance Since 1850*. New York: Oxford University Press.
Nee, Victor. 1992. "Organizational Dynamics of Market Transition: Hybrid Forms, Property Rights, and Mixed Economy in China." *Administrative Science Quarterly* 37:1-27.
Newfarmer, Richard S., and Willard F. Mueller. 1975. *Report to the Subcommittee on Multinational Corporations of the Committee on Foreign Relations United States Senate*. Washington, D.C.: U.S. Government Printing Office.
Newman, David. 1995. "Mexico Opens Up: Deregulation Brings Free Trade to Telecom." Data Communications on the Web, Global Networks. Online. Available: http://www.data.com./Global_Networks/Mexico_Opens.html.
North, Douglass C. 1981. *Structure and Change in Economic History*. New York: W. W. Norton.
———. 1990. *Institutions, Institutional Change, and Economic Performance*. Cambridge: Cambridge University Press.
Novedades. 1988. "Se concretó la venta de Mexicana de Aviación." March 3, p. 1F.
Novelo, Victoria, and Augusto Urteaga. 1979. *La industria en los magueyales: Trabajo y sindicatos en Ciudad Sahagún*. Mexico City: Centro de Investigaciones Superiores del Instituto Nacional de Antropología e Historia, Editorial Nueva Imagen.
O'Brien, Patrick. 1977. *The New Economic History of the Railways*. London: Croom Helm.
Oppenheimer, Andres. 1996. *México: En la frontera del caos: La crisis de los noventa y la esperanza del nuevo milenio*. Buenos Aires: Javier Vergara Editor.
Ortega, Max. 1988. *Estado y movimiento ferrocarrilero: 1958-1959*. Mexico City: Ediciones Quinto Sol.
Ortega Pizarro, Fernando. 1988. "A trasmano, poco a poco, el gobierno se va deshaciendo de Mexicana." *Proceso*, no. 601 (May 9): 10.
———. 1989. "Los soldados, para anunciar que Cananea había sido declarada en quiebra." *Proceso*, no. 669 (August 28): 6-12.
Osterberg, William P. 1997. "The Hidden Costs of Mexican Banking Reform." *Economic Commentary* (Federal Reserve Bank of Cleveland). Online. Available: http://www.clev.frb.org/research/com97/.
Ortiz, Rosario, and Rodolfo García. 1990. "Concertación en Teléfonos de México." In *Negociación y conflicto laboral en México*, ed. Graciela Bensusán and Samuel León. Mexico City: Fundación Friedrich Ebert Stiftung and FLACSO.
Ortiz Martínez, Guillermo. 1994. *La reforma financiera y la desincorporación bancaria*. Mexico City: Fondo de Cultura Económico.
Otero, Gerardo. 2000. "Rural Mexico After the 'Perfect Dictatorship.'" *Latin American Studies Association Forum* 31, no. 3:4-7.
Pardo, María del Carmen. 1986. "La ley federal de entidades paraestatales: Un nuevo intento para regular el sector paraestatal." *Foro Internacional* 27, no. 2:234-46.

Pellicer de Brody, Olga, and José Luis Reyna. 1978. *Historia de la revolución mexicana, 1952–1960*. Vol. 22, *El afianzamiento de la estabilidad política*. Mexico City: Colegio de México.
Pérez de Mendoza, Alfredo. 1989. "Teléfonos de México: Development and Perspectives." In *Changing Networks: Mexico's Telecommunications Options*, ed. Peter F. Cowhey, Jonathan D. Aronson, and Gabriel Székely. Monograph Series 32. Center for U.S.-Mexican Studies, University of California, San Diego.
Pérez Escamilla, Juan Ricardo. 1989. "Telephone Policy in Mexico: Rates and Investment." In *Changing Networks: Mexico's Telecommunications Options*, ed. Peter F. Cowhey, Jonathan D. Aronson, and Gabriel Székely. Monograph Series 32. Center for U.S.-Mexican Studies, University of California, San Diego.
Pérez Escobedo, María Antonieta. 1995. "Aviación civil: En alas de una nueva legislación." *Comercio Exterior*, July, 518–21.
Pérez-Moreno, Lucía. 2002. "Bancos bajo la lupa." *Expansión*, July 10–24, pp. 67–69.
Pérez Sandi, Juan Manuel Villanueva. 1996. "Las Administradoras de Fondos de Retiro (Afores): Resumen técnico." *Entorno* (Coparmex), año 7, no. 93:22–26.
Perrow, Charles. 1990. "Economic Theories of Organization." In *Structures of Capital: The Social Organization of the Economy*, ed. Sharon Zukin and Paul DiMaggio. Cambridge, Eng.: Cambridge University Press.
Peters, Phillip. 1998. "Mexico's Telmex: All Alone on the Telephone." *Wall Street Journal*, February 13, p. A15.
Petrazzini, Ben A. 1995. *The Political Economy of Telecommunications Reform in Developing Countries: Privatization and Liberalization in a Comparative Perspective*. Westport, Conn.: Praeger.
———. 1996. *Global Telecom Talks: A Trillion Dollar Deal*. Washington, D.C.: Institute for International Economics.
Petricioli, Gustavo. 1988. "Economía mixta." In *México: 75 años de revolución*. Vol. 2, *Desarrollo económico*. Mexico City: Fondo de Cultura Económica.
Philip, George. 1984. "Public Enterprise in Mexico." In *Public Enterprise and the Developing World*, ed. V. V. Ramanadham. London: Croom Helm.
Phillips, Don. 1996. "On the Track Back to Regulation?" *Washington Post*, October 17, pp. D1, D2.
Pichardo Pagaza, Ignacio. 1988. "El proceso de desincorporación de entidades paraestatales (el caso de México)." Discursos del Secretario no. 5, Dirección General de Comunicación Social. Mexico City: Secretaría de la Contraloría General de la Federación.
Piore, Michael J. 1996. "Review of *The Handbook of Economic Sociology*." *Journal of Economic Literature* 34 (June): 741–54.
Polanyi, Karl. 1957. *The Great Transformation: The Political and Economic Origins of Our Time*. 1st Beacon Paperback ed. Boston: Beacon Press.
———. 1957. "The Economy as Instituted Process." In *Trade and Markets in the Early Empires*, ed. Karl C. Polanyi, Conrad M. Arensberg, and Harry W. Pearson. New York: Free Press.

Polanyi, Karl, Conrad M. Arensberg, and Harry W. Pearson, eds. 1957. *Trade and Markets in the Early Empires.* New York: Free Press.
Portes, Alejandro. 1995a. "Economic Sociology and the Sociology of Immigration: A Conceptual Overview." In *The Economic Sociology of Immigration: Essays on Networks, Ethnicity, and Entrepreneurship,* ed. Alejandro Portes. New York: Russell Sage Foundation.
———. 1995b. "On Grand Surprises and Modest Certainties: Comment on Kuran, Collins, and Tilly." *American Journal of Sociology* 100, no. 6:1620–26.
———. 1997. " Neoliberalism and the Sociology of Development: Emerging Trends and Unanticipated Facts." *Population and Development Review* 23, no. 2:229–59.
———. 2000. "The Hidden Abode: Sociology as Analysis of the Unexpected." *American Sociological Review* 65 (February):1–18.
Portes, Alejandro, and Julia Sensenbrenner. 1993. "Embeddedness and Immigration: Notes on the Social Determinants of Economic Action." *American Journal of Sociology* 98, no. 6:1320–50.
Poulantzas, Nicos. 1972. "The Problem of the Capitalist State." In *Ideology in Social Science: Readings in Critical Social Theory,* ed. Robin Blackburn. Suffolk, Eng.: Chaucer Press.
———. 1973. *Political Power and Social Classes.* London: NLB and Sheed and Ward.
———. 1976. "The Capitalist State: A Reply to Miliband and Laclau." *New Left Review* 95 (January–February): 63–83.
Powell, Walter W. 1990. "The Transformation of Organizational Forms: How Useful Is Organizational Theory in Accounting for Social Change?" In *Beyond the Marketplace: Rethinking Economy and Society,* ed. Roger Friedland and A. F. Robertson. New York: Aldine de Gruyter.
Preston, Julia. 1998. "$62 Billion Bank Bailout Plan in Mexico Incites Outrage as Critics Say It Helps the Rich." *New York Times,* July 31, p. A6.
Privatisation International: The Monthly Intelligence Report on Privatisation and Private Infrastructure Worldwide. 1998. "Privatisation Review 1997: Telecom Sales Dominate Record Year." January, 22, 24.
Przeworski, Adam. 1980. "Material Basis of Consent: Economics and Politics in a Hegemonic System." In *Political Power and Social Theory.* Vol. 1, 21–66.
———. 1990. *The State and the Economy Under Capitalism.* Chur, Switzerland: Harwood Academic Publishers.
———. 1991. *Democracy and the Market: Political and Economic Reforms in Eastern Europe and Latin America.* Cambridge: Cambridge University Press.
Ramamurti, Ravi. 1996. "The New Frontier of Privatization." In *Privatizing Monopolies: Lessons from the Telecommunications and Transport Sectors in Latin America,* ed. Ravi Ramamurti. Baltimore: Johns Hopkins University Press.
Ramírez, Miguel D. 1994. "Privatization and the Role of the State in Post-ISI Mexico." In *Privatization in Latin America: New Roles for the Public and Private Sectors,* ed. Werner Baer and Melissa H. Birch. Westport, Conn.: Praeger.
Reforma. 1997. "Gasto Publicitario." August 4, p. 4A, Negocios.

Rey, Luis. 1995. "Competition Comes to Mexican Telecoms." *OECD Observer,* no. 194:26–29.
Reynolds, Clark. 1978. "Why Mexico's 'Stabilizing Development' Was Actually Destabilizing (with Some Implications for the Future)." *World Development* 6:1005–18.
Rodríguez, Luis Alberto. 1988. "Pidió el CEESP se acelere la desincorporación de paraestatales." *La Jornada,* March 19, p. 25.
Rogozinski, Jacques. 1993. *La privatización de empresas paraestatales: Una visión de la modernización de México.* Mexico City: Fondo de Cultura Económica.
———. 1997. *La privatización en México: Razones e impactos.* Mexico City: Editorial Trillas.
Rubio, Luis F. 1989. "El sector privado en el pasado y en el futuro de México." In *La economía mexicana y sus empresarios,* ed. Javier Elguea Solís. Mexico City: Universidad Anáhuac del Sur.
Ruggie, John Gerard. 1983. *The Antinomies of Interdependence: The Political Economy of International Change.* New York: Columbia University Press.
———. 1995. "At Home Abroad, Abroad at Home: International Liberalization and Domestic Stability in the New World Economy." Jean Monnet Chair Paper no. 5, Robert Schuman Centre at the European University Institute, San Domenico, Italy.
Ruiz Dueñas, Jorge. 1988. *Empresa pública: Elementos para el examen comparado.* Mexico City: Fondo de Cultura Económica.
Ruster, Jeff. 1997. "A Retrospective on the Mexican Toll Road Program (1989–94)." In *Public Policy for the Private Sector* (pamphlet). N.p.: World Bank Group.
Sales, Carlos, Enrique Sclar, and Luis Videgaray. 1999. "El desarrollo de la infraestructura carretera en México." In *Regulación de los sectores de infraestructura y energéticos en México,* ed. Pablo T. Spiller and Carlos Sales. Mexico City: Instituto Tecnológico Autónomo de México.
Salinas de Gortari, Carlos. 1987. "Comunicaciones y transportes: Modernización y decentralización." *Diálogo nacional: Revista de la consulta popular,* Instituto de Estudios Políticos Económicos y Sociales del PRI, no. 12 (December 30).
———. 1989. *Primer informe de gobierno, anexo estadístico.* Mexico City: Poder Ejecutivo Federal.
———. 1993. "Prólogo." In *Economía y telecomunicaciones.* (Ponencias presentadas en el seminario internacional "Las telecomunicaciones como factor de desarrollo y modernización económica.") Mexico City: SCT with Colegio Nacional de Economistas, A.C.
———. 1994. *Sexto informe de gobierno, anexo estadístico.* Mexico City: Poder Ejecutivo Federal.
Salmon, Agustín. 1988. "Beneficiadas las aerolíneas mexicanas con el convenio bilateral México-EU: Camacho G." *Excelsior,* February 17, p. 44.
Salpukas, Agis. 1999. "Conrail Chugs off into the Sunset: SCX and Norfolk Southern Take Over." *New York Times,* June 1, pp. C1, C8.
Sánchez, Isabel. 2001a. "La historia repetida." *Expansión,* October 31–November 14.
———. 2001b. "El saldo necrobancario." *Expansión,* May 2–16.

Sánchez, Manuel, and Rossana Corona, eds. 1993. *Privatization in Latin America.* Washington, D.C.: Inter-American Development Bank.

Sánchez, Manuel, Rossana Corona, Luis Fernando Herrera, and Otoniel Ochoa. 1993a. "A Comparison of Privatization Experiences: Chile, Mexico, Colombia, and Argentina." In *Privatization in Latin America,* ed. Manuel Sánchez and Rossana Corona. Washington, D.C.: Inter-American Development Bank.

Sánchez, Manuel, Rossana Corona, Otoniel Ochoa, Luis Fernando Herrera, Arturo Olvera, and Ernesto Sepúlveda. 1993b. "The Privatization Process in Mexico: Five Case Studies." In *Privatization in Latin America,* ed. Manuel Sánchez and Rossana Corona. Washington, D.C.: Inter-American Development Bank.

Sanderson, Steven. 1992. *The Politics of Trade in Latin American Development.* Palo Alto: Stanford University Press.

Saunders, Robert J., Jeremy J. Warford, and Björn Wellenius. 1994. *Telecommunications and Economic Development.* Published for the World Bank. Baltimore: Johns Hopkins University Press.

Schmitter, Philippe C. 1974. "Still the Century of Corporatism?" In *The New Corporatism: Social-Political Structures in the Iberian World,* ed. Fredrick B. Pike and Thomas Strich. South Bend: University of Notre Dame Press.

Schneider, Ben Ross. 1999. "The Politics of Administrative Reform: Intractable Dilemmas and Improbable Solutions." In *Sustainable Public Sector Finance in Latin America.* (Papers from a conference presented by the Latin American Research Group, Research Department, Federal Reserve Bank of Atlanta, http://www.frbatlanta.org.) Atlanta: Federal Reserve Bank of Atlanta.

Schoenberger, Erica. 1991. "The Corporate Interview as a Research Method in Economic Geography." *Professional Geographer* 43, no. 2:180–89.

Schumpeter, Joseph A. 1942. *Capitalism, Socialism, and Democracy.* New York: Harper and Row Publishers.

———. 1991. "Can Capitalism Survive?" In *The Economics and Sociology of Capitalism,* ed. Richard Swedberg. Princeton: Princeton University Press.

SCT (Secretaría de Comunicaciones y Transportes). 1970. *Las comunicaciones, medio de integración nacional y de desarrollo económico.* Mexico City: SCT.

———. 1987. *La aviación mexicana en cifras, 1975–1986.* Mexico City: SCT.

———. 1989. *El transporte en México.* Mexico City: Dirección General de Aeronáutica Civil.

———. 1990a. *La aviacion mexicana en cifras, 1980–1989.* Mexico City: Dirección General de Aeronáutica Civil.

———. 1990b. "Modificación al título de concesión de Teléfonos de México, S.A. de C.V." *Diario Oficial,* December 10.

———. 1992. *Secretaría de Comunicaciones y Transportes, anuario estadístico del sector comunicaciones y transportes.* Mexico City: Comité de Estadística del Sector Comunicaciones y Transportes.

———. 1993. *Economía y telecomunicaciones.* (Ponencias presentadas en el seminario internacional "Las telecomunicaciones como factor de desarrollo y modernización económica," SCT with Colegio Nacional de Economistas, A.C.) Mexico City: Poder Ejecutivo Federal.

———. 1994a. *Memoria 1988–1994*. Mexico City: SCT.

———. 1994b. *Programa nacional de autopistas 1989–1994*. Mexico City: Dirección General de Carreteras Federales.

———. 1994c. *El sector comunicaciones y transportes*. Oficialía Mayor, Coordinación de Asesores, no. 3.

———. 1995a. "Investment Opportunities in the Mexican Railroad System." Mexico City: Ferrocarriles Nacionales de México.

———. 1995b. "Ley federal de telecomunicaciones." *Diario Oficial*, June 7.

———. 1996a. *La aviación mexicana en cifras, 1989–1995*. Mexico City: Dirección General de Aeronáutica Civil.

———. 1996b. "Decreto por el que se crea la Comisión Federal de Telecomunicaciones." *Diario Oficial*, August 9.

———. 1996c. "Reglas para prestar el servicio de larga distancia internacional que deberán aplicar los concesionarios de redes públicas de telecomunicaciones autorizados para prestar este servicio." *Diario Oficial*, December 11.

———. 1996d. "Reglas del servicio de larga distancia." *Diario Oficial*, June 21.

———. 1996e. "Resolución administrativa por la que la Secretaría de Comunicaciones y Transportes establece la regulación tarifaria aplicable a los servicios de interconexión de redes públicas de telecomunicaciones, autorizadas para prestar servicios de larga distancia." Forma CG-1A, April 26, Mexico City.

———. 1997a. *Los puertos mexicanos en cifras: 1990–1996*. Mexico City: Coordinación General de Puertos y Marina Mercante.

———. 1997b. "Resumen de los procesos de apertura a la inversión concluidos y por realizar." SCT, Mexico City, July. Photocopy.

———. 2000. *El sector comunicaciones y transportes, 1994–2000*, Mexico City.

———. 2002. *La aviación mexicana en cifras, 1994–2000 (Preliminares)*. Mexico City: Dirección General de Aeronáutica Civil.

SHCP (Secretaría de Hacienda y Crédito Público). 1994a. *Desincorporación de entidades paraestatales: Información básica de los procesos del 1º de diciembre de 1988 al 31 de diciembre de 1993*. Mexico City: Fondo de Cultura Económica.

———. 1994b. *El proceso de enajenación de entidades paraestatales*. Mexico City: Unidad de Desincorporación de Entidades Paraestatales de la Secretaría de Hacienda y Crédito Público.

———. 1995a. "Acuerdo que crea la Comisión Intersecretarial de Desincorporación." *Diario Oficial*, April 7.

———. 1995b. *Plan Nacional de Desarrollo, 1995–2000*, Mexico City: SHCP.

———. n.d. "Enajenación de entidades paraestatales, participaciones accionarias, venta de activos y concesiones realizadas a partir del 1ro de diciembre de 1994 al 30 de noviembre del 2000." Mexico City: SHCP.

Segura, José García. 1988. "Mexicana restructurará su deuda de 291 milliones de dólares: Sosa." *El Día*, January 10, p. 5.

Sen, Amartya. 1999. *Development as Freedom*. New York: Anchor Books,.

Shelton, David H. 1964. "The Banking System: Money and the Goal of Growth." In *Public Policy and Private Enterprise in Mexico*, ed. Raymond Vernon. Cambridge: Harvard University Press.

Sindicatura. 1988. "Sindicatura de Aeroméxico: Aviso al público." *La Jornada,* May 17, p. 14.
Skocpol, Theda. 1979. *States and Social Revolutions: A Comparative Analysis of France, Russia, and China.* Cambridge: Cambridge University Press.
Smith, Adam. 1981. *An Inquiry into the Nature and Causes of the Wealth of Nations.* Indianapolis: Liberty Fund.
Smith, Peter H. 1979. *Labyrinths of Power: Political Recruitment in Twentieth-Century Mexico.* Princeton: Princeton University Press.
Smithies, Richard. 1998. "The Liberalization of Air Transport Services and the General Agreement on Trade in Services." In *Handbook of Airline Marketing,* ed. Gail F. Butler and Martin R. Keller. New York: Aviation Week Group, McGraw-Hill.
Spiller, Pablo T., and Carlos Sales, eds. 1999. *Regulación de los sectores de infraestructura y energéticos en México.* Mexico City: Instituto Tecnológico Autónomo de México.
SPP (Secretaría de Programación y Presupuesto). 1983. *Plan Nacional de Desarrollo, 1983-1988.* Mexico City: Poder Ejecutivo Federal.
———. 1986. "Ley federal de las entidades paraestatales." *Diario Oficial,* May 14.
———. 1989. *Plan Nacional de Desarrollo, 1989-1984.* Mexico City: Poder Ejecutivo Federal.
Stark, David. 1996. "Recombinant Property in East European Capitalism." *American Journal of Sociology* 101, no. 4:993-1027.
Stearns, Linda Brewster, and Kenneth D. Allan. 1996. "Institutional Environments: The Corporate Merger Wave of the 1980s." *American Sociological Review* 61, no. 4:699-718.
Streeck, Wolfgang, and Philippe C. Schmitter, 1991. "From National Corporatism to Transnational Pluralism: Organized Interests in the Single European Market." *Politics and Society* 19, no. 2:133-64.
Summers, Lawrence H. 1994. "A Changing Course Toward Privaization." In *Does Privatization Deliver? Highlights from a World Bank Conference,* ed. Ahmed Galal and Mary Shirley. Washington D.C.: World Bank.
Swedberg, Richard, Ulf Himmelstrand, and Göran Brulin. 1990. "The Paradigm of Economic Sociology." In *Structures of Capital: The Social Organization of the Economy,* ed. Sharon Zukin and Paul DiMaggio. Cambridge: Cambridge University Press.
Székely, Gabriel, and Jaime del Palacio. 1995. *Teléfonos de México: Una empresa privada.* Mexico City: Grupo Editorial Planeta.
Tamayo, Jorge. 1988a. "Las entidades paraestatales en México." In *México: 75 años de revolución.* Vol. 2, *Desarrollo económico.* Mexico City: Fondo de Cultura Económica.
———. 1988b. "Relación gobierno federal: Entidades paraestatales en México." In *Relación gobierno central: Empresas públicas,* ed. Nuria Cunill and Juan Martin. Caracas: Instituto Latinoamericano de Planificación Económico y Social y Centro de Administración para el Desarrollo.
Tandon, Pankaj. 1992a. "Aeroméxico." In *World Bank Conference on the Welfare Consequences of Selling Public Enterprises: Case Studies from Chile, Malaysia, Mexico, and the U.K.* Washington, D.C.: Country Economics

Department, Public Sector Management and Private Sector Development Division.

———. 1992b. "Mexicana de Aviación." In *World Bank Conference on the Welfare Consequences of Selling Public Enterprises: Case Studies from Chile, Malaysia, Mexico, and the U.K.* Washington, D.C.: Country Economics Department, Public Sector Management and Private Sector Development Division.

———. 1992c. "Vol. 1: Background, Telmex." Paper presented at the World Bank Conference on the Welfare Consequences of Selling Public Enterprises: Case Studies from Chile, Malaysia, Mexico and the U.K., Washington D.C., June 11–12.

———. 1994. "Mexico." In *Does Privatization Deliver? Highlights from a World Bank Conference,* ed. Ahmed Galal and Mary Shirley. Washington D.C.: World Bank.

———. 1996. "Divestiture and Deregulation: The Case of Mexico's Airlines." In *Privatizing Monopolies: Lessons from the Telecommunications and Transport Sectors in Latin America,* ed. Ravi Ramamurti. Baltimore: Johns Hopkins University Press.

Teichman, Judith A. 1995. *Privatization and Political Change in Mexico.* Pittsburgh: University of Pittsburgh Press.

Tello, Carlos. 1979. *La política económica en México, 1970–1976.* Mexico City: Siglo Veintiuno Editores.

———. 1984. *La nacionalización de la banca en México.* Mexico City: Siglo Veintiuno Editores.

Telmex (Teléfonos de México). 1991. *Historia de la telefonía en México: 1878–1991.* Mexico City: Subdirección de Comunicación Social.

———. 1991–2001. *Annual Reports.* Teléfonos de México: Mexico City.

———. 1997. Table with Rate Structure for 1997. Telmex. Photocopy.

———. n.d.a. "Densidad telefónica a 1996." Teléfonos de México, Mexico City. Photocopy.

———. n.d.b. "La desincorporación de teléfonos de México." Teléfonos de México, Mexico City. Photocopy.

———. n.d.c. "Privatización de las telecomunicaciones el caso de México." Telmex. Photocopy.

———. n.d.d. "Tarifas de liquidación." Teléfonos de México, Mexico City. Photocopy.

TMM (Transportación Marítima Mexicana). 1996a. *Annual Report.* Mexico City, TMM.

———. 1996b. "Form 20-F: Annual Report Pursuant to Section 13 or 15(d) of the Securities Exchange Act of 1934 for the Fiscal Year Ended December 31, 1996." Mexico City, Telmex.

Torres, Blanca. 1984. *Historia de la revolución mexicana 1940–1952.* Vol. 21, *Hacia la utopía industrial.* Mexico City: El Colegio de México.

Trabajo y Democracia Hoy. 1997. "Información y datos de los sindicatos del Foro." No. 35, Especial, 86–92.

Trejo Delarbre, Raúl. 1990. *Crónica del sindicalismo en México, 1976–1988.* Mexico City: Síglo Veintiuno Editores.

Tye, William B. 1990. *The Theory of Contestable Markets: Applications of Regulatory and Antitrust Problems in the Rail Industry.* New York: Greenwood Press.
United States Office of Federal Coordination of Transportation. 1934. *Foreign Experience with Transportation Control.* Vol. 5. Washington, D.C.: United States Office of Federal Coordination of Transportation.
Universiteit Utrecht. n.d. "Mexico: Historical Demographical Data of the Whole Country." Online. Available: http://www.library.uu.nl/wesp/populstat/ Americas/mexicoc.htm
Uno Más Uno. 1988a. "Aerovías de México, Vendida en 770 Mil Millones." October 24, p. 1.
———. 1988b. "Elogia la iniciativa privada la decisión de vender compañía mexicana de aviación." April 13, p. 13.
———. 1988c. "Solicitan compañías aéreas estatales y privadas a la SCT permiso para cubrir nuevas rutas nacionales." April 20, p. 14.
———. 1988d. "Teléfonos no será vendida: SPP y STPS." June 22, p. 1.
Uzzi, Brian. 1996. "Embeddedness and Economic Performance: The Network Effect." *American Sociological Review* 61, no. 4:674–98.
Valencia, Guillermo. 1988. "Desconocen Aeroméxico y CMA que haya un plan de reprivatización." *El Universal,* January 15, p. 21.
Valero, Jesús Rivera. 1988. "Sólo con diez paraestatales debe quedarse el gobierno: Ardavín." *Excelsior,* February 11, p. 1.
Vargas Medina, Agustín. 2001. "Los documentos ocultos del fobaproa." *Proceso,* no. 1303 (October 21).
Vázquez Rubio, Pilar. 1989. "Los telefonistas cruzaron el pantano: Concertaron con Telmex." *El Cotidiano,* no. 31 (September–October): 59, 60.
———. 1990. "El telefonista sostiene su apuesto." *El Cotidiano,* no. 35 (May–June): 66–71.
Vázquez Talavera, César. 1990. "Hacia una reestructuración de las relaciones laborales en la aviación comercial." In *Relaciones laborales en las empresas paraestatales,* ed. Graciela Bensusán and Carlos García. Mexico City: Fundación Friedrich Ebert.
Vera, Rodrigo. 1988. "Magnates y extranjeros se reparten lo que dejó Aeroméxico." *Proceso,* no. 607 (June 20): 14.
Vernon, Raymond. 1963. *The Dilemma of Mexico's Development.* Cambridge: Harvard University Press.
———. 1965. *The Dilemma of Mexico's Development.* Cambridge: Harvard University Press.
Vickers, John, and George Yarrow. 1988. *Privatization: An Economic Analysis.* Cambridge: MIT Press.
Vieyra, Alberto V. 1988. "La sindicatura de Aeroméxico obtuvo $5 mil millones en utilidades." *El Nacional,* October 2, p. 3.
Villarreal, René. 1977. "The Policy of Import-Substituting Industrialization, 1929–1975." In *Authoritarianism in Mexico,* ed. José Luis Reyna and Richard S. Weinert. Philadelphia: Institute for the Study of Human Issues.
———. 1988. *Mitos y realidades de la empresa pública: ¿Racionalización o privatización?* Mexico City: Editorial Diana.

Violland, Michel. 1996. "Whither Railways?" *OECD Observer*, no. 198 (February–March): 33–36.

Wade, Robert. 1990. *Governing the Market: Economic Theory and the Role of Government in East Asian Industrialization*. Princeton: Princeton University Press.

Walder, Andrew G. 1994. "Corporate Organization and Local Government Property Rights in China." In *Changing Political Economies: Privatization in Post-Communist and Reforming Communist States*, ed. Vedat Milor. Boulder, Colo.: Lynne Rienner.

———. 1996. "Markets and Inequality in Transitional Economies: Toward Testable Theories." *American Journal of Sociology* 101, no. 4:1060–73.

Wallerstein, Immanuel. 1974. *The Modern World-System: Capitalist Agriculture and the Origins of the European World Economy in the Sixteenth Century*. Vol. 1. New York: Academic Press.

———. 1979. *The Capitalist World-Economy*. Cambridge: Cambridge University Press.

Waterbury, John. 1993. *Exposed to Innumerable Delusions: Public Enterprise and State Power in Egypt, India, Mexico, and Turkey*. Cambridge: Cambridge University Press.

Williamson, Oliver E. 1985. *The Economic Institutions of Capitalism*. Free Press: New York.

Weber, Max. 1968. *Economy and Society*. Ed. Guenter Roth and Claus Wittich. New York: Bedminster Press.

White, Harrison C. 1981. "Where Do Markets Come From?" *American Journal of Sociology* 87, no. 3:517–47.

Wiles, Peter. 1962. *Communist International Economics*. Oxford: Basil Blackwell.

Wilkie, James W., Eduardo Alemán, and José Guadalupe Ortega, eds. 2000. *Statistical Abstract of Latin America*. Vol. 36. Los Angeles: Latin American Center Publications, University of California.

Williamson, Oliver E. 1975. *Markets and Hierarchies*. New York: Free Press.

———. 1985. *The Economic Institutions of Capitalism: Firms, Markets, Relational Contracting*. New York: Free Press.

Wills, Rick. 1999. "Agency Confronts Mexico Airline Company as Monopoly." *New York Times*, October 6, p. C4.

World Bank. 1985. *Annual Report*. Oxford, Eng.: World Bank.

———. 1988. *Annual Report*. Oxford, Eng.: World Bank.

———. 1989. *Annual Report*. Oxford, Eng.: World Bank.

———. 1995a. *Bureaucrats in Business: The Economics and Politics of Government Ownership*. Published for the World Bank. Oxford: Oxford University Press.

———. 1995b. *World Tables. 1995*. Baltimore: Johns Hopkins University Press.

———. 1997. *World Development Report: The State in a Changing World*. Oxford: Oxford University Press.

Wood, Ellen Meiksins. 1981. "The Separation of the Economic and the Political in Capitalism." *New Left Review*, May/June, pp. 66–95.

Xelhuantzi López, María. 1988. *El sindicato de telefonistas de la república mexicana: Doce años (1976–1988)*. Mexico City: Sindicato de Telefonistas de la República Mexicana.

Yergin, Daniel, and Joseph Stanislaw. 1998. *The Commanding Heights: The Battle Between Government and the Marketplace That Is Remaking the Modern World.* New York: Touchstone.
Zamora, José Gasca. 1989. "Fuentes para el estudio de las empresas paraestatales de México y su privatización, 1983–1988." *Comercio Exterior* 39, no. 2:151–75.
Zapata, Francisco. 1995. *El sindicalismo mexicano frente a la restructuración.* Mexico City: El Colegio de México.
Zavala, Oswaldo. 1999. "En EU, denuncia de sobrecargos." *Proceso,* no. 1203 (November 21).
Zedillo, Ernesto Ponce de Leon. 1996. *Segundo informe de gobierno, anexo estadístico.* Mexico City: Poder Ejecutivo Federal.
———. 1997. *Tercer informe de gobierno, anexo estadístico.* Mexico City: Poder Ejecutivo Federal.
———. 1998. *Cuarto informe de gobierno, anexo estadístico.* Mexico City: Poder Ejecutivo Federal.
Žižek, Slavoj. 1989. *The Sublime Object of Ideology.* London: Verso.
Zuckerman, Laurence. 2001. "U.S. Takes Big Role in Airlines' Crisis." *New York Times,* October 4.
Zukin, Sharon, and Paul DiMaggio, eds. 1990. *Structures of Capital: The Social Organization of the Economy.* Cambridge: Cambridge University Press.

INDEX

access code dialing, 172, 181, 243
Acle Tomasini, Alfredo, 62
Aerocalifornia, 143, 144
Aeroméxico, 42
   Aeroperu acquisition, 145
   backward linkages and, 122
   bankruptcy of, 79, 133, 134
   collective bargaining agreement, 134, 135
   employment levels, 123, 124
   fleet, 120, 122, 124, 125, 133
   foreign debt of, 131, 132, 148
   founding of, 120
   incorporated into federal budget, 43
   investment and maintenance, 123
   labor, 49, 121
   management and turnover at, 121
   Mexicana de Aviación acquisition, 145
   nationalization of, 121
   operation in receivership, 134, 135
   Pan American participation in, 120
   politicization of service, 121, 123
   renaming of firm, 111 n. 1, 135
   restructuring, 133, 134
   revenue, 124, 125, 126
   routes, 123, 124, 126
   sale of, 136
   strike against, 133
AFL-CIO, 247
Afore. *See* Retirement Fund Administrator
agent banks. *See* privatization in Mexico
Agreement on Actions to Modernize the Mexican Rail System, 214
Agreement on Assumption of Liabilities, 216
Air Commerce Act, 115

Air France, 138
Airports and Auxiliary Services, 42
   creation, 118, 119
   revenues, 119
Air Transportation Stabilization Board, 116
Alemán, Miguel
   Mexican National Railroad, 200, 204
   parastate sector growth, 38, 39
   private sector and, 38
Alemán Velasco, Miguel, 136
Alestra, 174, 181, 183
   fibre-optic network, 184, 185
Alfa. *See* Grupo Alfa
Alien Land Law, 55
Altos Hornos
   backward linkages in railroads, 200
   creation of, 42
API. *See* Complete Port Administration
Aronson, Jonathan, 149
Arrow, Kenneth, 22
Article 25 of Constitution, 36
   ammendment to, 74
Article 26 of Constitution, 36
Article 27 of Constitution, 36, 55
Article 28 of Constitution, 36
   ammendment to, 74
   railroads added to, 211
   railroads removed from, 219
   reformed prior to bank privatization, 80
Article 90 of Constitution, 36
Article 123 of Constitution, 36 n. 1, 220
ASA. *See* Airports and Auxiliary Services
ASPA. *See* Union Association of Pilot Aviators
Aspe, Pedro, 41, 109
   Mexicana de Aviación restructuring, 140

Aspe, Pedro *(cont'd)*
  Unit for the Divestiture of Parastate Entities, 81
ASSA. *See* Union Association of Flight Attendants
asset specificity, 22
AT&T, 166, 174, 175, 178, 180, 181, 182, 238
atomization
  appearance of in market, 11, 13
  as foundation of neoclassical theory, 7, 8, 9
  conception of control and, 10, 11
  in neoclassical theory, 3
  social construction of, 10
autonomy, of non-state actors, 29. *See also* state autonomy
Avantel, 174, 175, 178, 181–83
  fibre-optic network, 184, 185
aviation
  code sharing, 117
  competition in, 115, 116
  cooperation among carriers, 116, 117
  deregulation in United States, 116
  economies of scale in, 114
  equipment and service, 116
  European Union, 115
  interdependence with other sectors, 113, 115
  international routes, 115
  loan guarantees, 116
  niche pricing, 116
  role of state in, 114–15
  subsidies to, 115
  terrorist attacks of September 11, 116
aviation in Mexico
  Aeroméxico nationalization, 120
  backward linkages in, 118, 122, 131, 132, 137
  bilateral agreements, 118, 141
  carriers competing in market, 127
  charter service, 142
  competition from other forms of transportation, 127
  consolidation of industry, 145, 146
  Constitution of 1917 and, 117
  demand for, 127, 128, 129
  deregulation of, 131, 141
  economic crisis, 127
  fares, 118, 143, 145
  federal investment in, 119, 120
  Gulf War and, 141
  foreign debt and, 132
  foreign influence over regulation of, 118
  international alliances, 146
  international competition, 127, 130, 141, 142, 146
  international routes, 118
  jet fuel prices, 137
  labor, 118, 122
  Ministry of Communications and Transportation, 117, 118
  North American Free Trade Agreement and, 103
  parastate firms, 119
  price wars, 143
  profit motive, 147
  restrictions on foreign ownership, 117
  share of total transportation, 127
  sovereignty over, 117
  role of state in, 117–19
  subsidies to, 119
Ávila Camacho, Manuel
  labor at Mexican National Railroad and, 203
  Mexican Natioal Railroad and, 200
  parastate sector growth, 38, 39
  private sector and, 38

balance of trade, 57
Banamex, 54
  long-distance market, 174
  Mexicana de Aviación sale, 137
Banco Internacional, Telmex privatization, 169, 170
Bancomer, 54
  Aeroméxico sale, 136
Banco Serfin, 221
Bank Divestment Committee
  private sector links to, 105
  structure of, 82, 87
Bank Fund for Savings Protection
  airlines, 146
  creation of, 85
  financial sector, 85
  Grupo Cintra, 146
Bank of America, 102
Bank of Mexico
  Bank Divestment Committee, 82
  Banker's Alliance and, 50, 51
  creation of, 36
Banker's Alliance, 50, 51
  SHCP and, 63
  state dependence on, 51, 52
  struggle over control of parastate firms, 60

banking and banks. *See* financial sector
Banobras
    Aeroméxico bankruptcy, 134, 135
    Aeroméxico sale, 135 n. 9
Barclays, 102
Bell Atlantic, 174
Black-McKeller Act, 116
Block, Fred, 233, 257
Bolaños, Sergio, 139
Booz Allen and Hamilton, 102
bounded rationality, 22
Bowles, Samuel, 1
Brandenburg, Frank, 33
Brener, Pablo and Israel, 140
Burlington Northern Santa Fe, 193, 214
business. *See* firms; private sector
Business Coodinating Council, 244

C.S. First Boston, 221
Calles, Plutarco Elías
    foreign debt, 195
    foreign investment, restrictions on, 55
    Lázaro Cárdenas and, 37
Canacintra. *See* National Chamber of Manufacturers
Canadian National Railway, 193
Capital. *See* private sector
capital flight, 51, 52, 59, 95, 104, 244. *See also* peso crisis
capitalism
    defined, 12, 13
    distinct from market, 12, 13
    role of firms in, 14
    role of market in, 14
Cárdenas Coalition
    composition of, 50, 51, 52
    struggle over control of parastate firms, 60
Cárdenas, Cuauhtémoc, 97, 249
Cárdenas, Lázaro
    economic policies, 38
    foreign investment, 38
    petroleum nationalization, 56
    petroleum workers, 49
    Plutarco Elías Calles, 37
    political base, 37
    railroad nationalization, 196
Carranza, Venustiano, 195
Carso. *See* Grupo Carso
CCE. *See* Business Coodinating Council
CDB. *See* Bank Divestment Committee
Cemex. *See* Grupo Cemex
CFC. *See* Federal Competition Commission

CFE. *See* Federal Electric Commission
charrazo, 204
charros, 47
Chartist Movement, 250
Chase Manhattan Bank, 83
    Grupo Falcón, 139
    Mexican National Railroad, 228
Chihuahua-Pacific Railroad, 189
    privatization of, 227
CIGF. *See* Interministerial Committee on Spending and Finance
Cintra. *See* Grupo Cintra
Civil Aeronautics Act, 115
Civil Aeronautics Board, 115
Cofetel. *See* Federal Telecommunications Commission
collective action problem, state autonomy and, 26, 243
Colosio, Luis Donaldo, 249, 250
Commission to Finance and Strengthen the Endowment of the PRI, 97
Committee for the Promotion of State Socioeconomic Development, 63
Committee of Operators, 252. *See also* Telecommunications in Mexico, Committee of Operators
commodity fetishism. *See* fetishism of commodities
communications and transportation
    difficulty of privatization in, 110
    monopoly characteristics, 110
    political value, 110, 111
    positive externalities, 111
    role of state in, 30, 110
    theoretical importance, 110
    use value of, 110, 111
communications and transportation in Mexico
    as focus of privatization efforts, 85, 86
    graphic depiction of, 109
    National Development Plans, 212, 213
    stabilizing development and, 109
Communications Workers of America, 184, 247
Compañía Telefónica de España, 170
Compañía Telefónica y Telegráfica Mexicana, 153
companies. *See* firms; private sector
competition
    as component of ideal-type market, 11
    disembedded exchange, 16, 17
    economic sociology, 11
    social construction of, 11, 16

Complete Port Administration, 87, 89
　private sector and, 89
Conasupo. *See* National Company of Staple Goods
Concamin. *See* National Confederation of Chambers of Industry
Concanaco. *See* National Confederation of Chambers of Commerce
conception of control, 10
　neoclassical paradigm as, 10, 27
Consolidated Rail Corporation, 192
Constitution of 1917, 236
　ammendments to, 74, 80
　Articles of, 36 n. 1, 55, 211, 219, 220
　commodity fiction, 36
　interpretations of, 36
　provisions of, 36
　reform under President Zedillo, 85
　role of state in economy and, 36
Coparmex. *See* Employer's Confederation of the Mexican Republic
Córdova, Arnaldo, 35
corporatism, 46
Council of Advisors, 77-78
　financial sector consolidation, 82
　private sector links, 104
Cowhey, Peter, 149
Credit Commerciale of France, 102
CSX, 193
CTM. *See* Mexican Confederation of Labor
CWA. *See* Communications Workers of America

de la Huerta-Lamont Pact, 195
de la Madrid, Miguel, 69
　constitutional amendments, 94
　financial sector, 94, 95
　private sector and, 94-96
　privatization under, 71, 74-78
　railroad consolidation, 212
　railroad investment, 212
　railroads in National Development Plan, 212
　role of state in National Development Plan, 74
debt crisis, 74
democracy
　in neoclassical theory, 232
　neoliberal reforms and, 247-50
　of capital, 239
Democratic Railroad Workers Movement, 217, 218
Democratic Revolutionary Party, 249
Department of Railroads, Transit and Tariffs, 196
Díaz, Porfirio, 194, 195, 236
　Mexican National Railroad and, 195
Dictum. *See* Grupo Dictum
Disel Nacional, 42
disembedded exchange. *See also* embeddedness
　as exchange value, 16, 17
　capitalism, 13
　indicators of, 16, 17
　market, 16
Donaldson Lufkin and Jenrette, 102
Drexel Burnham Lambert, 139

Echeverría, Luis, 39
　Mexican Railroad Workers Union dissidents, 206
　parastate sector investment, 50
　private sector and, 54
　student movement of 1968, 49
　tax reform, 50
economic crises
　as source of change, 30, 243, 244
　social struggle and, 30
economic sociology
　conception of control, 10
　contrasted with Marxist theory, 13
　critique of, 10
　distinction between public and private sectors, 9, 10
　embeddedness contrasted with atomization, 8
　limits to market exchange, 9
　neoclassical economics, 8-12
　neoliberal reforms, 9
　new and old economic sociology, 7 n. 3
　property rights, 18 n. 9
　role of state in, 9
　social relations in market, 9
　stability in markets, 11
　state autonomy, 25
　utility maximization, 8, 9
ejido. *See* land reform
embeddedness. *See also* disembedded exchange
　as use value, 16, 17
　autonomy, 27, 28
　contrasted with atomization, 8
　critique of, 10
　defined, 8

economic sociology, 3
indicators of, 16, 17
Karl Polanyi, 15
of Mexican state in international economy, 69, 70, 237
state's relationship with private sector, 28
Employer's Confederation of the Mexican Republic, 244
response to privatization of Mexicana de Aviación, 138
response to privatization of Telmex, 162
Ericsson, 152, 159
Evans, Peter, 25, 27, 233, 258
exchange value
as disembedded exchange, 16, 17
defined, 15
forms of rationality and, 15
ideal types of exchange, 15
operationalized, 16, 17
relation to use value in Marxist theory, 12, 13
export markets, volatility of, 55
EZLN. *See* Zapatista Army of National Liberation

Falcón. *See* Grupo Falcón
FCC. *See* Federal Communications Commission
Federal Aviation Act, 115
Federal Communications Commission, 177, 182
Federal Competition Commission, 240
lawsuit against Grupo Cintra, 147
Federal Electric Commission, 42, 48
incorporated into federal budget, 43
rates, 43
World Bank loans, 100
Federal Law of Parastate Entities
provisions of, 75, 76, 213
rationality, 80
regulations governing, 80
Federal Reserve Bank of the United States, 45, 46
Federal Telecommunications Commission, 177, 181, 182
Federal Telecommunications Law, 172, 174 n. 7, 175, 177
Federation of Unions of Goods and Services Workers
Carlos Salinas de Gortari, 92, 166
composition of, 92
Francisco Hernández Juárez, 92, 166

international solidarity, 93
Mexican Telephone Workers Union, 166
Fernández Varela, Félix Vélez, 36 n. 4
Fertimex, 200
Fesebes. *See* Federation of Unions of Goods and Services Workers
fetishism of commodities
appearance of atomization in market, 12, 14
fictitious commodities, 15
Ficorca. *See* Fund to Cover Exchange Rate Risk
financial sector
added to constitution, 74
bailout of, 99
Bank Fund for Savings Protection, 85
concentration of resources in, 53, 54
consolidation and restructuring, 82
coordination with industry, 53
foreign investment in, 103, 104 Table 5
NAFTA, 103
nationalization of, 45, 46, 59, 74
recapitalization of, 59
universal banking legislation, 53
World Bank adjustment loan to, 102
firms
as locus of social struggle, 23
in institutional economics, 22, 23
Marxist theory and, 14
organization of compared to market, 14
overcoming transaction costs, 23
First Boston, 102
Fligstein, Neil, 11
FNM. *See* Mexican National Railroad
Fobaproa. *See* Bank Fund for Savings Protection
foreign debt, 58, 59
revenue from privatization, 83
foreign investment
concentration of investment, 55
constitutional restrictions on, 55
finance capital, 70
financial sector, 58, 103, 104 Table 5
limits to in telecommunications, 237
manufacturing, 58
petroleum boom, 58
petroleum nationalization, 56
priority sectors of economy, 56
railroads, 241
reparations from Revolution, 54
repatriation of profits, 55
restrictions in NAFTA, 103
stabilizing development, 53

Fox, Vicente, 248, 250
France Telecom, 83, 168, 170
Friedman, Milton, 231
Fund to Cover Exchange Rate Risk, 84
　　Fig. 5, 99

Galiani, Fernando, 13
Gasca Neri, Rogelio, 132, 133
General Agreement on Trade in Services, 179
General Director of Maritime Work, 87 n. 8
General Director of Port Operation, 87 n. 8
General Directorate of Fares, 143
General Law of Means of Communication and Transportation, 118
　　interconnection requirements in, 153, 156
　　labor in, 156
　　replaced by Federal Telecommunications Law, 175
Gintis, Herbert, 1
Glade, William P., 33
Goldman Sachs, 102
　　Telmex privatization, 169, 170
Goldsmith, Sir James, 139
Governing Project for the National Air Transportation System, 143
Grupo Alfa, 174, 175
　　bailout of, 95, 245
Grupo Carso
　　purchase of Telmex, 168
　　relationship of Carso firms with Telmex, 181
Grupo Cemex, bailout of, 95, 245
Grupo Cintra, 240
　　companies owned by, 146
　　competition and Federal Competition Commission, 147
　　consolidation of aviation under, 146
　　creation of, 146
　　debt of Aeroméxico and Mexicana de Aviación, 148
　　national versus international competition, 148
Grupo Dictum, purchase of Aeroméxico, 136
Grupo Falcón, purchase of Mexicana de Aviación, 139
Grupo ICA, 227
Grupo Iusa, 174
Grupo México, 227
Grupo Serbo, attempt to buy Mexicana de Aviación, 138

Grupo Visa, bailout of, 95, 245
Grupo Vitro, bailout of, 95, 245
Grupos, 53
GTE Telephone Corporation, 170
Guadarrama, Enrique Rojas, 136
Guanos y Fertilizantes de México, 42

Haggard, Stephan, 25, 253
Hamilton, Nora, 30
hard budget constraints, 106, 232
　　arbitrary use of property and, 20
　　autonomy of actors in market, 29
　　balance of public sector, 106
　　barriers to cooperation in market, 4
　　defined, 19
　　disembedded exchange, 16, 17
　　institutional economics, 19, 20
　　paradox of, 20
　　private firms and, 107
　　public firms, 106
　　softened by international market, 106
　　state as enforcer of, 20
　　state autonomy and, 24
Harp Helú, Alfredo, 100
Harvey, David, 12
Hayek, Friedrich A., 232, 233, 248, 254
Heilbroner, Robert, 231, 233, 253
Hernández Galicia, Joaquín, 139
　　Carlos Salinas de Gortari and, 91
Hernández Juárez, Francisco, 92
　　Labor Congress, 164
　　Mexican Confederation of Labor, 164
　　Mexican Telephone Workers Union militance, 157, 158
　　modernization, labor flexibility, 92
　　Telmex privatization, 164
　　Telmex union democracy movement, 157
Herzog, Jesús Silva, 76, 77

ICA. *See* Grupo ICA
IMF. *See* International Monetary Fund
import substituting industrialization, role of state-owned firms in, 2, 55
industrial relations. *See* labor
institutional economics
　　hard budget constraints, 19, 20
　　market failure, 20–22
　　property rights, 18, 19
　　transaction costs, 22, 23
interest rates, 45
Interministerial Committee for Divestment, 228

Interministerial Committee on Spending and Finance, 63
Mexicana de Aviación sale, 79
international economy
  Mexico in, 236–38
  states in, 14 n. 7
International Monetary Fund, 245
  austerity programs, 45, 50, 96, 101
  convergence with World Bank, 101
International Telecommunications Union, 177
ITT, 153
ITU. *See* International Telecommunications Union
Iusa. *See* Grupo Iusa
Iusacell, 174

Kahler, Miles, 25
Kansas City Southern Railroad, 226
Kaufman, Robert, 25, 253
key informant interviews, 4, 257–64
Kingsley of Mexico, 226
Kornai, János, 1, 18, 19, 22

La Quina. See Hernández Galicia, Joaquín
labor
  Aeroméxico, 120, 134, 240
  austerity, 90, 91
  Carlos Salinas de Gortari, 91, 92
  commodity fiction of, 15
  competition in telecommunications, 184
  corporatism, 46
  Democratic Revolutionary Party, 94
  Disel Nacional, 91
  Federal Law of Parastate Entities, 75
  flight attendants strike, 91
  foreign investment, 200
  foreign investment, 47
  international solidarity, 93, 247
  limits to commodification of, 13
  Mexican National Railroad, 49, 91, 220, 239, 240
  Mexicana de Aviación privatization, 140
  modernization and labor flexibility, 90
  NAFTA, 247
  national industry unions, 46
  organization in aviation, 122
  parastate sector compensation, 48, 49
  parastate sector employment, 77, 239
  profit maximizationi in telecommunications, 156
  property rights, 19
  requisa, 156, 220, 239
  restructuring in privatization, 82, 238–40
  Revolutionary Institutional Party, 46
  student movement of 1968, 49
  Telmex, 91
  Telmex restructuring, 165, 240
  use value in telecommunications, 156
  wage labor and capitalism, 12
land reform, 246
Law for Federal Government Control of Decentralized Organisms and Firms with State Participation, 60, 62
Law of Budgeting, Accountability and Public Expenditures, 64
Law of State Ministries and Departments, 61
Legorreta, Gerardo de Prevoisin, 136, 145
LFEP. *See* Federal Law of Parastate Entities
Limantour, José Yves, 194
López Mateos, Adolfo, 189
  Aeroméxico nationalization, 48
  capital flight, 52
  labor conflict at the Mexican National Railroad, 205
López Portillo, José, 67
Lufthansa, 138

Manufacturers Hanover, 102
market failure, 233
  defined, 20, 21
  embeddedness, 16, 17
  hard budget constraints, 21
  justification for state intervention in economy, 21
  limits to market exchange, 4
  property rights, 21
  redistribution, 21
markets
  appearance of atomization in, 11, 13
  commodity fiction, 15
  competition, 11
  democracy and, 232
  distinct from capitalism, 12, 13
  economic sociology, 9
  hierarchies and, 5 n. 2, 10
  ideal type, 11
  interdependence with non-market institutions, 24
  juridical equality in, 16
  Karl Polanyi, 15
  limits to, 24, 253–55
  Marxist theory, 13
  neoclassical theory, 7, 8, 234
  political freedom and, 9, 232

markets *(cont'd)*
    role of state in, 11, 12
    social fragmentation necessary to, 11, 16
    social relations in, 14
Martínez, Guillermo Ortiz
    financial sector privatization, 86
    Zedillo administration, 87
Marx, Karl, 12, 13
MCI, 174, 182
McKinsey Corporation, 102
    Telmex privatization and, 169, 170
M-C-M', 13
Mercedes Benz, 138
Mercer Management Consulting, 102, 221
Merrill Lynch, 102, 228
Mexican Confederation of Labor
    Carlos Salinas de Gortari, 92
    composition of, 46
    Mexican Telephone Workers Union, 92
Mexican Light and Power Company, 47
Mexican Maritime Transport
    containerized shipping, 213, 225
    export market, 225
    investment in Mexican National Railroad, 213
    multimodal shipping, 225
    ports privatization, 225
Mexican National Railroad
    backward linkages, 199, 200
    cargo versus passenger transportation, 197, 198
    commodity prices, 208
    competition, 208, 209
    composition of cargo transportation, 214, 215, 221
    concessions, 219, 220
    containerized shipping, 213
    coordination with foreign firms, 214
    creation of, 42, 195
    de la Huerta-Lamont Pact and Pani amendment, 195
    domination of transportation, 195
    foreign control over, 195
    foreign debt, 195, 196, 216
    foreign investment, 220
    forward linkages, 200, 214, 215, 221
    imports, 208
    incorporated into federal budget, 43
    investment in, 206, 212
    labor, 48. *See also* Mexican Railroad Workers Union
    labor restructuring, 217, 225
    labor struggle at, 49, 220
    Mexican Maritime Transport, 213
    Mexican Revolution, 195
    military intervention in, 49
    multimodal shipping, 213
    nationalization of, 196
    organizational restructuring, 218, 219
    private investment in, 213, 214
    profitability, 199, 216
    property rights specified, 220
    rates for shipping, 43, 196, 199, 213, 216
    requisa, 220
    restructuring, 212, 241
    routes and structure of privatized railroad, 221-24
    share of cargo transportation, 214
    subsidies to, 197, 199, 206
    World War II, 196
Mexican Rail Transportation, 226, 227
Mexican Railroad Workers Union
    charrazo, 204
    collective bargaining agreement, 203, 204, 217
    corruption, 206
    creation of, 201
    employment levels, 201, 202, 204, 217, 228
    foreign ownership of railroads, 200
    labor solidarity, 201
    pension fund, 228
    rates for shipping, 201, 202, 203, 205, 206
    restructuring and labor flexibility, 217, 218
    restructuring proposal of dissidents, 205
    sabotage, 206
    state intervention in, 201, 203, 204
    strike, 205
    support for nationalization of railroads, 200
    worker administration at the Mexican National Railroad, 201-3
    World War II, 203
Mexican Revolution, 236
    interpretations of, 35
    popular sectors and, 35
    populism, 35
Mexican Telephone Workers Union, 166
    concessions and guarantees in Telmex privatization, 165
    international solidarity, 184

labor struggle at Telmex, 156
long-distance market, 184
Mexican Confederation of Labor, 156
privatization of Telmex, 164
property rights restricted at Telmex, 157
purchase of shares of Telmex, 165
response to privatization, 165
restructuring and labor flexibility, 165
state manipulation of, 156
technology, 157, 165
Telmex management, 184
Telmex Title of Concession, 167
Union Democracy Restoration Movement, 156
worker retraining, 184
Mexicana de Aviación, 240
acquisition by Aeroméxico, 145
acquisition by Mexican investors, 122
Aeroméxico bankruptcy, 138
backward linkages, 122, 131, 132
coordination with Aeroméxico, 145
employment levels, 123, 124, 125
fleet, 122
foreign debt, 131, 132, 137, 145, 148
founding of, 121
Grupo Cintra, 146
Grupo Falcón, 83, 139
labor concessions, 145
nationalization of, 122, 132
Pan American participation in, 121
privatization of, 79, 136–39
profitability, 132
restructuring, 137, 138
revenue, 124, 125, 126
routes, 121, 123, 124, 126
Star Alliance, 146
Turborreactores purchase, 145
Ministry of Agrarian Reform, 50
Ministry of Agriculture, 50
Ministry of Commerce, Interministerial Committee for Divestment and, 86
Ministry of Communication and Public Works. *See* Ministry of Communications and Transportation
Ministry of Communications and Transportation
Aeroméxico bankruptcy, 133
Agreement on Actions to Modernize the Mexican Rail System, 214
aviation concessions, 79, 117
Cárdenas Coalition, 50
General Directorate of Fares, 118

Interconnection rates, 175
long-distance market opening, 181
Mexicana de Aviación sale proposal, 136
Ministry of Communication and Public Works, 147 n. 2
privatization under President Zedillo, 87
railroad privatization, 219
rate regulation in trucking, 209
telecommunications five-year plan, 153
Telmex operations, 154
Telmex restructuring, 165
Telmex Title of Concession, 167
Unit for the Support of Structural Change, 87
Ministry of Control and Administrative Development, Interministerial Committee for Divestment and, 86
Ministry of Energy, 86
Ministry of Energy, Mines, and Parastate Industry, 61–63
Ministry of Hydraulic Resources, 50
Ministry of Labor and Social Provision, 167
Interministerial Committee for Divestment, 86
Telmex restructuring, 167
Ministry of Planning and Budget
Ernesto Zedillo, 98
Federal Law of Parastate Entities, 75
Mexican National Railroad oversight of, 213
relations with private sector, 245
telecommunications, 154, 155
Telmex restructuring, 165
Telmex Title of Concession, 167
Ministry of the Treasury and Public Credit
autonomy, outsourcing, 105
Bank Divestment Committee, 82
Banker's Alliance, 51, 63
Federal Law of Parastate Entities, 75
Interministerial Committee for Divestment, 86
parastate sector oversight and control, 60, 61, 63
privatization and centralization of control over parastate sector, 71
railroads, 213
relations with private sector, 245
telecommunications, 154, 155
Minstry of the President, 61–63
money
capitalism and, 12
commodity fiction of, 15

money *(cont'd)*
  limits to commodification of, 13
Montalvo, Melchor, 133 n. 8
multimodal transportation, 241. *See also* Mexican Maritime Transport, railroad transportation, and Mexican National Railroad

Nafinsa, 51
  Aeroméxico bankruptcy, 134
  World Bank loans, 56
NAFTA. *See* North American Free Trade Agreement
National Action Party, 95, 96, 248
  alliance with President Salinas, 97, 98, 249
National Air Transportation System Guidelines, 142
National Association of Exporters, 214
National Bank for Agricultural Credit, 36
National Banking Commission, 61
National Chamber of Manufacturers
  and Cárdenas Coalition, 50
  and foreign investment, 58
  reaction to neoliberal reforms, 96
  reaction to privatization of Telmex, 163
National Commission on Foreign Investment, 226
National Commission of Investment, 61
National Company of Staple Goods, 42, 79, 200
  incorporated into federal budget, 43
National Confederation of Chambers of Commerce, 214
  and foreign investment, 58
  response to privatization of Mexicana de Aviación, 138
National Confederation of Chambers of Industry, foreign investment and, 58
National Council for Foreign Trade, 214
National Development Plan
  of President de la Madrid, 74
  of President Salinas, 79, 80
  profitability in, 79, 80
National Export and Import Company, 42
National Insurance Commission, 61
National Irrigation Commission, 36
National Lottery, incorporated into federal budget, 43
National Port Coordinating Commission, 87 n. 8
National Rail Car Manufacturer, 207
  backward linkages in railroads, 200
  imports, 207
National Rail System Modernization Program, 215
National Revolutionary Party. *See* Revolutionary Institutional Party
National Roads Commission, 36
National Union of Aeroméxico Technicians and Workers
  Aeroméxico bankruptcy, 134
  Aeroméxico restructuring, 133
  dissolution of, 135
  strike against Aeroméxico, 133
National Union of Workers of Airlines and Related Companies, 135
National Workers Union, 93
NCS International, 173, 181, 182
neoclassical economics
  as conception of control, 10, 27
  atomization, 7, 8
  contrasted with economic sociology, 8–12
  critique of, 3
  orthodox paradox, 25
  role of state in, 9, 25, 254
  self-interest, 8, 9, 232
neoliberal reforms
  democratization and, 247–50
  economic sociology, 9
  limits to, 253–56
  political reform and, 7
  poverty and, 243
  state autonomy and, 25
Norfolk Southern, 193
North American Free Trade Agreement, 103, 225
  anti-competitive practices, 179
  civil aviation, 141
  labor side agreement, 184
  labor solidarity, 247
  national control of economy, 103
  opposition to, 246, 247
  ports, 103
  railroads, 225
  support for, 247
  telecommunications, 179
North, Douglass, 18, 22
Northeastern Railroad, 241
  characteristics of, 224 Table 16
  routes, 222 Fig. 15
Northern Pacific Railroad
  characteristics of, 224 Table 16
  routes, 222 Fig. 15

opportunism, 22
Organic Law of Federal Public Administration, 74
Organic Law of Public Administration, 63
Organic Law of the Mexican National Railroad, 212, 213
orthodox paradox, 25

PAN. See National Action Party
parastate sector
  agriculture, 41
  bank nationalization, 45, 74
  characteristics of, 40
  defined, 33 n. 1, 40
  educational institutions, 40, 41
  employment levels, 45, 77, 93
  federal budget, 43, 44, 45
  foreign debt financing, 45
  foreign exchange, 45
  governance of sectorization, 63
  governance of triangle of efficiency, 61, 62
  governance of, 59–64
  heavy industry, 41
  infrastructure, 49
  labor/popular sectors, 46–50
  number of firms, 39, 71, 72, 76
  presidentialism, 60
  prices of goods produced by, 43, 50, 80
  private sector, 50–54
  private sector, subsidies to, 49, 51
  privatization of, 71
  resistance to privatization, 75, 76
  revenue, 99
  sectors of activity, 78
  stabilizing development, 53
  surplus channelled to debt repayment, 100
  surplus, 43, 44, 45
  tax revenue, 45
  value of, 43, 82
  World Bank, 56, 101, 102
Pemex, 200, 254. See also petroleum
  financing, 45
  foreign debt, 45
  foreign exchange, 45
  incorporated into federal budget, 43
  Joaquín Hernández Galicia, 91
  labor, 49, 91
  nationalization of, 41, 56
  North American Free Trade Agreement, 103
  prices, 43
  restructuring, 83
  size, 41
Penn Central Railroad, 192
Pérez de Mendoza, Alfredo, 169
peso crisis. See also capital flight
  balance of trade, 225
  devaluation of 1976, 50
  of 1982, 46
  of 1994, 85, 99
Petricioli, Gustavo, 77
petroleum. See also Pemex
  foreign control over, 48
  nationalization of, 56
  North American Free Trade Agreement, 103
  world prices, 45, 50, 96
  World War II, 56
petroleum boom, 45
Petroleum Law, 55
Polanyi, Karl, 244, 245, 246, 250, 255, 256
  fictitious commodities, 15
  ideal types of exchange, 15
  social struggle, 30, 31
populism, 35
Portfolio Purchase and Capitalization Program, 85. See also Bank Fund for Savings Protection
Ports, 103
poverty, neoliberal reforms and, 243
PRD. See Democratic Revolutionary Party
PRI. See Revolutionary Institutional Party
price
  component of ideal-type market, 11
  disembedded exchange, 16, 17
  economic sociology, 9
  neoclassical economics, 8
  parastate firms, 82
  social construction of, 16
Price Waterhouse, 102, 221
principal-agent dilemma, 62
private property. See also property rights
  capitalism and, 12
  distinguished from state ownership, 9, 10
  economic sociology, 9, 10
  state as enforcer of, 30
private sector
  Business Coodinating Council, 54
  Carlos Salinas de Gortari, 96–98
  composition of, 50, 51
  concentration of resources, 53, 54, 99, 100
  Ernesto Zedillo, 98–100
  Miguel de la Madrid, 94–96

private sector *(cont'd)*
   privatization, 94–100
   representation on Bank Divestment Committee, 82
   Revolutionary Institutional Party, 52
   state dependence on, 51
   subsidies from parastate firms, 49, 51
   transformation during stabilizing development, 53
privatization
   developing economies, 1, 2
   economic sociology and, 9
   private property, 10
   worldwide, 1
privatization in Mexico
   resectorization, 81, 86
   agent banks, 78, 252
   banks, 80, 82
   Carlos Salinas de Gortari, 79–83
   Ernesto Zedillo, 84–89
   foreign debt repayment, 84 Fig. 6
   innovations in process, 76–78, 86
   international economy, 100–104
   labor, 82, 90–94
   Miguel de la Madrid, 74–78
   ports, 87, 89
   private sector and, 94–100
   revenue from, 73, 84 Fig. 6, 88
   sectors privatized, 78
   social security, 85
   World Bank loans, 101, 102
profit motive. *See* self-interest
profitability
   component of ideal-type market, 11
   disembedded exchange, 16, 17
   exchange value, 21
   hard budget constraints, 19, 20
   market failure and, 21
   Marxist theory, 12, 13
   National Development Plan, 79, 80
   public firms, 29, 30
property rights, 232, 237
   arbitrary use of property, 18
   Article 27 of Mexican Constitution, 55
   as exclusion, 18
   barriers to cooperation in market, 4
   disembedded exchange, 16, 17
   economic sociology, 18 n. 9
   institutional economics, 18, 19
   labor and, 90
   limits to, 13
   market failure and, 21
   paradoxes of, 18, 19
   restrictions in Mexican Constitution, 74
   state as enforcer of, 18, 19
   state autonomy, 24
   transaction costs and limits to, 23
   United States response to Mexican Revolution, 54, 55
protectionism. *See* tariffs
public goods
   contrasted with market exchange, 18
   institutional economics, 22
   market failure and, 20, 21
   property rights and, 18
public ownership. *See* state-owned firms; parastate sector
Puche, Jaime Serra, 87

railroad transportation
   competition with trucking, 192
   concessions by state to operate, 190
   consolidation of industry, 193
   containerized shipping, 192, 193, 213
   diffusion of technology, 190
   economic development and contribution to, 190, 191
   financing of in United States, 190
   multimodal shipping, 192, 193, 213
   new economic history, 191
   state regulation of, 191, 192
   state subsidies to, 192
   structure of ownership, 190
   technical characteristics of, 191, 192
railroad transportation in Mexico
   concessions by state to operate, 193
   Constitutional ammendment, 74
   export markets, 194, 208
   federal investment in, 120, 197, 208
   foreign ownership of, 193
   growth of, 193
   labor, 200–207
   legal restructuring, 219, 220
   rate regulation and rate structure, 194
   requirements of private firms, 194
   strategic classification, 211
   structure and routes of privatized railroads, 221–24
   subsidies to, 194
   tax exemptions, 194
rationality
   arbitrary use of property, 18
   bounded, 22
   exchange value, 15

Federal Law of Parastate Entities, 80
ideal types of exchange, 15
instrumental values, 18, 21
National Development Plan, 79
rational calculation in market, 14
ultimate values, 21
use value, 15
rectorship of state
  in constitution, 74
  National Development Plan, 213
  railroads, 219
  relationship to private sector, 213
  Telmex, 163
Regional Confederation of Mexican Workers, railroads and, 201
Regulatory Law of Public Service, Banking and Credit, 74
requisa, 156, 220, 239
Resolution Plan on Long Distance Public Network Interconnection, 175
Restructuring Committee for Railroads, 220, 221, 225
Retirement Fund Administrator, 85
Revolutionary Institutional Party
  composition of, 37, 97
  foreign investment, 56
  founding of, 37
  interest representation, 37–38
  National Revolutionary Party, 37, 236
  opposition to neoliberal reforms, 97
  parastate firms and, 34
  popular sectors, 37–38, 46
  presidentialism, 59
  private sector, 38, 50, 51, 97, 98
Rogozinski, Jacques, 41, 69, 77, 81
Ruiz Cortines, Adolfo, capital flight and, 52

Sacristán, Carlos Ruiz, 220
  Salinas administration, 86, 87
  Zedillo administration, 87
Salazar, José Chapa, 148
Salgado, Salustio, 157
Salinas de Gortari, Carlos
  communications and transportation, 109
  federal labor law, 92
  labor, 91, 92
  legal reforms, 80
  Mexican Confederation of Labor, 92
  Mexican National Railroad workers, 92
  mine workers, 92
  National Action Party, 97
  National Development Plan, 79

petroleum workers, 91
political reform under, 248, 249
port workers, 92
private sector, 96–98
privatization under, 71, 79–83
railroads in National Development Plan, 213
Telmex workers, 91
Saro, 144
satellite communications, added to constitution, 74
Schumpeter, Joseph A., 235, 256
SCT. *See* Ministry of Communications and Transportation
self-interest, 232
  among managers of public firms, 62, 63
  foundation of neoclassical theory, 7, 8, 9
  limits to, 255
  market exchange and, 14, 29
  market failure and, 21
  motivation for neoliberal reform, 26
  property rights and, 18
  transaction costs and limits to, 23
SEMIP. *See* Ministry of Energy, Mines, and Parastate Industry
Sen, Amartya, 248
Serbo. *See* Grupo Serbo
Serrano Segovia, José, 136
SHCP. *See* Ministry of the Treasury and Public Credit
Sidermex, 200
Silberstein, Jorge, 87, 227
Singapore Telecommunications, 170
Skocpol, Theda, 27, 29
Slim, Carlos, 100
Smith, Adam, 8, 20, 21, 232, 254
social struggle, 244
socialization of loss. *See* Agreement on Assumption of Liabilities; Bank Fund for Savings Protection; Fund to Cover Exchange Rate Risk
soft budget constraints. *See* hard budget constraints
Solomon Brothers, 102
Sosa de la Vega, Manuel, 139
Southeastern Railroad
  characteristics of, 224 Table 16
  routes, 222 Fig. 15
Southern Pacific, 226, 193
South Orient Railroad, 214
Southwestern Bell, 83, 168, 170, 227
sovereignty, 54–59, 233

SPP. *See* Ministry of Planning and Budget
Sprint, 186
stabilizing development, parastate firms and, 53
Staggers Rail Act, 192
state
    delgation of authority, 11, 12
    dependence on market, 21
    enforcement of property rights, 18, 19
    intervention in economy, 20, 21
    legitimacy of, 19, 22
    locus of social struggle, 23
    necessary losses in economy, 22
    neoliberal reforms, 25
    orthodox paradox, 25
    role in creating markets, 16, 25
state autonomy
    as capacity and insulation, 26
    business groups, 252
    changing meaning of, 27, 28
    collective action problem and, 26
    economic reform, 104–6
    economic sociology, 25
    embeddedness, 27, 28
    hard budget constraints, 24, 29
    internal and external dimensions of, 27
    Marxist theory, 27
    neoliberal reforms, 25
    popular sectors, 28
    potential autonomy of state, 27
    property rights, 24, 29
    social struggle, 4
state ownership. *See* parastate sector
state-owned enterprises. *See* parastate sector; state-owned firms
state-owned firms, 30. *See also* parastate sector
    economic sociology, 9
    import substituting industrialization, 2
    role in economic development, 1
    social obligations of, 18
STFRM. *See* Mexican Railroad Workers Union
STRM. *See* Mexican Telephone Workers Union
student movement of 1968, 49
Sugar Mills, 42, 43

TAESA, 143, 144
tariffs, 52, 53, 57, 58
    reduction under Miguel de la Madrid, 96
taxes
    market failure and, 21

    petroleum boom and, 70
    rates in Mexico, 49
    reform effort of 1970s in Mexico, 50
telecommunications
    access code dialing and transaction costs, 172
    competition, 179
    computers, 151
    coordination among firms, 151
    cross subsidization, 179
    deregulation of Chilean long distance market, 171, 172
    International Telecommunications Union, 151
    market segments, 150
    price wars in Chile, 172
    role of private sector in, 150
    role of state in, 150, 151
    technology, 151
telecommunications in Mexico
    access code dialing and transaction costs, 172, 181, 243
    average long-term incremental costs, 167, 177
    backward linkages, 153, 154
    Chilean long-distance market and, 171, 172
    Committee of Operators and collective action problem, 178, 181, 172, 173, 252
    competition, 153
    competition in long-distance market, 181, 182
    competition, limits to, 173
    concessions, 152
    cross subsidization, 160
    currency devaluations, 153
    effects of inflation, 159
    fibre-optic networks, 183–85
    Five Year plans, 153, 154
    foreign debt, 160
    foreign technology, 159
    interconnection, 153, 172, 175, 179
    international settlement rate, 160, 176
    labor and strategic nature of, 156
    labor militance, 156
    long-distance market, 173, 179–82, 237
    Ministry of Communications and Transportation, 154
    modernization, 162
    nondiscrimination, 175, 179
    phone service tax, 153
    proportional return, 180

rate setting and regulation, 152, 155, 159, 160
reciprocity, 175, 179
requisa, 156
role of state in, 152
separation of regulation from operation, 178
subsidies to, 154
telephone density, 162, 154, 183
third-party mediation, 173, 177, 178
trade surplus in long distance, 160
Teléfonos del Noroeste, 166
Tello, Carlos, 67
Telmex
  cellular service, 167
  competition and limits to, 166, 242
  conditions of sale, 163, 169
  conditions of sale, 167, 170, 175
  creation of, 153
  cross subsidization, 167, 171
  discrimination in pricing, 179
  equipment providers, 180
  external financing, 160
  financial restructuring, 168–70
  Five Year plans, 153, 154
  interconnection rates and rules, 174, 175
  international settlement rate, 176
  international surplus and foreign exchange, 160, 161
  justification of sale, 163, 164
  labor, 155–58
  labor restructuring, 164–66
  Mexicanization of, 153
  monopoly in local service, 174
  monopoly in long distance, 167
  nationalization of, 154
  network growth, 154, 155
  privatization of and private sector, 163
  rate regulation, 167, 171
  rate structure, 160, 161, 169, 171, 242
  relationship of Grupo Carso firms with Telmex, 181
  restructuring, 162–64, 166–68, 242
  size of, 162
  Sprint alliance, 184
  structural separation of, 179, 180
  taxes, 161
  Title of Concession, 167
  trade surplus in long distance, 161
  use value in service, 170
TFM. *See* Mexican Rail Transportation
TMM. *See* Mexican Maritime Transport
transaction costs, 233
  as friction, 22
  defined, 22
  embeddedness, 16, 17
  hard budget constraints, 23
  limits to market exchange, 4
  properties of, 22
  property rights and, 23
transportation sector. *See* communications and transportation
trucking, 208
  characteristics of, 209
  costs, 209
  federal investment in, 208, 209

U.S. Metals Reserve Company, 196
UACE. *See* Unit for the Support of Structural Change
UDEP. *See* Unit for the Divestiture of Parastate Entities
uncertainty. *See* transaction costs
Union Association of Flight Attendants
  Aeroméxico bankruptcy, 134
  Aeroméxico restructuring, 133
  concessions, 135
Union Association of Pilot Aviators
  Aeroméxico bankruptcy, 134
  Aeroméxico restructuring, 133
  concessions, 135
  salaries and working conditions, 122
  uniqueness as union, 122
Union Pacific, 227, 228
  Southern Pacific Purchase, 193
Unit for the Divestiture of Parastate Entities, 71
  creation of, 81
  private sector links, 104
  renaming of, 85, 86
  resectorization, 81
  structure of, 81
  Telmex privatization, 169
Unit for the Support of Structural Change
  Chihuahua-Pacific privatization, 227
  Northeastern Railroad privatization, 227
  privatization in communications and transportation, 87
  railroads, 220
United States
  Mexican Revolution, 55
  petroleum nationalization, 56
Unpasa, 200
use value
  as embeddedness, 16, 17
  defined, 15

use value *(cont'd)*
  forms of rationality and, 15
  ideal types of exchange, 15
  operationalized, 16, 17
  relation to exchange value in Marxist theory, 12, 13
utility maximization. *See* self-interest

Vallejo, Demetrio, 205, 206
value, as social relation, 13, 14. *See also* use value and exchange value
Vega Hutchison, Juan Manuel, 62
Velázquez, Fidel, 93
Visa. *See* Grupo Visa
Vitro. *See* Grupo Vitro

Walder, Andrew, 9
Weber, Max, 15
White, Harrison, 11
Williamson, Oliver, 5 n. 2, 22
World Bank, 231, 245, 254
  convergence with International Monetary Fund, 101
  Federal Electric Commission, 100
  railroad restructuring loan, 216
  structure of lending, 56, 57
  support for parastate firms, 56, 100
  support for privatization, 101, 102
  telecommunications, 110
world systems theory, 14 n. 7
workers. *See* labor

Zapatista Army of National Liberation, 246, 249
Zedillo, Ernesto
  bailout of bankrupt firms, 251, 252
  cabinet changes, 86
  Fund to Cover Exchange Rate Risk, 99
  legal reforms, 85
  Ministry of Planning and Budget
  political reform under, 248, 250
  private sector, 98–100
  privatization under, 71, 73, 84–89
  railroad privatization, 219

www.ingramcontent.com/pod-product-compliance
Lightning Source LLC
Chambersburg PA
CBHW031545300426
44111CB00006BA/184